Peterson's

MASTER THE™
GED:
MATHEMATICS

David Herzog

3rd Edition

PETERSON'S

A **nelnet** COMPANY

PETERSON'S

A ***nelnet*** COMPANY

An ARCO Book

ARCO is a registered trademark of Peterson's, and is used herein under license by Peterson's.

About Peterson's, a Nelnet company
Peterson's (www.petersons.com) is a leading provider of education information and advice, with books and online resources focusing on education search, test preparation, and financial aid. Its Web site offers searchable databases and interactive tools for contacting educational institutions, online practice tests and instruction, and planning tools for securing financial aid. Peterson's serves 110 million education consumers annually.

For more information, contact Peterson's, 2000 Lenox Drive, Lawrenceville, NJ 08648; 800-338-3282; or find us on the World Wide Web at www.petersons.com.

ISBN-13: 978-0-7689-2516-6
ISBN-10: 0-7689-2516-9

Printed in the United States of America

10 9 8 7 6 5 4 3 2 1   09 08 07

Third Edition

Petersons.com/publishing

Check out our Web site at www.petersons.com/publishing to see if there is any new information regarding the test and any revisions or corrections to the content of this book. We've made sure the information in this book is accurate and up-to-date; however, the test format or content may have changed since the time of publication.

OTHER RECOMMENDED TITLES:

Peterson's 30 Days to the GED
Peterson's Master the GED
Peterson's Master the GED: Language Arts, Reading
Peterson's Master the GED: Languge Arts, Writing
Peterson's Master the GED: Social Studies
Peterson's Master the GED: Science

Contents

Contents

Contents

PART IV: TWO PRACTICE TESTS

Before You Begin

HOW THIS BOOK IS ORGANIZED

Congratulations! You've just picked up the best preparation guide for the GED Mathematics Test you can buy. This book serves two purposes: first, to prepare you for the GED examination in mathematics, and second, to help you learn mathematics. Here's how you can use it to get your best GED Mathematics score.

- **Top 10 Strategies to Raise Your Score** gives you tried-and-true test-taking strategies.

- **Part I** contains answers to all your questions about the GED Mathematics Test. You'll learn what kinds of questions to expect, what the questions look like, and how you can keep your cool on test day.

- **Part II** gives you your first chance to try your hand at practice questions. Practice Test 1: Diagnostic can show you where your skills are strong—and where they need some improvement.

- **Part III** leads you through the mathematics review. Here you will review the arithmetic of whole numbers; algebra as a structure means for translating word problems into mathematical shorthand and then solving them; and geometry as a way of reasoning. Skim, scan, or study these reviews, depending on your *own* needs. You'll also get plenty of chances to test your skills with a "Test Yourself" quiz for each topic. To get the most out of the quizzes, you should cover the answers that directly follow them. In addition, you'll learn about each question type you will see on your exam.

- **Part IV** contains two practice tests. Try to take both of them if time allows. Remember: Practice makes perfect!

SPECIAL STUDY FEATURES

Peterson's Master the GED: Mathematics was designed to be as user-friendly as it is complete. It includes several features to make your preparation easier.

Overview

Each chapter begins with a bulleted overview listing the topics that will be covered in the chapter. You know immediately where to look for a topic that you need to work on.

Summing It Up

Each strategy chapter ends with a point-by-point summary that captures the most important points.

Test Yourself Quizzes and Exercises

Each chapter offers Test Yourself Quizzes throughout and Exercises at the end of the chapter. Take as many quizzes and exercises as you can. Use the results to determine where you still need work.

Bonus Information

In addition, be sure to look in the page margins of your book for the following test prep tools:

Note

Notes highlight critical information about the format of the GED Mathematics Test.

Tip

Tips draw your attention to valuable concepts, advice, and shortcuts for tackling the tests.

Alert!

Whenever you need to be careful of a common pitfall or test-taker trap, you'll find an *Alert!* This information reveals and eliminates the wrong turns many people take on the exam.

YOU'RE WELL ON YOUR WAY TO SUCCESS

Remember that knowledge is power. By using *Peterson's Master the GED: Mathematics, 3rd Edition,* you'll be studying with the most comprehensive GED mathematics preparation guide available and you'll become extremely knowledgeable about the GED. We look forward to helping you obtain your GED. Good luck!

GIVE US YOUR FEEDBACK

Peterson's publishes a full line of resources for your educational and career needs. Peterson's publications can be found at your local book store, library, and guidance office, and you can access us online at www.petersons.com.

We welcome any comments or suggestions you may have about this publication and invite you to complete our online survey at www.petersons.com/booksurvey. Or you can fill out the survey at the back of this book, tear it out, and mail it to us at:

Publishing Department
Peterson's, a Nelnet company
2000 Lenox Drive
Lawrenceville, NJ 08648

Your feedback will help us make your educational dreams possible.

TOP 10 STRATEGIES TO RAISE YOUR SCORE

1. **Always be aware that you are being timed on this test.** Since each question counts the same as every other question, if you are not able to answer a question in one or two minutes, leave that question, at least temporarily, and move on to the others. If you have time later, you can return to any skipped questions.

2. **Questions usually go from easiest to most difficult—you should, too!** So, work your way through the earlier, easier questions as quickly as you can.

3. **The easy answer isn't always the best.** Remember the hardest questions are usually at the end of a section and that also means the answers are more complex. Look carefully at the choices and really think about what the question is asking.

4. **An educated guess is always best.** The process of elimination is the best way to improve your guessing odds. Check out the answer choices and try to cross out any that you definitely know are wrong. If you're answering a question with 5 answer choices and you're able to knock out two choices, the odds go up to 33 percent that you will be correct.

5. **Check frequently to make sure that you are marking the correct section and the correct problem numbers on your answer sheet.** If you accidentally skip one problem number, you might end up marking the wrong answers to all of the questions that follow.

6. **Be certain to mark your answers on the grid on the answer sheet.** You will get no credit for answers written in the test booklet or on scrap paper.

7. **Make certain to answer the question being asked.** Sometimes test-takers get a problem wrong because, after solving for x, they choose that value as the answer to the question, when the question actually asked for the value of $x + 2$.

8. **If you are stuck, try looking at the multiple-choice answers.** Since one of the answers has to be right, the choices may give you an idea of how to proceed.

9. **Remember that from time to time (but not that often) one of the multiple-choice answers may be, "There is not enough information."** When you see such an answer choice, bear in mind that there is a real possibility that there may *not* be enough information to answer the question.

10. **By all means, relax—and good luck!**

PART I

GED MATHEMATICS BASICS

All About the GED Mathematics Test

OVERVIEW

- What is the GED?
- What is the GED Mathematics Test?
- Preparing to take the mathematics test
- Using the calculator
- Summing it up

WHAT IS THE GED?

The General Educational Development (GED) tests are a series of examinations designed to determine whether the person taking them has the literacy and computational skills equivalent to those of the upper two thirds of the students currently graduating from high schools in the United States. The tests are sponsored by the American Council on Education, a nonprofit educational organization located in Washington, D.C.

Since 1942, millions of adults have earned their high school credentials by passing the GED. More than 1 million adults take the GED test each year, and more than 600,000 of them are awarded high school equivalency diplomas. Although passing rates vary widely from state to state, about 75 percent of all test takers pass the five-part exam. All fifty states, the District of Columbia, nine United States territories and possessions, and ten Canadian provinces use GED results as the basis for issuing high school equivalency diplomas. All tests are administered under the supervision of state or (in Canada) provincial offices at designated GED testing centers (there are about 2900 such centers), and the standards for a passing grade are set by each state or province. In addition, federal and state correctional and health institutions and the military services also administer the tests to people in their institutions.

What do the tests measure? According to the American Council on Education, the tests measure "broad concepts and general knowledge, not how well they (test-takers) remember details, precise definitions, or historical facts. Thus, the tests do not penalize test-takers who lack recent academic or classroom experience or who have acquired their education informally."

chapter 1

WHAT IS THE GED MATHEMATICS TEST?

The **Mathematics Test** is usually the part of the exam most universally feared by test-takers. In reality, it shouldn't be. Many of the questions can be solved by using the basic arithmetic operations of addition, subtraction, multiplication, and division. In addition, there are some questions involving elementary algebra and plane geometry, as well as questions based on data interpretation, basic statistics, and probability. Many of the questions are presented as word problems, involving real-life situations. Others ask you to interpret information presented in graphs, tables, charts, or diagrams. You will be given a sheet of important formulas to help you solve the problems on the test. All scrap paper will be collected at the conclusion of the test.

The test itself lasts for 90 minutes and consists of two different sections. Each section contains 25 questions, but the concepts tested in the two different sections vary substantially. In the first section, you will be allowed to use a hand-held scientific calculator. In fact, before the test begins, you will be given a calculator to use and a practice worksheet to help make you more comfortable with the calculator. In this section, as you might guess, you will not have any questions that ask you to simply perform mathematical computations. The questions in this section emphasize mathematical understanding and application, and the calculator enables the test makers to ask you word problems with realistic numbers. Since you will not have to spend a lot a time performing computations on this section, completing this section should take you a good deal less time than completing the second section.

In the second section, you must answer your questions without the aid of a calculator, and this means that you will have to perform all of your computations by hand. In spite of this, the emphasis in this section is on mental math and estimation. For example, some of the questions will ask you how you would go about solving a particular problem, but not require you to actually solve it!

Each of the two sections on the test contains a number of "Alternate Format" questions. These questions do not contain the usual multiple choices; instead, they require you to figure out your answer and code it into a special grid. Later on, there will be a section that will show you how to code your answers into these grids.

Test	Content Areas	No. of Questions	Time Limit (minutes)
Mathematics	Numbers, Number Sense,		
Booklet One:	Operations – 25%		
Calculator	Data, Statistics, Probability – 25%		
Booklet Two:	Geometry and Measurement – 25%		
No Calculator	Algebra, Functions, Patterns – 25%	50	90

PREPARING TO TAKE THE GED MATHEMATICS TEST

Multiple-Choice Questions

The GED Mathematics Test contains 50 math questions. Of these, 40 are in the standard multiple-choice format; that is, you are asked a question and then need to choose between five possible answer choices. These eight tips will help you answer multiple-choice questions.

❶ On the "non-calculator" portion of the test, try not to waste time doing unnecessary computations. Remember that one of the answer choices must be the correct one. Estimate as much as possible as you attempt to determine the correct answer.

❷ Be careful (especially when solving geometry questions involving measurement) to express your answer in the same units of measure as given in the multiple choice answers. For example, if you determine that the answer to a question is 3 feet, and all of the answer choices are in inches, convert your answer to 36 inches. Don't work so quickly that you mistakenly select the answer 3 inches.

❸ If the answer you obtain doesn't match one of the choices given, it might still be right. Try to write it in a different form, and then see if it matches. For example, the answer $x^2 + 6x$ can also be written as $x(x + 6)$.

❹ In the same way, most fractional and square root answers on the test will be expressed in simplified form. Therefore, if you obtain $\frac{9}{15}$ as the answer to a question, and don't see it among the multiple choices, don't panic. See if the fraction you obtained can be simplified. For example, write $\frac{9}{15}$ as $\frac{3}{5}$ and see if you can find $\frac{3}{5}$ as an answer choice. Similarly, you may not see $\sqrt{60}$ among the answer choices, but you might find it in its simplified form, $2\sqrt{15}$.

❺ If you are stuck, try looking at the multiple choice answers. Since one of the answers has to be right, the choices may give you an idea of how to proceed.

❻ Some of the questions you may be able to answer correctly by simply estimating. For example, if the five answer choices vary greatly in size, you may be able to tell which one is correct by estimating the size of the correct answer.

❼ Remember that, from time to time, one of the multiple choice answers may be "There is not enough information." When you see such an answer choice, bear in mind that there is a real possibility that there may not be enough information to answer the question.

❽ Never do any more work than you are being asked to do. Some of the questions on the test ask you how to solve the problem, but do not actually ask you to solve it. In such questions, pick the answer choice that expresses the way to solve the problem, but do not waste any time trying to determine the actual numerical solution to the problem.

ALERT!

As you read through the answer choices, beware of answer choices that simply repeat numbers that are presented in the problem, especially if the problem is a word problem.

Alternate Format Questions

As we have already discussed, the new GED contains 50 math questions. Of these, 40 will be in the usual standard multiple choice format, that is, there will be five answer choices and you will be asked to select the correct one. The remaining ten questions are special, alternate format questions. In these questions, you will not be given any answer choices, but instead will have to determine your own answer, and code it on the answer grid in the correct way.

There are two different types of alternate format questions. In one type, you will be asked a geometry question, the answer to which will be a point on the coordinate plane. Instead of simply writing the point down, you will be asked to shade it in on a coordinate plane grid. As an example, if you determine the answer to a particular problem to be (2, 3), you will need to shade that point in on a grid similar to the one below:

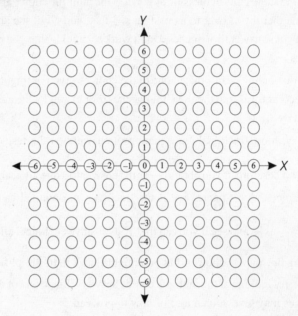

That is, in order to answer the question, you will need to shade in the grid as shown below:

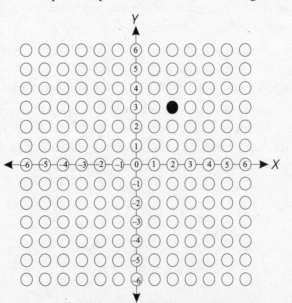

There are only two of these "coordinate plane grid" questions on the test, one in the first (calculator) part, and one in the second (non-calculator) part. Notice that, since the blank grid only extends between the numbers –6 and +6 in both the x and y directions, the only permissible answers are ones in which the values for the x- and y-coordinates are between –6 and 6. If, in working the problem, you obtain the answer (7, –10), it is time to try it again. In the same way, fractional values for the coordinates are not permitted. Should you get an answer of $\left(2\frac{1}{2}, 5\frac{1}{2}\right)$, it is time to try again!

As you can see, there is nothing particularly difficult about coding the answers to these types of questions. Begin by simply solving the problem as you normally would, and then, instead of looking for the correct answer from among five multiple choices, make a dot on the correct circle on the grid. Try the two examples below to make certain that you understand how to answer this type of question.

Q What is the midpoint of the line segment whose endpoints are (–3, –4) and (3, 6)? Mark the midpoint on the coordinate plane below.

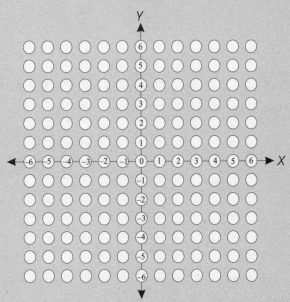

A The x-coordinate of the midpoint of a line segment is the average of the x-coordinates of the endpoints; similarly, the y-coordinate of the midpoint of a line segment is the average of the y-coordinates of the endpoints. Therefore, the x-coordinate of the midpoint is $\frac{-3+3}{2} = \frac{0}{2} = 0$. In the same way, the y-coordinate of the midpoint is $\frac{6+(-4)}{2} = \frac{2}{2} = 1$. Therefore, the midpoint is (0, 1). You should shade the grid as shown on the following page.

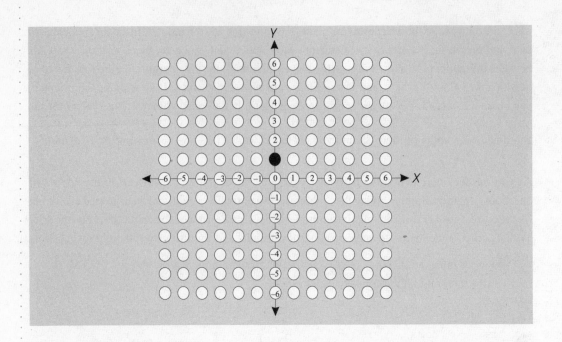

Q What is the *y*-intercept of the line whose equation is $2y = 6x + 10$? Mark the answer on the coordinate plane below.

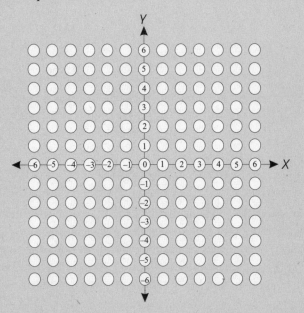

A Recall that the slope-and-intercept from for the equation of a line is given by $y = mx + b$, where *m* represents the slope, and *b* represents the *y*-intercept. The equation that we are given, $2y = 6x + 10$, is almost in this form. In fact, if we isolate the *y* on the left-hand side of the equation by dividing both sides of the equation by 2, we will have put the equation in the slope-and-intercept form: $y = 3x + 5$.

From this, we can easily see that the *y*-intercept of the equation is 5, and this means that the graph contains the point (0, 5). This point should be shaded in on the grid below, as shown:

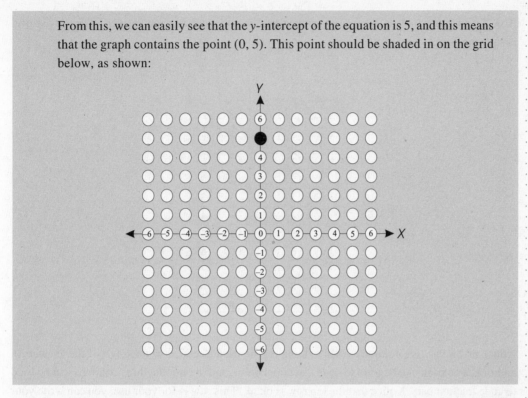

The other type of alternate format question on the GED is more important, at least to the extent that there are 8 such questions on the test, six of which appear on the first (calculator) part of the test, and the remaining two of which appear on the second (non-calculator) part of the test. These questions will have numerical answers, but instead of looking for the correct answer from among a group of five possible multiple choices, you will need to "code" the answer in a special grid.

Technically, these questions should be a bit more challenging than the multiple choice ones, since if you get confused on a multiple choice question, you can always look at the possible answer choices to get some idea of how to proceed. On these questions, however, you are on your own.

Before we look at this alternate format grid, let's consider some specific aspects of this type of question—aspects that will probably make more sense once you start to get used to coding answers on the grid. First of all, for many of the questions, there will be more than once correct way to code an answer. This should not concern you; as long as the way you have coded your answer is one of the possible correct ways, you will receive credit for your answer. When you look at the grid, you will also see that there is no way to code a negative answer. This is also not a problem. The answers to these types of questions will always be positive, so you will not have to worry about needing to code a negative answer. In fact, you can use this fact to your advantage. If, in solving one of these questions, you should obtain a negative answer, you know that you must of made a mistake, and you should try the problem again! Finally, there is no way to code certain types of numbers, in particular, mixed numbers. So, say, for example, that you are working a problem and end up with an answer of $7\frac{1}{4}$. What are you supposed to do with this answer? Well, you simply need to re-write it in an equivalent form that can be coded, such as the decimal form 7.25.

Now, let us take a look at the answer grid and see how to use it to code answers. The answer grid is shown below:

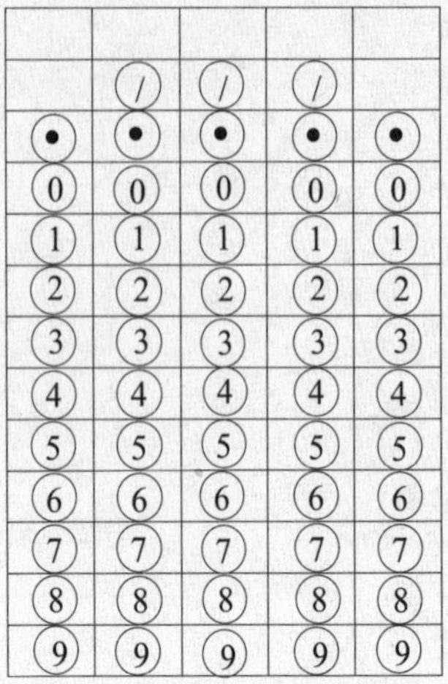

Notice that it has five columns. Each column can be used to code a *single digit* of the answer. In addition, a column can be used to code a decimal point, and the middle three columns can be used to code fraction bars. Notice that the top row is blank. This row is for your use; you can write your answer in this row to help in coding it below. However, remember that the answer sheet is machine read. Therefore, you must code your answer in the grid. You will not get any credit if you have the correct answer hand-written in the first row, but you have coded the grid incorrectly.

Now, let's see how to fill in the grid. Suppose that you are working a problem, and you determine the answer to be 765. Since this answer contains three digits, you will need to use three columns to code it. But which three? The answer to this question is simple: use any three consecutive columns that you wish. Therefore, there are three correct ways to code the number 765. You can start filling in your answer all the way to the left, you can center it, or you can place it all the way to the right. The following three codings would represent correct answers:

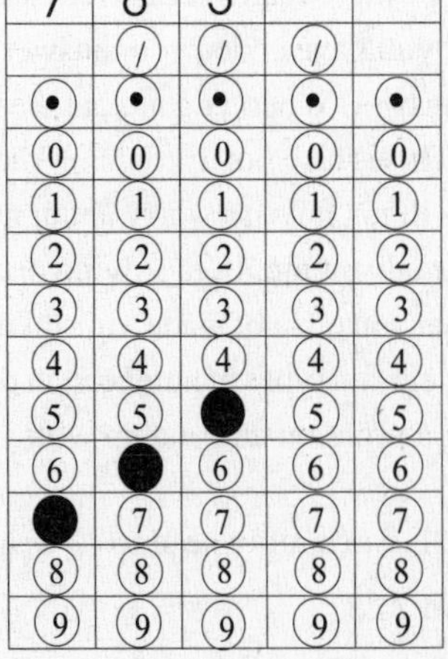

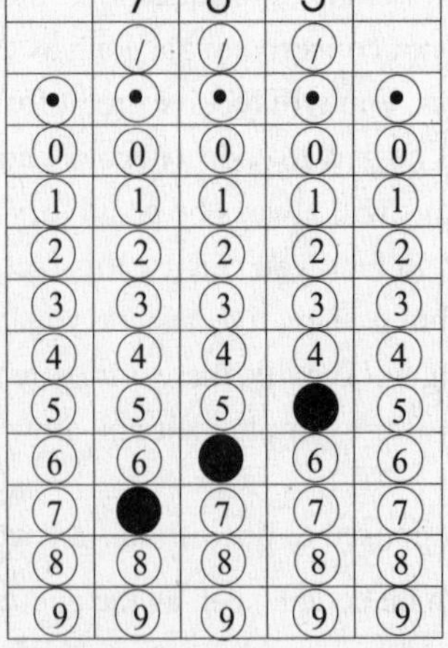

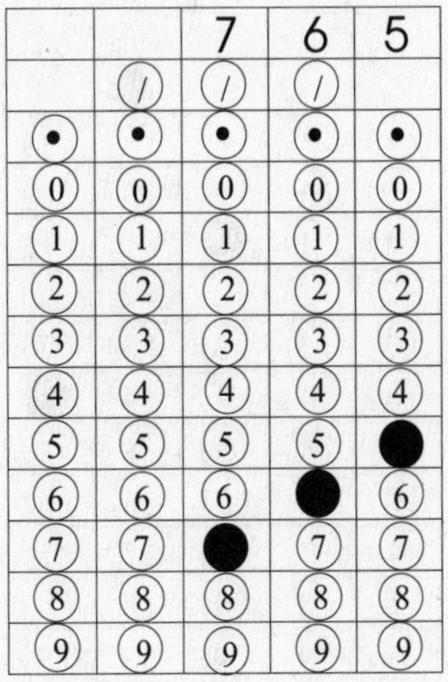

Suppose your answer comes out to be the fraction $\frac{1}{4}$? There are a lot of different ways that this answer could be coded. First of all, you could use the fraction bar that appears in columns two, three, or four to enter $\frac{1}{4}$ as a fraction. There would be three ways to do this, as shown below. Any one of the three would be correct.

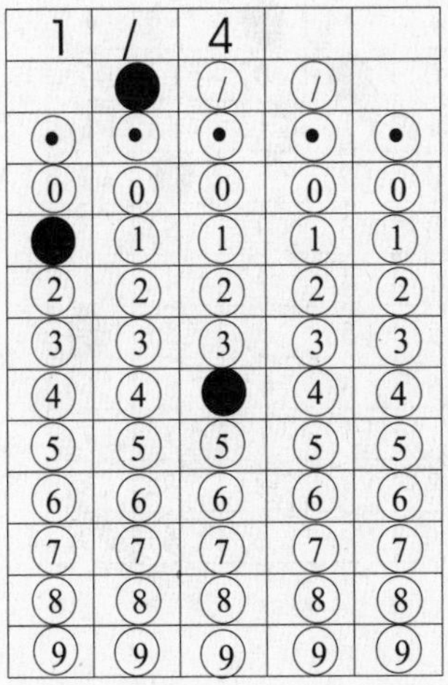

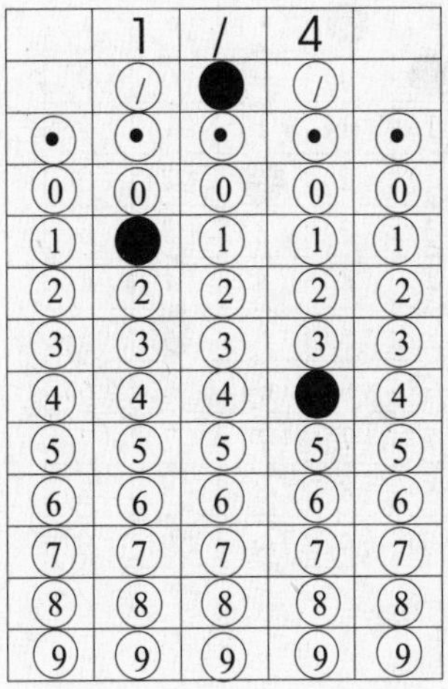

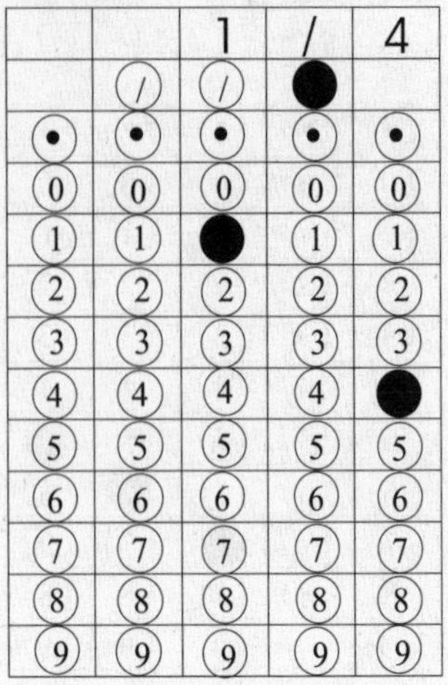

The solution $\frac{1}{4}$ could also be changed to a decimal, 0.25, and coded that way. There are several ways that 0.25 could be coded: four of them are shown below. Note that displaying the 0 to the left of the decimal point is optional.

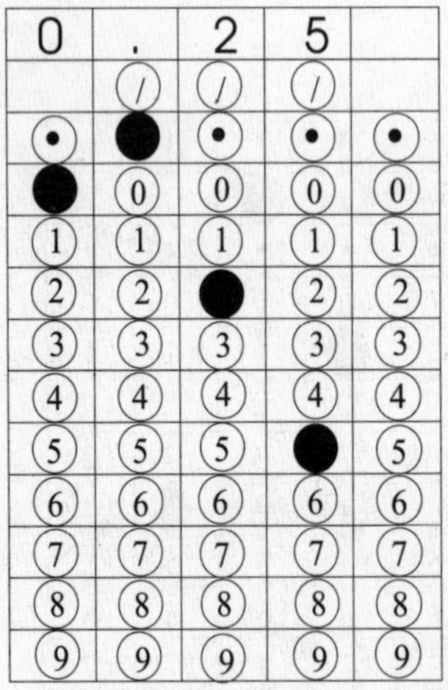

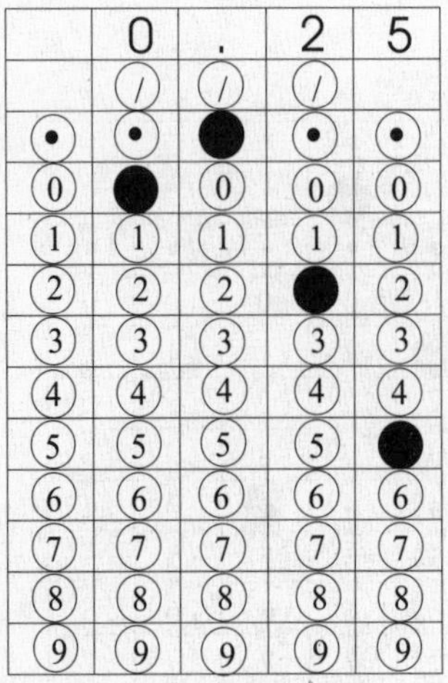

Note that, should the answer to the question turn out to be a fraction such as $\frac{2}{3}$, this would need to be coded as a fraction since the decimal representation repeats infinitely.

Now, how would you code a mixed number answer, such as $7\frac{1}{2}$? Using the grid, there is no way to code a mixed number, so it would be necessary to rewrite the $7\frac{1}{2}$ as either a decimal number, 7.5, or an improper fraction, $\frac{15}{2}$. Below are four different ways to code the answer $7\frac{1}{2}$.

And that is all that there is to it. Simply remember that when you code your answers, work slowly and carefully so as to be certain that you are coding them correctly. Remember that the boxes in the first row are there for you to write your answers in, but only as an aid in gridding them correctly. Since the answers are machine read, no one will be looking at what you write in the top row. Therefore, correct gridding is essential. Remember that there are several ways to code any answer, and they are all equally correct. There is also no way to code a mixed number, so if you obtain a mixed number as an answer to one of the problems, you must rewrite it as an improper fraction or as a decimal number. And that is all there is to it!

USING THE CALCULATOR

As we have already discussed, you will be allowed to use a calculator to help you answer the first 25 questions on the math test. Your initial thought upon hearing this may have been, "This is great. With a calculator, any math test will be very easy!" Not so fast.

The majority of the math questions on the calculator section emphasize problem solving and the application of mathematical ideas. This means that you are not going to see any questions in this section that ask you to simply perform a mathematical calculation, that, of course, could easily be done using a calculator. The stated reason for allowing you to use a calculator on the test is "to eliminate the tedium of complex calculations in realistic settings." What this means is that, while the calculator will certainly help you to "crank out the numbers" in these problems, you are going to have to begin by doing a lot of thinking as to how to approach and set up the problem before you can even begin to think about using the calculator. As a simple example, consider a geometry problem that gives you the lengths of the two legs of a right triangle as 10 and 12, and asks you for the length of the hypotenuse. Further, let us suppose that the five answer choices are

(1) 14.9
(2) 15.1
(3) 15.5
(4) 15.6
(5) 15.9

Now, your calculator will be extremely useful, after you obtain the answer $\sqrt{244}$, as a way of quickly computing the value of this square root to the nearest tenth. But, unless you understand the purpose of, and how to use, the Pythagorean Theorem, there is no way that you will be able to get this far. Unless you understand that $a^2 + b^2 = c^2$, and that therefore the missing length is given by the formula $\sqrt{10^2 + 12^2} = \sqrt{100 + 144} = \sqrt{244}$, your calculator will not be of very much help.

That being said, the calculator will still be of tremendous use to you on the test, and therefore it is crucial that you know how to use it quickly and accurately. Everyone taking the test will be given the same type of calculator to use, the Casio FX-260.

Now, as you no doubt know, every calculator is set up a little bit differently. For example, the clear keys tend to be in different locations on different calculators, and perhaps they are labeled differently. The Casio FX-260 looks like this:

Even though you will be given time before the test begins to practice with the calculator, it may take you a little while to get used to it, especially if you are already used to a different calculator made by a different company. Therefore, a very helpful thing to do before the test date would be this: Go to a local stationery or electronics store and purchase a Casio FX-260. The calculator is readily available (specifically because the manufacturer knows that this is the calculator that is going to be used on the GED), and should cost between $10 and $15. If you spend a lot of time practicing with it before the test, you will know your way around it like a pro by the time you take the test. You won't have to waste time looking for a particular key, or trying to figure out how to use the "shift" key.

By this time, you are probably very familiar with calculators, and can easily use them to add, subtract, multiply, and divide. Still, there are six strategies that you should remember.

❶ When you are first given the calculator, turn it ON by pressing the $\boxed{\text{ON}}$ key, that is located in the upper right-hand corner.

❷ After you complete a question, in order to prepare for the next question, be certain to clear the calculator by either pressing the $\boxed{\text{ON}}$ key, or the red $\boxed{\text{AC}}$ key. AC stands for "All Clear." If you do not clear your calculator between problems, you run the risk of having your answer to problem 4 add into your answer to problem 5.

❸ Any arithmetic operation can be entered in the order in which it is written. The calculator is already programmed to adhere to the algebraic order of operations. Therefore, if you wish to compute, say $5 + 2 \times 3$, simply key in:

$$\boxed{5}\ \boxed{+}\ \boxed{2}\ \boxed{\times}\ \boxed{3}\ \boxed{=}$$

The answer will be 11. Note that the calculator has performed the multiplication before the addition, as is appropriate according to the Order of Operations.

❹ Should you ever need to multiply an expression in parentheses by another number, be certain to press the $\boxed{\times}$ key on the calculator when appropriate. For example, suppose that you need to compute $4(3 + 2)$. Even though, in algebra, it is not necessary to write a times sign after the 4, when you do the computation on the calculator, it will be necessary. Therefore, to compute $4(3 + 2)$, key in

$$\boxed{4}\ \boxed{\times}\ \boxed{(}\ \boxed{3}\ \boxed{+}\ \boxed{2}\ \boxed{)}\ \boxed{=}$$

The calculator will show the correct answer of 20.

❺ To enter a negative number using the calculator, enter the number without a sign, and then press the "change sign" key, that looks like this" $\boxed{+/-}$. Therefore, to perform the subtraction $-9 - -4$, you must key in

$$\boxed{9}\ \boxed{+/-}\ \boxed{-}\ \boxed{4}\ \boxed{+/-}\ \boxed{=}$$

The calculator will show the correct answer, –5.

❻ One of the trickier things to do with the calculator is to use it to compute a square root. To do this, you must use the $\boxed{\text{SHIFT}}$ key and the $\boxed{x^2}$. Say, for example, we need to find $\sqrt{244}$, as required in the sample problem above. You would begin by entering the number 244 in the calculator, then press the "shift" key, and then the $\boxed{X^2}$ key. By

shifting the $\boxed{X^2}$ key, you access its second function, that is the square root. Conclude

by pressing $\boxed{=}$. To summarize, to determine $\sqrt{244}$, key in the following:

$$\boxed{2}\ \boxed{4}\ \boxed{4}\ \boxed{\text{SHIFT}}\ \boxed{X^2}\ \boxed{=}$$

SUMMING IT UP

- The GED Mathematics Test actually contains two parts, or booklets. Each booklet contains 25 questions and you will have 90 minutes to answer the 50 questions.

  ❶ On the first 25 questions, you will be allowed to use a calculator—the Casio FX260 solar calculator.

  ❷ On the second 25 questions, you are not allowed to use a calculator.

- Most of the questions are multiple-choice. A few questions require you to mark the correct answer in an answer grid or on a plane.

- 25 percent of the questions will measure skills in numbers, number sense, and operations; 25 percent will measure skills in algebra; and 25 percent will measure skills in geometry.

- Many of the questions will be presented as word problems or practical everyday situations.

- An important point to remember is that there is no penalty for a wrong answer. With this in mind, you should not spend too much time on any one question.

- If you have extra time at the end of the test, go back and check your work. The most common errors that are made on math tests result from careless mistakes.

PART II

DIAGNOSING STRENGTHS AND WEAKNESSES

CHAPTER 2   Practice Test 1: Diagnostic

ANSWER SHEET PRACTICE TEST 1

Part 1

1. ① ② ③ ④ ⑤   7. ① ② ③ ④ ⑤   14. ① ② ③ ④ ⑤   20. ① ② ③ ④ ⑤
2. ① ② ③ ④ ⑤   8. ① ② ③ ④ ⑤   15. ① ② ③ ④ ⑤   24. ① ② ③ ④ ⑤
3. ① ② ③ ④ ⑤   11. ① ② ③ ④ ⑤   17. ① ② ③ ④ ⑤   25. ① ② ③ ④ ⑤
5. ① ② ③ ④ ⑤   12. ① ② ③ ④ ⑤   18. ① ② ③ ④ ⑤
6. ① ② ③ ④ ⑤   13. ① ② ③ ④ ⑤   19. ① ② ③ ④ ⑤

4.

9.

10.

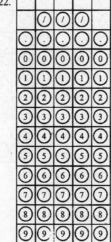

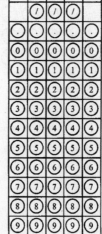

16.

21.

22.

23.

answer sheet

Part 2

1. ① ② ③ ④ ⑤
2. ① ② ③ ④ ⑤
3. ① ② ③ ④ ⑤
4. ① ② ③ ④ ⑤
5. ① ② ③ ④ ⑤
6. ① ② ③ ④ ⑤
7. ① ② ③ ④ ⑤
8. ① ② ③ ④ ⑤
9. ① ② ③ ④ ⑤
10. ① ② ③ ④ ⑤
12. ① ② ③ ④ ⑤
13. ① ② ③ ④ ⑤
14. ① ② ③ ④ ⑤
16. ① ② ③ ④ ⑤
18. ① ② ③ ④ ⑤
19. ① ② ③ ④ ⑤
20. ① ② ③ ④ ⑤
21. ① ② ③ ④ ⑤
22. ① ② ③ ④ ⑤
23. ① ② ③ ④ ⑤
24. ① ② ③ ④ ⑤
25. ① ② ③ ④ ⑤

11.

15.

17.

PRACTICE TEST 1: DIAGNOSTIC

90 Minutes • 50 Questions Total

PART 1—25 QUESTIONS (A CALCULATOR IS PERMITTED): 45 MINUTES

PART 2—25 QUESTIONS (A CALCULATOR IS NOT PERMITTED): 45 MINUTES

> **Directions:** The Mathematics Test consists of questions intended to measure general mathematics skills and problem-solving ability. The questions are based on short readings that often include a graph, chart, or figure. Work carefully, but do not spend too much time on any one question. Be sure you answer every question. You will not be penalized for incorrect answers.

Formulas you may need are given on the following pages. Only some of the questions will require you to use a formula. Record your answers on the separate answer sheet. Be sure that all information is properly recorded.

There are three types of answers found on the answer sheet:

❶ Type 1 is a regular format answer that is the solution to a multiple-choice question. It requires shading in 1 of 5 bubble choices.

❷ Type 2 is an alternate format answer that is the solution to the standard grid "fill-in" type question. It requires shading in bubbles representing the actual numbers, including a decimal or division sign where applicable.

❸ Type 3 is an alternate format answer that is the solution to a coordinate plane grid problem. It requires shading in the bubble representing the correct coordinate of a graph.

Type 1: Regular Format, Multiple-Choice Question

To record your answers for multiple-choice questions, fill in the numbered circle on the answer sheet that corresponds to the answer you select for each question in the test booklet.

> **Q** Jill's drug store bill totals $8.68. How much change should she get if she pays with a $10.00 bill?
>
>   **(1)** $2.32
>
>   **(2)** $1.42
>
>   **(3)** $1.32
>
>   **(4)** $1.28
>
>   **(5)** $1.22
>
> **A** **The correct answer is (3).** Therefore, you should mark answer space (3) on your answer sheet.

practice test 1

Type 2: Alternate Format, Standard Grid Question

To record the answer to the previous example, "1.32," using the Alternate Format, Standard Grid, see below:

Standard Grid

Type 3: Alternate Format, Coordinate Plane Grid Question

To record your answer, fill in the numbered circle on the answer sheet that corresponds to the correct coordinate in the graph. For example:

Q A system of two linear equations is given below.

$$x = -3y$$
$$x + y = 4$$

What point represents the common solution for the system of equations?

A **The correct answer is (6, –2).** The answer should be gridded as shown below.

Coordinate Plane Grid

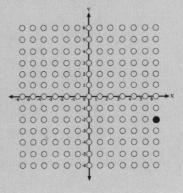

FORMULAS

Description	Formula
AREA (A) of a:	
square	$A = s^2$; where s = side
rectangle	$A = lw$; where l = length, w = width
parallelogram	$A = bh$; where b = base, h = height
triangle	$A = \frac{1}{2} bh$; where b = base, h = height
circle	$A = \pi r^2$; where π = 3.14, r = radius
PERIMETER (P) of a:	
square	$P = 4s$; where s = side
rectangle	$P = 2l + 2w$; where l = length, w = width
triangle	$P = a + b + c$; where a, b, and c are the sides
Circumference (C) of a circle	$C = \pi d$; where π = 3.14, d = diameter
VOLUME (V) of a:	
cube	$V = s^3$; where s = side
rectangular container	$V = lwh$; where l = length, w = width, h = height
cylinder	$V = \pi r^2 h$; where π = 3.14, r = radius, h = height
square pyramid	Volume $= \frac{1}{3} \times (\text{base edge})^2 \times \text{height}$
cone	Volume $= \frac{1}{3} \times \pi \times \text{radius}^2 \times \text{height}$; π is approximately equal to 3.14.
Pythagorean theorem	$c^2 = a^2 + b^2$; where c = hypotenuse, a and b are legs of a right triangle
distance (d) between two points in a plane	$d = \sqrt{(x_2 - x_1)^2 + (y_2 - y_1)^2}$; where (x_1, y_1) and (x_2, y_2) are two points in a plane
slope of a line (m)	$m = \dfrac{y_2 - y_1}{x_2 - x_1}$ where (x_1, y_1) and (x_2, y_2) are two points in a plane
trigonometric ratios	given an acute angle with measure x of a right triangle, $\sin x = \dfrac{\text{opposite}}{\text{hypotenuse}}$, $\cos x = \dfrac{\text{adjacent}}{\text{hypotenuse}}$, $\tan x = \dfrac{\text{opposite}}{\text{adjacent}}$
mean	mean $= \dfrac{x_1 + x_2 + \cdots + x_n}{n}$; where the x's are the values for which a mean is desired, and n = number of values in the series
median	median = the point in an ordered set of numbers at which half of the numbers are above and half of the numbers are below this value
simple interest (i)	$i = prt$; where p = principal, r = rate, t = time
distance (d) as function of rate and time	$d = rt$; where r = rate, t = time
total cost (c)	$c = nr$; where n = number of units, r = cost per unit

You may use a scientific calculator for Part 1. (A Casio FX-260 Scientific Calculator will be provided at your Official GED Testing Center.)

CALCULATOR DIRECTIONS

You may practice with your calculator, using the following directions.

CALCULATOR DIRECTIONS

To prepare the calculator for use the *first* time, press the [ON] (upper-rightmost) key. "DEG" will appear at the top-center of the screen and "0" at the right. This indicates the calculator is in the proper format for all your calculations.

To prepare the calculator for *another* question, press the [ON] or the red [AC] key. This clears any entries made previously.

To do any arithmetic, enter the expression as it is written. Press [ON] (equals sign) when finished.

EXAMPLE A: $8 - 3 + 9$

First press [ON] or [AC].
Enter the following:

[8] [−] [3] [+] [9] [=]

The correct answer is 14.

If an expression in parentheses is to be multiplied by a number, press [x] (multiplication sign) between the number and the parenthesis sign.

EXAMPLE B: $6(8 + 5)$

First press [ON] or [AC].
Enter the following:

[6] [x] [(] [8] [+] [5] [)] [=]

The correct answer is 78.

To find the square root of a number
- enter the number;
- press the [SHIFT] (upper-leftmost) key ("SHIFT" appears at top-left of the screen);
- press [x²] (third from the left on top row) to access its second function: square root.
DO NOT press [SHIFT] and [x²] at the same time.

EXAMPLE C: $\sqrt{64}$

First press [ON] or [AC].
Enter the following:

[6] [4] [SHIFT] [x²] [=]

The correct answer is 8.

To enter a negative number such as -8,
- enter the number without the negative sign (enter 8);
- press the "change sign" ([+/−]) key which is directly above the [7] key.
All arithmetic can be done with positive and/or negative numbers.

EXAMPLE D: $-8 - -5$

First press [ON] or [AC].
Enter the following:

[8] [+/−] [−] [5] [+/−] [=]

The correct answer is -3.

Part 1

Directions: You may now begin Part 1 of the Mathematics Test. Bubble in the correct response to each question on Part 1 of your answer sheet.

1. David packed a total of 40 boxes in 5 hours. How many boxes would Samuel pack in 3 hours if he packed the same number of boxes per hour?

   (1) 24

   (2) 25

   (3) 45

   (4) 15

   (5) 30

2. If 24 pencils cost $4.82, how much would 4 pencils cost?

   (1) $2.40

   (2) $.81

   (3) $.60

   (4) $1.20

   (5) $1.40

3. The prices of a gallon of milk at different grocery stores are $1.39, $1.22, $1.29, and $1.30. What is the average price of a gallon of milk?

   (1) $1.32

   (2) $1.31

   (3) $1.25

   (4) $1.29

   (5) $1.30

4. Mrs. Gabaway wants to telephone her friend in Boston. The day rate is $.48 for the first minute and $.34 for each additional minute. The evening rate discounts the day rate by 35 percent. If Mrs. Gabaway is planning a 45-minute chat, to the nearest penny, how much would she save if she took advantage of the evening rate by calling after 5 p.m.?

   Mark your answer in the circles in the grid on your answer sheet.

5. $\sqrt{64} + 16 =$

   (1) 80

   (2) 48

   (3) 24

   (4) 32

   (5) 26

6. Melissa had $500 in her checking account. She wrote checks in the amounts of $35.75, $120.50, $98.25, and $350. She then deposited $375. How much money did she have in her account after her deposit?

   (1) $375

   (2) $270.50

   (3) $604.50

   (4) $200

   (5) Insufficient data is given to solve the problem.

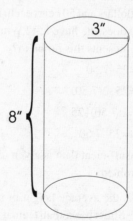

7. How many cubic inches of liquid can the cylindrical can above hold?

   (1) 48π

   (2) 64π

   (3) 96π

   (4) 72π

   (5) 108π

8. Find the value of $-6a + a^2 + 3(a + 6)$ if $a = 3$

    (1) −18

    (2) 27

    (3) 28

    (4) 18

    (5) 9

9. Jonathan drove his 66-year-old grandmother and 9-year-old little brother to the movie theater on Saturday afternoon. He treated all three of them to the matinee. The prices read as follows: Adults, $3.50; Children Under 12, $1.50; Senior Citizens, 20 percent discount. How much change did Jonathan receive from his $10.00 bill?

    Mark your answer in the circles in the grid on your answer sheet.

10. Quadrilateral ABCD is a square. The coordinates of point A are (3,2), the coordinates of point B are (−3,2), and the coordinates of point C are (−3,−4). On the coordinate plane on your answer sheet, mark the location of coordinate D.

11. Sarah has 25 dollars. She buys 2 books that cost 7 dollars and 50 cents each. How much money does she have left? Which expression represents this problem?

    (1) $x = 25 - 7.50$

    (2) $x = 25 - 2(7.50)$

    (3) $x = 2(7.50) - 25$

    (4) $2x = 25 - 7.50$

    (5) Insufficient data is given to solve this problem.

12. What is the average height of a player on the 10th Street basketball team if the heights of the individual players are 5'8", 6'1", 5'10", 6'3", and 5'9"?

    (1) 5'8"

    (2) 5'9"

    (3) 5'11"

    (4) 5'10"

    (5) 6'

13. Sam returned a pair of shoes and was credited with $54.99 in his account. At the same store, he bought a shirt for $22.50, a tie for $14.95, and socks for $3.98. How much credit was remaining in his account?

    (1) $13.56

    (2) $13.00

    (3) $11.43

    (4) $12.50

    (5) $15.99

14. Ed bought his gasoline at a station that recently converted its pumps to measuring gasoline in liters. If his tank took 34.0 liters, approximately how many gallons did it take? One gallon is equal to 3.785 liters.

    (1) 9

    (2) 8

    (3) 9.25

    (4) 10

    (5) 11.5

15. At midnight, the temperature was −15°. By dawn, the temperature had risen 25°. What was the temperature at dawn?

    (1) 25°

    (2) −10°

    (3) 10°

    (4) −5°

    (5) −25°

16. Tom and Johnny leave the house at 6 a.m. to go camping. Colin decides to go with them, but to his dismay, when he reaches Tom's house, he learns that he and Johnny had left an hour ago. If Colin drives 65 MPH, how many hours will it take him to overtake his friends who are traveling at 45 MPH?

    Mark your answer in the circles in the grid on your answer sheet.

17. A college football team carried the ball 5 times in one quarter. During those 5 times, they gained 4 yards, lost 6 yards, lost 4 yards, gained 8 yards, and lost 5 yards. How many yards did they lose in all?

(1) 11

(2) 10

(3) 5

(4) 15

(5) 3

18. Two ships leave the same harbor at the same time and travel in opposite directions, one at 30 km/hr and the other at 50 km/hr. After how many hours will they be 360 kilometers apart?

(1) $2\frac{1}{2}$

(2) $3\frac{1}{2}$

(3) $5\frac{1}{2}$

(4) $4\frac{1}{2}$

(5) $6\frac{1}{2}$

19. In 1990, the average salary of a New Jersey teacher was $30,588. Ten years later, teachers in New Jersey were averaging $61,008. Find the nearest whole percentage of increase in salary from 1990 to 2000.

(1) 99 percent

(2) 100 percent

(3) 50 percent

(4) 86 percent

(5) 75 percent

20. Juanita walked 1 hour 20 minutes every day for one full week. At the end of the week, how many hours did she walk altogether?

(1) 7 hr. 20 min.

(2) 8 hr.

(3) 9 hr. 20 min.

(4) 10 hr. 10 min.

(5) Insufficient data is given to solve the problem.

21. In order to determine the expected mileage for a particular car, an automobile manufacturer conducts a factory test on five of these cars. The results, in miles per gallon, are 25.3, 23.6, 24.8, 23.0, and 24.3. What is the median mileage?

Mark your answer in the circles in the grid on your answer sheet.

22. Stock in North American Electric fluctuated in price with a high of $67\frac{3}{4}$ and a low of $63\frac{5}{8}$. Find the difference between the high and the low price.

Mark the answer in the circles in the grid on your answer sheet.

23. A woman bought $4\frac{1}{2}$ yards of ribbon to decorate curtains. If she cut the ribbon into 8 equal pieces, how long was each piece?

Mark your answer in the circles in the grid on your answer sheet.

24. Thomas purchased 1 lb. 4 oz. bananas, 2 lb. 8 oz. apples, 3 lb. peaches, and 3 lb. 6 oz. plums. How much fruit did he buy?

(1) 12 lb. 8 oz.

(2) 10 lb. 2 oz.

(3) 9 lb. 8 oz.

(4) 10 lb. 8 oz.

(5) 9 lb. 2 oz.

25. A discount toy store generally takes $\frac{1}{5}$ off the list price of their merchandise. During the holiday season, an additional 15 percent is taken off the list price. Mrs. Johnson bought a sled listed at $62.00 for Joey and a dollhouse listed at $54.00 for Jill. To the nearest penny, how much did Mrs. Johnson pay?

(1) $92.80
(2) $78.88
(3) $82.64
(4) $75.40
(5) $112.52

Part 2

Directions: You may not return to Part 1 or use your calculator for this part. Bubble in the correct response to each question on Part 2 of your answer sheet.

1. Mona bought 2 packages of ground beef, each weighing 3 pounds 4 ounces, and a package of pork chops weighing 2 pounds 9 ounces. How many pounds of meat did she buy?

(1) 5 pounds 13 ounces
(2) 9 pounds
(3) 9 pounds 1 ounce
(4) 6 pounds 9 ounces
(5) 8 pounds 9 ounces

2. Mr. Jackson has a pine board measuring 8 ft. 5 in. and a redwood board measuring 6 ft. 8 in. How much larger is the pine board than the redwood board?

(1) 1 ft. 9 in.
(2) 2 ft. 3 in.
(3) 2 ft. 9 in.
(4) 2 ft.
(5) 1 ft. 3 in.

3. Find the value of x in the equation $x + 3x + (x+3) = 18$.

(1) 4
(2) 3
(3) 2
(4) 5
(5) 6

4. There were only 8 inches of snowfall in 1993 in Laketown and 24 inches the following year. What is the ratio of snowfall from 1993 to 1994?

(1) 8:24
(2) 24:8
(3) 3:1
(4) 8:32
(5) 1:3

5. A plumber completed five jobs yesterday. On the first job, she earned $36.45, the second $52.80, the third $42.81, the fourth $49.54, and the fifth $48.90. What was the average amount she earned for each job?

- **(1)** $38.50
- **(2)** $39.75
- **(3)** $40.80
- **(4)** $46.10
- **(5)** $42.50

6. Henry's VW Rabbit Diesel gets 50 mpg. Diesel fuel costs an average of $1.10 per gallon. This summer, Henry and his family drove 1,200 miles to Niagara Falls for vacation. How much did Henry pay for fuel?

- **(1)** $52.80
- **(2)** $26.40
- **(3)** $105.60
- **(4)** $132.20
- **(5)** $264.00

7. Lili bought 3 quarts of soda to make punch for a party. If she had 3 pints left over, how much did she use?

- **(1)** 1 quart
- **(2)** 2 quarts
- **(3)** 3 pints
- **(4)** 2 pints
- **(5)** 1 pint

diagnostic test

Top Purchasers of U.S. Exports

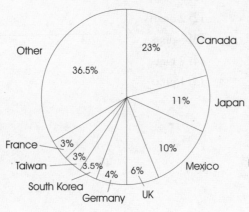

2004
Total U.S. Exports—$510 Billion

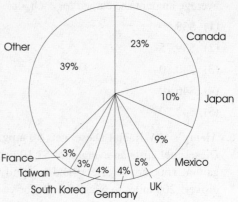

2006
Total U.S. Exports—$575 Billion

8. How much more did Germany spend on U.S. exports than Taiwan in 2006?

    (1) $5.75 billion

    (2) $17.25 billion

    (3) $23 billion

    (4) $2.3 billion

    (5) $575,000,000

9. Approximately how much money did the United Kingdom spend on the purchase of U.S. exports in 2004?

    (1) $31 billion

    (2) $3.1 billion

    (3) $29 billion

    (4) $2.9 billion

    (5) $33 billion

10. How many countries spent more than $25 billion for U.S. exports in 2006?

    (1) 2

    (2) 3

    (3) 8

    (4) 6

    (5) 4

11. Consider the equation $y = 7x - 3$. On the grid on your answer page, mark the y-intercept of this equation.

12. Jackie is leaving from New York on a flight to Los Angeles at 1:30 p.m.. The flight takes 4 hours 30 minutes, and she will gain 3 hours of time by traveling to the West Coast. What time will she arrive in Los Angeles?

    (1) 2:30 p.m.

    (2) 6:00 p.m.

    (3) 5:00 p.m.

    (4) 3:00 p.m.

    (5) 5:30 p.m.

13. On a map, 1 inch represents 3 miles. How many inches are needed to represent a road that is actually 171 miles long?

    (1) 513 inches

    (2) 57 inches

    (3) 121 inches

    (4) 3 inches

    (5) 17 inches

14. A coat that lists for $240 is on sale for $180. By what percent has the coat been discounted?

(1) 40 percent

(2) 35 percent

(3) 25 percent

(4) 30 percent

(5) 20 percent

15. The snail can creep at speeds up to 0.03 mph, but the snail has also been observed to travel as slowly as 0.0036 mph. Find the difference between the snail's fastest and slowest speeds.

Mark your answer in the circles in the grid on your answer sheet.

16. The diameter of a circular playground is 60 feet. If Anthony walks around the playground once, how many feet will he have walked?

(1) 376.8 ft.

(2) 120 ft.

(3) 300.8 ft.

(4) 188.4 ft.

(5) 326.7 ft.

17. In right triangle *DEF* below, what is the value of tan *D*?

Mark your answer in the circles in the grid on your answer sheet.

18. $527(316 + 274)$ has the same value as which of the following?

(1) $316(527) + (274)$

(2) $316 + 527 + 274$

(3) $527(274) + 316(274)$

(4) $316(527) + 316(274)$

(5) $527(316) + 527(274)$

19. Twice the sum of 3 and a number is 1 less than 3 times the number. If the letter N is used to represent the number, which of the following equations could be solved in order to determine the number?

(1) $2(3) + N = 3N - 1$

(2) $2(3 + N) = 3N - 1$

(3) $2(3N) = 3N - 1$

(4) $2(3 + N) - 1 = 3N$

(5) $2(3 + N) = 1 - 3N$

20. Over a six-month period, a certain stock rose 6 points, fell 4 points, fell another 8 points, rose 5 points for each of the next two months, and ended the period by rising another 3 points. What was the total gain for this stock?

(1) 4 points

(2) 2 points

(3) 19 points

(4) 7 points

(5) 6 points

21. The sailfish is built for speed and can swim through the water at speeds of 68 mph. Approximately how many kilometers can it travel in an hour? 1 kilometer = .62 miles.

(1) 109 km/hr.

(2) 68 km/hr.

(3) 96 km/hr.

(4) 110 km/hr.

(5) 42 km/hr.

22. If $x^2 + 5x + 6 = 0$ then $x =$

    (1) +3 and +2

    (2) −3 only

    (3) −2 and −3

    (4) +5 only

    (5) +2 only

23. James plans to cut 58 meters of fencing into 8 pieces of equal length. How long will each piece be?

    (1) 464 meters

    (2) 7.5 meters

    (3) 8 meters

    (4) 64 meters

    (5) 7.25 meters

24. There are 155 children signed up for a class field trip. The number of girls exceeds the number of boys by 17. If B represents the number of boys, which of the following equations could be solved to determine the number of boys signed up for the class trip?

    (1) B + (B − 17) = 155

    (2) B + (B + 17) = 155

    (3) 155 + B = B + 17

    (4) 155 + B = B − 17

    (5) B + (B − 17) = 138

25. A vending machine contains $21 in dimes and nickels. Altogether, there are 305 coins. If N represents the number of nickels, which of the following equations could be solved in order to determine the value of N?

    (1) .10(N − 305) + .5N = 21

    (2) 10(305 − N) + 5N = 21

    (3) .10N + .05(305 − N) = 21

    (4) .10(305 − N) + .05N = 21

    (5) .10(N − 21) + 5N = 305

ANSWER KEY AND EXPLANATIONS

Part 1

1. 1	**6.** 2	**11.** 2	**16.** 2.25	**21.** 24.3
2. 2	**7.** 4	**12.** 3	**17.** 5	**22.** 4.125
3. 5	**8.** 4	**13.** 1	**18.** 4	**23.** .5625
4. 5.40	**9.** 2.20	**14.** 1	**19.** 1	**24.** 2
5. 3	**10.** 3, −4	**15.** 3	**20.** 3	**25.** 4

1. **The correct answer is (1).** Set up a grid to solve the problem. Multiply diagonally with known numbers, and then divide by the third known number.

boxes	40	?
hours	5	3

   $40 \times 3 = 120$

   $120 \div 5 = 24$

2. **The correct answer is (2).** Set up a grid to solve the problem. Multiply diagonally with known numbers, and then divide by the third known number.

pencils	24	4
cost	$4.82	?

   $\$4.82 \times 4 = \$19.28$

   $\$19.28 \div 24 = \$0.81$

3. **The correct answer is (5).** Add $1.39, $1.22, $1.29, and $1.30, then divide by 4 (the number of items you added).

4. A day call would cost $0.34 × 44 minutes + $0.48, or $15.44. If the evening rate discounts the day rates by 35%, an evening call would cost 65% of $15.44, or 0.65 × $15.44 = $10.04. The savings is $15.44 − $10.04, or $5.40. Therefore, the number $5.40 should be coded on your answer sheet, as shown below.

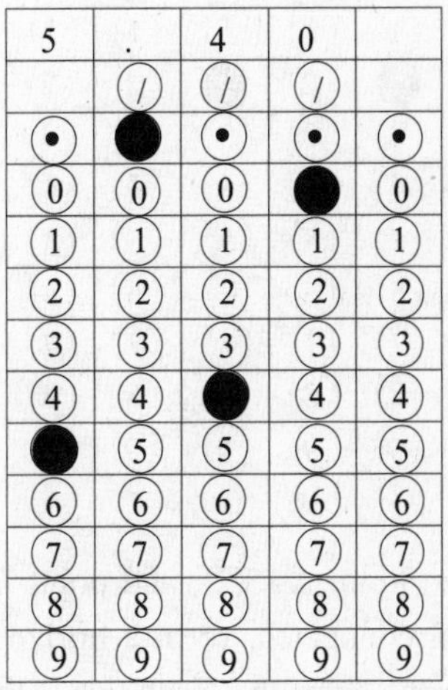

5. **The correct answer is (3).** The square root of 64 is 8 and 8 + 16 = 24.

6. **The correct answer is (2).** Add the amounts of the check written, subtract that number from the original $500.00 in the account, then add the deposit.

7. **The correct answer is (4).** The formula for the volume of a cylinder is $V = \pi r^2 h$. Thus, $V = \pi(3)^2(8) = \pi(9)(8) = 72\pi$.

8. **The correct answer is (4).**

$$-6a + a^2 + 3(a + 6) =$$

$$-6(3) + 9 + 3(9) =$$

$$-18 + 9 + 27 = 18$$

9. If Jonathan drove, he must be an adult, so he paid $3.50 for his ticket and $1.50 for his brother's. His grandmother's ticket cost 80% of $3.50 = $.8 \times 3.50 = \$2.80$. Add the three ticket prices together and get $7.80. Jonathan got $2.20 in change. Therefore, the number $2.20 must be coded on your answer sheet.

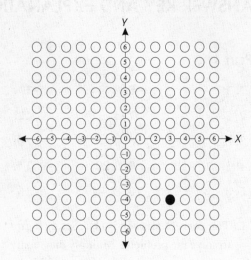

10. When Point A (3,2), and point B (–3,2), when connected, form a horizontal line segment of length 6, each side of the square must be length 6. The missing corner is 6 units below (3,2), which puts it at (3,–4). Therefore, the point (3,–4) must be entered on your answer sheet as shown below.

11. **The correct answer is (2).** $x = 25 - 2(7.50)$. 25 dollars is the original amount. Subtract the amount she spent on the books, $(2)(7.50)$.

12. **The correct answer is (3).** To find the average, add all the values given, then divide by the number of values. To add the values in this problem, first change all the heights to inches:

5'8" = 68"

6'1" = 73"

5'10" = 70

6'3" = 75"

5'9" = 69"

355" ÷ 5" = 71"

71" ÷ 12" = 5'11"

13. **The correct answer is (1).** Add the amounts he spent. $22.50 + $14.95 + $3.98 = $41.43. Subtract this amount from the total credit.

$54.99 – $41.43 = $13.56.

14. **The correct answer is (1).** Divide 34.0 by 3.875 to get 8.98, which is closest to 9 liters.

15. **The correct answer is (3).** Add +25 (the amount the temperature rose) to –15 (the temperature at midnight).

16. Recall that $D = rt$.

Colin and his friends travel the same distance, but Colin travels for 1 hour less than his friends.

Let t = the number of hours Tom and Johnny travel.

Then, $t - 1$ = the number of hours Colin travels.

$45t = 65(t - 1)$

$45t = 65t - 65$

$-20t = -65$

$t = 3\frac{1}{4}$ hours

$t - 1 = 2\frac{1}{4}$ hours. Thus, it will take Colin 2.25 hours to catch up to his friends. The answer must be coded on the answer sheet as shown below:

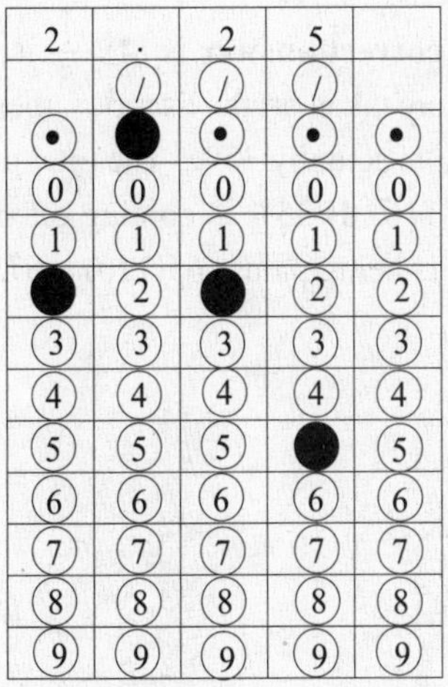

17. The correct answer is (5). Add the yardage gained $4 + 8 = 12$. Add the yardage lost $(-6) + (-4) + (-5) = -15$. Add the gain and the loss, $12 - 15 = -3$. Therefore, they lost 3 yards.

18. The correct answer is (4). Recall that $D = rt$. Since the two ships are moving in opposite directions, we may imagine one ship to be standing still and the other ship moving away from it at their combined rates of 80km/hr. In that case,

$D = rt$

$360 = 80t$

$t = 4\frac{1}{2}$ hours

19. The correct answer is (1). The percent of increase in salary is computed by

$$\frac{61,008 - 30,588}{30,588} = \frac{30,420}{30,588} = 0.9945 =$$
99.45% or 99%.

20. The correct answer is (3). Multiply 1 hour 20 minutes times $7 = 7$ hours 140 minutes. Convert 140 minutes into hours. $140 \div 60 = 2$ hours 20 minutes. Add the converted hours to the answer = 9 hours 20 minutes.

21. The median mileage is simply the mileage in the middle when the 5 mileages are written in numerical order. This number is 24.3. Therefore, 24.3 must be coded on the answer sheet as shown below:

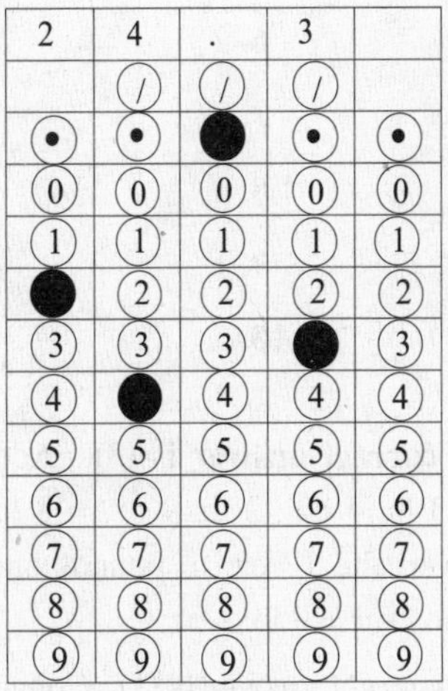

22. $67\frac{3}{4} - 63\frac{5}{8} = 67\frac{6}{8} - 63\frac{5}{8} = 4\frac{1}{8} = 4.125.$

Therefore, the number 4.125 must be coded on the answer sheet as shown below:

4	.	1	2	5
	/	/	/	
•	**/**	•	•	•
0	0	0	0	0
1	1	**1**	1	1
2	2	2	**2**	2
3	3	3	3	3
4	4	4	4	4
5	5	5	5	**5**
6	6	6	6	6
7	7	7	7	7
8	8	8	8	8
9	9	9	9	9

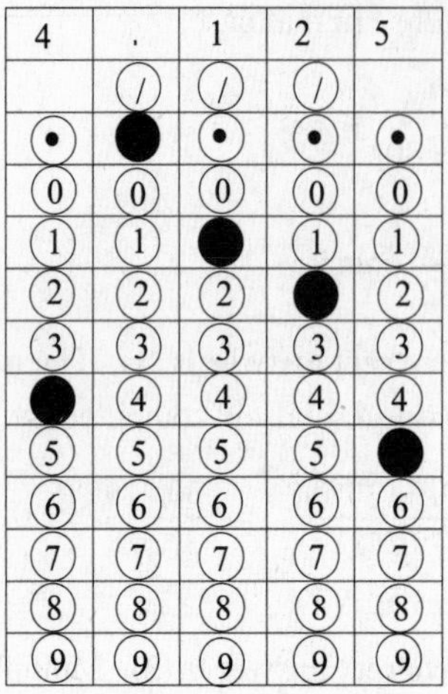

23. Divide $4\frac{1}{2}$ by 8: $4.5 \div 8 = 0.5625$. This answer must be coded onto your answer sheet as shown below.

.	5	6	2	5
	/	/	/	
•	•	•	•	•
0	0	0	0	0
1	1	1	1	1
2	2	2	**2**	2
3	3	3	3	3
4	4	4	4	4
5	**5**	5	5	**5**
6	6	**6**	6	6
7	7	7	7	7
8	8	8	8	8
9	9	9	9	9

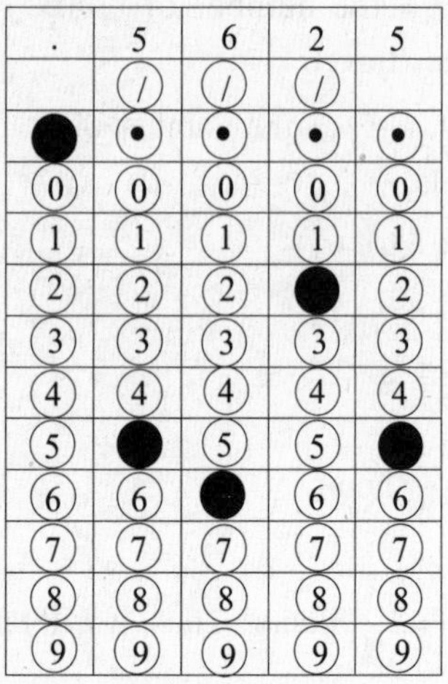

24. The correct answer is (2). 1 lb. 4 oz. + 2 lb. 8 oz. + 3 lb. 6 oz. = 9 lb. 18 oz. Convert to 10 lb. 2 oz.

25. The correct answer is (4). $\frac{1}{5}$ is 20%. If both are taken off the list price, then the discount is actually 35%. The total purchases are $116. If 35% is coming off that, then 65% is being paid. .65(116) = $75.40.

Part 2

1. 3	**6.** 2	**11.** 0, –3	**16.** 4	**21.** 4
2. 1	**7.** 3	**12.** 4	**17.** 2.4	**22.** 3
3. 2	**8.** 1	**13.** 2	**18.** 5	**23.** 5
4. 5	**9.** 1	**14.** 3	**19.** 2	**24.** 2
5. 4	**10.** 5	**15.** .0264	**20.** 4	**25.** 4

1. The correct answer is (3). Add 3 lb. 4 oz., 3 lb. 4 oz., 2 lb. 9 oz. = 8 lb. 17 oz. Convert the 17 oz. to pounds and add to the answer = 9 lb. 1 oz.

2. The correct answer is (1). Convert 8 ft. 5 in. to 7 ft. 17 in. 7 ft. 17 in. – 6 ft. 8 in. = 1 ft. 9 in.

3. The correct answer is (2).

$x + 3x + (x + 3) = 18$

$5x + 3 = 18$

$5x = 18 - 3$

$5x = 15$

$x = \frac{15}{5}$

$x = 3$

4. **The correct answer is (5).** The ratio of snowfall from 1993 to 1994 is 8:24. Reduced to the lowest terms, the ratio is 1:3.

5. **The correct answer is (4).** To find the average, add all the values given, then divide by the number of values.

6. **The correct answer is (2).** 1200 miles ÷ 50 miles = 24 gallons used. Then multiply $1.10 by 24 to get a total cost of $26.40.

7. **The correct answer is (3).** Convert 3 quarts to 6 pints, then subtract 3 pints = 3 pints.

8. **The correct answer is (1).** Germany spent $575 billion × .04 = $23 billion. Taiwan spent $575 billion × .03 = 17.25 billion. The difference is 23 − 17.25 = $5.75 billion.

9. **The correct answer is (1).** The United Kingdom purchased 6% of $510 billion = $510 billion × .06 = $30.6 billion, which is closest to $31 billion.

10. **The correct answer is (5).** The quickest way to answer this problem is to find what percent $25 billion is of $575 billion. Since 25 ÷ 575 = 0.043 = 4.3%. So, any country that purchases more than 4.3% of the U.S. exports would have spent more than $25 billion. There are 4 such countries: Canada, Japan, Mexico, and the United Kingdom.

11. The given equation, $y = 7x - 3$, is already in the slope-intercept form. Therefore, the y-intercept is at −3, that is, at the point (0,−3). The grid should be filled in as shown.

12. **The correct answer is (4).** 1:30 + 4 hr. 30 min. = 6:00. Subtract 3 hours for the time difference. 6:00 − 3 hr. = 3:00.

13. **The correct answer is (2).** A proportion can be established between the scale of the map in inches and the actual distance in miles:

$$\frac{1}{3} = \frac{x}{171}$$
$$3x = 171$$

$x = 57$ inches

14. **The correct answer is (3).** Make a proportion:

$$\frac{180}{240} = \frac{x}{100}$$

$$\frac{3}{4} = \frac{x}{100}$$
$$4x = 300$$

$x = 75\%$

Because x is 75% of the original price, the original price has been reduced by 25%.

15. Since 0.03−0.0036 = 0.0264, this number must be coded on the answer grid, as shown below.

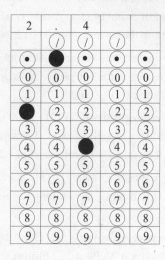

16. The correct answer is (4).

D = πd

C = 3.14 × 60

C = 188.4

17. To begin, find the length of the missing side *EF* by using the Pythagorean Theorem:

$(EF)^2 + 5^2 = 13^2$

$(EF)^2 + 25 = 169$

$(EF)^2 = 144$

$EF = 12$

Since the tangent of an angle is the ratio $\dfrac{opposite}{adjacent}$, compute tan D = $\dfrac{12}{5}$. Since this is equal to 2.4, this number must be coded onto your answer sheet as shown.

18. The correct answer is (5). This is an illustration of the distributive property.

527(316 + 274) = 527(316) + 527(274)

19. The correct answer is (2). The word *sum* indicates addition, so twice the sum of 3 and a number is represented by 2(3 + N). Then, *less than* indicates subtraction, so 1 less than 3 times the number is by 3N − 1.

20. The correct answer is (4). 6 + 5 + 5 + 3 = 19 represents the months in which the stock rose. (−4) + (−8) = (−12) represents the months in which the stock fell. (+19) + (−12) = +7.

21. The correct answer is (4). Since a kilometer is less than a mile, it will travel more than 68 kilometers. Multiplying by a fraction (.62) will give a smaller number, so we must divide $\dfrac{68}{.62}$ = 109.67 = 110 kilometers.

22. The correct answer is (3).

$x^2 + 5x + 6 = 0$

$(x+3)(x+2) = 0$

$x + 3 = 0$ or $x + 2 = 0$

$x = -3$ or $x = -2$

23. The correct answer is (5).

$\dfrac{58}{8}$ = 7.25 meters

24. **The correct answer is (2).** If B represents the number of boys, then the number of girls would be represented by B + 17. Since the number of girls plus the number of boys totals 155, it must be true that B + (B + 17) = 155.

25. **The correct answer is (4).** If N represents the number of nickels, then the number of dimes must be 305–N. Each nickel is worth $0.05, so the total value of the nickels in the machine is .05N. Similarly, the total value of the dimes in the machine is .10(305 – N). The value of the nickels plus the value of the dimes added together is $21, so .10(305 – N) + .05N = 21.

PART III

MATHEMATICS REVIEW

Whole Numbers

OVERVIEW

- Place-value numeration
- Addition of whole numbers
- Differences between whole numbers
- Multiplication of whole numbers
- Dividing whole numbers
- Summing it up

PLACE-VALUE NUMERATION

The system that we use for representing numbers is known as the decimal system of numeration. *Deci* means ten, so indeed our system of writing numerals—the symbols that represent numbers—is based on groupings of tens. A number is an idea of quantity. A numeral is the symbol that we use to represent that idea.

Look at the fingers on both your hands. Get an idea of how many fingers there are. You are now aware of a number. If you are like most people, the number that is in your head is the same as the number that is in mine. If you were Roman, you would represent that number by the symbol *X*. That would be your numeral for the number of fingers that you have. If you were a caveperson, you might lay out ten stones, and that would be your numeral for naming the number of fingers that you have. Most of us would write the symbol 10. That is the decimal numeral for the number of fingers that each of us has. You might think that 10 is a very natural way to represent the number ten. In fact, however, it took thousands of years to develop such a symbol. That symbol was worked on by the Hindus and then by the Arabs, until it came to be what we now recognize as standing for ten objects.

Decimal numeration has several aspects to it. Let us consider them one at a time.

Digits

There are ten digits available to us in the decimal system: Those digits are: 0, 1, 2, 3, 4, 5, 6, 7, 8, 9. A digit is the name given to a single-place numeral. It is not a coincidence that digit is also the name given to a finger or toe. Using the ten digits listed above, it is possible to write any numeral. That is, we may represent any number, no matter how small or how large, by using some combination of those ten digits.

chapter 3

Places

When we write a numeral in the decimal system, we consider more than just the face value of any digit. We must also consider what place the digit holds within the numeral. Look at the chart below, and examine numerals *A, B, C,* and *D.*

A				3
B		3		
C			3	
D	3			

You will notice that *A, B, C,* and *D* are each represented by a single digit 3. They represent four quite different numbers. Were they to be removed from the columns that they are in, zeros would have to be added to *B, C,* and *D* to indicate the place in each numeral where the digit 3 actually belongs. They would be written:

A	3
B	300
C	30
D	3000

There is quite a range between three and three thousand, yet if we do not consider the place in which each digit is written, they are each represented by the same digit, as are thirty and three hundred.

Once more, we turn to the fact that the decimal system is based upon the number ten. That fact tells the relationship of one place to another within any decimal numeral.

...	10×100	10×10	10×1	1	...
A				3	
B		3			
C			3		
D	3				

Notice that in the chart, the column on the right has 1 as its name. That means that any digit written in that column must be multiplied by one to find its true value. All right, there's a three in that column. How much is 3×1? Then numeral *A* is worth 3!

Now, move left one column heading. You will see that the next place is worth ten times as much as the column to its right. Since $10 \times 1 = 10$, any digit written in that place is worth 10 times its face value. What is numeral *C*'s real value?

Move left again. The third column from the right is worth ten times as much as the place to its right.

$10 \times 10 = 100$, so numeral *B*, which is written in that place, actually is worth 3 times 100. You can figure out the value of numeral *D* for yourself. The three dots to the left of the thousands place are there to indicate that as you continue to move to the left one place at a time, each place will be worth ten times as much as the place to its right. The three dots on the extreme right are there as a teaser. This system does not end at the ones place, but we are not going to worry about decimal fractions for a while.

Zero as a Placeholder

We do not normally go around writing numerals on charts that look like the one above. That is why we need to have some way of indicating what place a digit is in, even though it is not written in a column. That is where the idea of a placeholder comes in. A placeholder is a digit which does not have any value of its own, but will fill up the holes left in a numeral when that numeral is written without column headings. Zero does just that. If you look back at numeral *B*, you will notice that there are two empty places to the right of the digit 3. When we write numeral *B* without the columns, we use zeros to fill in, or hold the place of, those empty columns. That is why numeral *B* is written as 300. Two empty places, two placeholders!

Test Yourself

Directions: Look at the chart below. Write the numeral in place-value form (with the appropriate placeholders) on the corresponding line below the chart.

	1,000,000	100,000	10,000	1000	100	10	1
A						5	
B				4			
C			7				
D		6					
E					2		
F			3		5		
G	1		5			2	
H		4		8			1
I	7					3	

A _____

B _____

C _____

D _____

E _____

F _____

G _____

H _____

I _____

Answers

A. 50	**B.** 4000	**C.** 70,000	**D.** 600,000	**E.** 200
F. 30,500	**G.** 1,050,020	**H.** 408,001	**I.** 7,000,030	

Use of Commas

You may have noticed the commas that were used in the last section in the numerals that were greater than a thousand. The purpose of placing commas in large numerals is to make them easier to read. The traditional rule for placement of commas was really quite simple: If a numeral contains four digits or more, start at the right, and for each group of three digits that you count, place a comma, like so:

$$42369178$$

$$\underset{3\,2\,1\,3\,2\,1}{42369178}$$

$$42,369,178$$

"Oh, that's all very nice," you say, "but I notice that you did not write 4000 (above) as 4,000. It has four digits."

I'm glad you noticed that. I did say that the traditional method was quite simple. Unfortunately, there have been some changes lately, both in textbook writing and in thinking on the subject. Some educators are now saying that commas should only be included if the numeral contains five digits or more, although the rule of counting by threes from the right still applies. They would write 4000, but would still write 70,000. Then, there is the system of international units (SI, for short) which prefers that no commas be used at all. Instead, they would group numerals with four or more digits into threes, like so: 4 000 or 70 000 or, for the big one above, 42 369 178.

What to do; what to do?! Do not get upset. As noted before, the whole idea of grouping with commas is to make it easier to read the numeral. Be aware of the fact that there are three different ways in which commas may be used. All of them group the digits into groups of three starting at the right side, and all of them look similar. In this book, we'll follow the second method, but you should feel free to follow either the first or the second. The third method is not widely used in this country at present.

First method	Second method	Third method
5,000	5000	5 000
23,000	23,000	23 000
786	786	786
24,172,000	24,172,000	24 172 000

Reading Large Numerals

If you sometimes find it difficult to read numerals that contain many digits, you are not alone. Many people find large numerals confusing and difficult to read. You may, however, find it interesting to learn that, as is true in so much of mathematics, there is a pattern to the way in which numerals are written in the decimal system. That pattern is formed by the constant repetition of the first three place names: hundreds, tens, and units.

Look at the numeral 432. Read 432 aloud. It is read "four hundred thirty-two." Don't stick an "and" in there. We're saving the "and" for fractions. 432 is read "four hundred thirty-two." Try 687. Did you read it as "six hundred eighty-seven"? Now, notice how the places are arranged:

Hundreds	Tens	Units

We will abbreviate the headings as H, T, and U. (Units is just another way of saying "ones.") The places occupied by the 4, 3, and 2 in 432, and by the 6, 8, and 7 in 687 would look like this:

H	T	U
4	3	2
6	8	7

Now, stay with us for one more round, and you'll see the pattern. Remember, we said that there was going to be a pattern formed by repeating hundreds, tens, and units, or, simply, Hs, Ts, and Us. Now, look at the numeral 531,789. Let's see how it fits under the repeating place-headings:

H	T	U	H	T	U
5	3	1	7	8	9

If it follows the pattern of reading each digit from left to right and saying the name of the column head after it, we should read this numeral "five hundred thirty-one, seven hundred eighty-nine." What's that? You say, "That can't be right." Well, not exactly, but it's very close. There is just one thing missing. Those Hs, Ts, and Us repeat and repeat periodically. All that we need to do is to tell one period from another. Now, the rightmost period is called the ones period. We never bother to say its name. Every other period beside the ones, however, must be named. Now the next period after the ones is the thousands. If we put those period names into place, the numeral will look like this:

Thousands			Ones		
H	T	U	H	T	U
5	3	1 ,	7	8	9

Now try reading it from left to right, but after each period—except the ones—say the period's name. You should have read "five hundred thirty-one thousand, seven hundred eighty-nine." Did you get that? If you did not, try it again. (Psst! Just between us, did you notice how the comma separates the two periods?)

Now try these three:

Thousands			Ones			Write the numeral's name:
H	T	U	H	T	U	
	7	,	8	4	3	_____
3	5	,	6	5	7	_____
4	5	9	,	6	8	2	_____

The first was seven thousand, eight hundred forty-three. The second was thirty-five thousand, six hundred fifty-seven. And the last was four hundred fifty-nine thousand, six hundred eighty-two. If you had trouble with any of those, look them over again. Read the numerals in the leftmost period as if you were reading a three digit numeral; then say the name of the period. Read the next three digits as if it were a three digit numeral, and since it is the ones period, do not say the period name. Get it?

Still Bigger Numerals

Now that you're becoming such an expert at reading those big numerals, we're going to expand your horizon by two more period names. The next period to the left of the thousands is the millions. (You might notice that each period name is a thousand times the previous period's name. A thousand ones make a thousand, a thousand thousands make a million. Do you know what a thousand millions are?) To the left of the millions are the billions:

Billions			Millions			Thousands			Ones		
H	T	U	H	T	U	H	T	U	H	T	U
	4	0	,	0	0	0	,	0	0	0	,	0	0	0

Can you read that number? You might read it as forty billion, no million, no thousand, no ones. The numeral is forty billion.

Test Yourself

Directions: See whether you can identify and name the following numerals.

Billions			Millions			Thousands			Ones				
H	T	U	H	T	U	H	T	U	H	T	U		
					6	,	4	3	0	,	0	0	0	A	_____
			2	0	3	,	0	0	0	,	0	0	0	B	_____
	2	3	,	1	7	5	,	0	0	0	,	0	0	6	C	_____
3	5	7	,	0	0	0	,	2	3	0	,	0	1	7	D	_____
4	0	0	,	0	5	0	,	0	2	0	,	9	0	0	E	_____

Answers

A. 6 million, 430 thousand

B. 203 million

C. 23 billion, 175 million, 6

D. 357 billion, 230 thousand, 17

E. 400 billion, 50 million, 20 thousand, 9 hundred

Now, for the ultimate test of reading large numerals.

Test Yourself

Directions: Below are several large numerals written in decimal form. There is no chart of periods or places. See whether you can name these numerals without referring back to the chart. If you have difficulty, you may refer back to the chart, but first try them on your own.

A. 247,000   _____

B. 5,617,000   _____

C. 25,419,002   _____

D. 138,000,045   _____

E. 7,632,491,228   _____

F. 84,000,720,000   _____

G. 367,591   _____

H. 2,516,794   _____

I. 83,704,003   _____

J. 563,807,004,010   _____

Answers

A. 247 thousand

B. 5 million, 617 thousand

C. 25 million, 419 thousand, 2

D. 138 million, 45

E. 7 billion, 632 million, 491 thousand, 228

F. 84 billion, 720 thousand

G. 367 thousand, 591

H. 2 million, 516 thousand, 794

I. 83 million, 704 thousand, 3

J. 563 billion, 807 million, 4 thousand, 10

ADDITION OF WHOLE NUMBERS

Addition is a combining operation. When you add, you are putting together a number of quantities to get a larger quantity. There are key phrases and words to look for that let you know when a problem or a situation requires addition. Examine the following problem:

> Arthur brought eight guests to the party, Joan brought seven guests, and Michael brought four. Altogether, how many guests were brought to the party?

Reading the problem through once, it is apparent that you are being asked to find a total: the total number of guests brought to the party. That fact is underscored by the problem's use of the word "altogether." Ask yourself whether the total will be more or less than the number of guests any one person brings. It should be apparent that the total will be **more** from the fact that the number of guests that any one person brings is just a part of those that are brought **altogether.** That calls for a combining operation.

The next step is to recognize which combining operation is needed. Are the numbers to be combined the same? If they **are** all the same, then addition **may** be used, but there might also be an alternative.* **If the numbers to be combined are not all the same, then you must add.** Adding $8 + 7 + 4$, you will find that 19 guests were brought to the party.

Now examine the problem below and see whether you can find the key word(s) that tell you what to do:

> What is the sum of 9, 15, and 23?

There is really only one key word: sum. In case you do not know it, sum is the name given to the solution of an addition problem. To find the solution to the problem, therefore, you must add.

We will deal at much greater length with analyzing word problems later in this volume. Now, let us turn our attention to the actual operation of addition.

Finding Sums Less Than Twenty

If you can add up to 18, you can add any two numbers in the decimal system. That is because the largest unique addition that is possible to write in the decimal system is $9 + 9$. Are you skeptical? Then consider the following addition:

$$\begin{array}{r} 342 \\ +527 \\ \hline \end{array}$$

At first glance, it appears that you are being asked to add two numbers in the hundreds. However, if you add in columns, not one of the columns in the addition sums to more than 9:

$$\text{First: } \begin{array}{r} 342 \\ +527 \\ \hline 9 \end{array} \qquad \text{Next: } \begin{array}{r} 342 \\ +527 \\ \hline 69 \end{array} \qquad \text{Finally: } \begin{array}{r} 342 \\ +527 \\ \hline 869 \end{array}$$

So, you think that was pretty sneaky, don't you? "What about 'carrying'?" you ask. Well, I never said that there would never be a time when you had to find a sum greater than 9. I said that the sum would never be greater than 18. That allows for "carrying." (Notice that I keep putting quotation marks around

---

*The alternative is multiplication.

"carrying." That is because it is a very non-mathematical word and will not be used again in this book. It is being used here only because it probably has some meaning to you from the past.)

There are lots of clever shortcuts available in mathematics as well as alternative ways to do certain operations if you are having difficulty doing it the "standard" way—whatever that means. This book is full of such shortcuts and alternatives. However, there is no shortcut or alternative for learning the addition facts through 10. If you do not know every combination that adds up to 10 or less well enough to do the Test Yourself that follows without having to count on your fingers or think about them, you are in trouble. Try these. Time yourself. They should not take you more than one minute.

Test Yourself

Directions: Find the sums to the equations below.

1. $2 + 3 =$	**2.** $4 + 3 =$	**3.** $5 + 2 =$	**4.** $4 + 6 =$
5. $3 + 5 =$	**6.** $6 + 4 =$	**7.** $8 + 2 =$	**8.** $2 + 7 =$
9. $2 + 5 =$	**10.** $5 + 5 =$	**11.** $3 + 4 =$	**12.** $1 + 9 =$
13. $2 + 2 =$	**14.** $4 + 4 =$	**15.** $5 + 3 =$	**16.** $4 + 2 =$
17. $3 + 3 =$	**18.** $2 + 8 =$	**19.** $3 + 6 =$	**20.** $3 + 2 =$
21. $2 + 4 =$	**22.** $3 + 7 =$	**23.** $2 + 6 =$	**24.** $6 + 3 =$
25. $4 + 5 =$	**26.** $6 + 2 =$	**27.** $5 + 4 =$	**28.** $9 + 1 =$

Answers

1. 5	**2.** 7	**3.** 7	**4.** 10	**5.** 8	**6.** 10
7. 10	**8.** 9	**9.** 7	**10.** 10	**11.** 7	**12.** 10
13. 4	**14.** 8	**15.** 8	**16.** 6	**17.** 6	**18.** 10
19. 9	**20.** 5	**21.** 6	**22.** 10	**23.** 8	**24.** 9
25. 9	**26.** 8	**27.** 9	**28.** 10		

If you are thoroughly conversant with addition facts through 10, the exercise above should have taken less than 30 seconds. If you took more than a minute, or if you got any of the answers incorrect, do not go any farther in this book until you make yourself a set of flashcards with the exercises above on them. Learn them so that you can recite the answers in your sleep without hesitation. There are very few things in mathematics that need to be absolutely perfectly committed to memory, but this is one of them.

Once you are thoroughly versed with all the addition facts that sum to 10 or less, it is a relatively simple matter to deal with facts in the teens. Mainly, it is a matter of applying the facts that you already know to derive those that you do not. This is a theme that is repeated over and over again in mathematics. Consider the following sum:

$$8 + 9 = \rule{2cm}{0.4pt}$$

Now, even if you already know what the sum is, follow along with the logic that is involved:

a) I am familiar with all sums to 10, and this is not one of them. It therefore follows that this sum is more than 10.

b) If the sum is more than 10, then it must be 10 and something.

c) Looking at 8 + 9, I know that I must add 2 to the 8 to make 10. The 2 that I add to the 8 must come from the 9.

d) I can rewrite 8 + 9 as 10 + 7. 10 + 7 is 17:

$$\begin{array}{r} 10 \\ +7 \\ \hline 17 \end{array}$$

Therefore, 8 + 9 = 17.

The only tricky part was in step c. One of the numbers must be changed to a 10. It is changed to a 10 by adding whatever must be added to it to make it into a 10. In the case of 8, 2 more was required for it to become 10. Of course, the 2 to be added to the 8 must come from somewhere, and the only place for it to come from is the other number:

$$8 + 9 = (8 + 2) + 7 = 10 + 7 = 17$$

When the 2 is taken from the 9 to add to the 8, then only 7 of the 9 remains. 10 + 7 then makes a total of 17.

This form of addition is known as "grouping to ten." Below, you will find several exercises in grouping to 10.

Test Yourself

> **Directions:** Try the following problems.

1. $7 + 6 = 10 + \underline{\hspace{1cm}}$
2. $6 + 9 = 10 + \underline{\hspace{1cm}}$
3. $5 + 8 = 10 + \underline{\hspace{1cm}}$
4. $8 + 4 = 10 + \underline{\hspace{1cm}}$
5. $9 + 6 = 10 + \underline{\hspace{1cm}}$
6. $7 + 4 = 10 + \underline{\hspace{1cm}}$
7. $3 + 8 = 10 + \underline{\hspace{1cm}}$
8. $9 + 4 = 10 + \underline{\hspace{1cm}}$
9. $5 + 7 = 10 + \underline{\hspace{1cm}}$
10. $4 + 7 = 10 + \underline{\hspace{1cm}}$
11. $7 + 9 = 10 + \underline{\hspace{1cm}}$
12. $8 + 6 = 10 + \underline{\hspace{1cm}}$
13. $6 + 6 = 10 + \underline{\hspace{1cm}}$
14. $8 + 8 = 10 + \underline{\hspace{1cm}}$
15. $9 + 9 = 10 + \underline{\hspace{1cm}}$
16. $4 + 9 = 10 + \underline{\hspace{1cm}}$
17. $7 + 8 = 10 + \underline{\hspace{1cm}}$
18. $6 + 7 = 10 + \underline{\hspace{1cm}}$

Answers

1. 3   2. 5   3. 3   4. 2   5. 5   6. 1
7. 1   8. 3   9. 2   10. 1   11. 6   12. 4
13. 2   14. 6   15. 8   16. 3   17. 5   18. 3

Larger Sums

Have you ever stopped to consider what makes 20 different from 2, or 70 different from 7? For that matter, what makes 5 different from 50, different from 500, or different from 5000? It is only the place that the non-zero digit is in. There are additional placeholders in all these other numerals, but the computational part of all is the same. Here's an example:

$$
\begin{array}{r} 5 \\ + 4 \\ \hline 9 \end{array}
\qquad
\begin{array}{r} 50 \\ + 40 \\ \hline 90 \end{array}
\qquad
\begin{array}{r} 500 \\ + 400 \\ \hline 900 \end{array}
\qquad
\begin{array}{r} 5000 \\ + 4000 \\ \hline 9000 \end{array}
$$

Notice that, other than those extra placeholders, the additions and the sums are in all cases identical. But what happens when the sum is greater than ten? Do you think that the same idea will still apply?

$$
\begin{array}{r} 8 \\ + 9 \\ \hline 17 \end{array}
\qquad
\begin{array}{r} 80 \\ + 90 \\ \hline 170 \end{array}
\qquad
\begin{array}{r} 800 \\ + 900 \\ \hline 1700 \end{array}
\qquad
\begin{array}{r} 8000 \\ + 9000 \\ \hline 17,000 \end{array}
$$

Does that answer the question?

All right, that's all very interesting, but what does it have to do with practical addition? That is a question that will have to be answered in two parts. First, let us consider what is known as "expanded notation." Expanded notation is simply a way of writing numerals so as to spell out the value of the digit in each place.

98 written in expanded form would be 90 + 8.

76 written in expanded form would be 70 + 6.

235 written in expanded form would be 200 + 30 + 5.

In other words, each digit is multiplied by the value of the place that it is in, and then the result is written down. (If you just got lost, look back at the section on place-value at the beginning of this chapter.)

Test Yourself

Directions: Try writing the following place-value numerals in expanded form.

1. 38 _____
2. 59 _____
3. 135 _____
4. 246 _____
5. 1784 _____
6. 2497 _____

Answers

1. 30 + 8
2. 50 + 9
3. 100 + 30 + 5
4. 200 + 40 + 6
5. 1000 + 700 + 80 + 4
6. 2000 + 400 + 90 + 7

Now let us take a look at an addition involving two larger numbers. Suppose we wish to add 346 + 232. First, let us expand the two numerals:

$$346 \qquad\longrightarrow\qquad 300+40+6$$
$$+232 \qquad\qquad\qquad\qquad +200+30+2$$

Next we add the ones together, the tens together, and the hundreds together:

$$300+40+6$$
$$+\,200+30+2$$
$$\overline{500+70+8}$$

Finally, to get the answer back into place-value form (instead of expanded form), we stack up the numerals from the sum in column form and add them together. Since no digit gets added to anything but zeros, the solution is rather straightforward:

$$500$$
$$70$$
$$+\ \ 8$$
$$\overline{578}$$

Test Yourself

Directions: Try the following addition problems.

1. 526 $\longrightarrow$ $500 + 20 + 6$
     $+473$ $\qquad\qquad\qquad$ $+400 + 70 + 3$
     $\qquad\qquad\qquad\qquad$ $\_\_ + \_ + \_\_ = \_\_\_\_$

2. 435 $\longrightarrow$ $400 + 30 + 5$
     $+264$ $\qquad\qquad\qquad$ $+ \_\_ + \_ + \_$
     $\qquad\qquad\qquad\qquad$ $\_\_ + \_ + \_\_ = \_\_\_\_$

3. 173 $\longrightarrow$ $\_\_ + \_ + \_$
     $+514$ $\qquad\qquad\qquad$ $+ \_\_ + \_ + \_$
     $\qquad\qquad\qquad\qquad$ $\_\_ + \_ + \_\_ = \_\_\_\_$

Answers

1. $900 + 90 + 9 = 999$
2. $600 + 90 + 9 = 699$
3. $600 + 80 + 7 = 687$

Now let us try to handle a somewhat more complex addition in expanded form. How about the sum of 584 and 249?

$$\begin{array}{r} 584 \\ +249 \\ \hline \end{array} \longrightarrow \begin{array}{r} 500 + 80 + 4 \\ +200 + 40 + 9 \\ \hline \end{array}$$

First, we'll get those zeros into place, since they are only placeholders and are not involved in the actual addition:

$$\begin{array}{r} 500 + 80 + 4 \\ + 200 + 40 + 9 \\ \hline 00 + \ 0 + \end{array}$$

Next, we'll perform the actual addition of ones, tens, and hundreds:

$$\begin{array}{r} 500 + 80 + 4 \\ + 200 + 40 + 9 \\ \hline 700 + 120 + 13 \end{array}$$

Well now, that 700 looks fine, but the 120 and the 13.... 120 is a mixture of hundreds and tens. We only want tens in that position. The problem in the right hand column is similar. We want only ones in that position, but we have 13—a mixture of tens and ones. Do you see a solution?

Let's take the sum as it is currently written, and then expand the parts of it that are causing the difficulty:

$$\begin{array}{ccccc} 700 & + & 120 & + & 13 \\ \downarrow & & \swarrow\searrow & & \swarrow\searrow \\ 700 & + & (100 + 20) & + & (10 + 3) \end{array}$$

The parentheses above are used to indicate the numerals that were expanded. $120 = 100 + 20$, while $13 = 10 + 3$. Next, we will regroup those numerals so that the hundreds can be combined and the tens can be combined. This time parentheses are used to show the numbers that will be added together:

$$700 + 100 + 20 + 10 + 3$$

$$(700 + 100) + (20 + 10) + 3$$

Adding them, we get:

$$800 + 30 + 3 \text{ which equals } 833.$$

Does this seem to you to be a long and involved process for adding two numbers? It is. Furthermore, it is not a method that is recommended for everyday use. The primary purpose of this technique is, in fact, to give you a better understanding of how numbers can be grouped and regrouped in order to make them more convenient to work with. In other words, if you are not familiar with a particular way of solving a certain problem or doing a particular computation, you may be able to rework the problem to make it "friendlier."

Test Yourself

> **Directions:** Try solving these problems in expanded form.

1.

```
   365
 + 473
```

```
  ___ +  ___ +  ___
+ 400 +  70 +  3
  ___ +  ___ +  ___  = (700 + 100) +  ___ +  ___ = _____
```

2.

```
   486
 + 395
```

```
 ___ + ___ + ___
 ___ + ___ + ___
 ___ + ___ + ___  = _____ = _____
```

3.

```
   139
 + 386
```

```
 _____
 _____
 _____ = _____ = _____
```

4.

```
   574
 + 648
```

```
 _____
 _____
 _____ = _____ = _____
```

Answers

1. $700 + 130 + 8 = (700 + 100) + 30 + 8 = 838$
2. $700 + 170 + 11 = (700 + 100) + (70 + 10) + 1 = 881$
3. $400 + 110 + 15 = (400 + 100) + (10 + 10) + 5 = 525$
4. $1100 + 110 + 12 = 1000 + (100 + 100) + (10 + 10) + 2 = 1222$

Column Addition

No matter how many numbers you may wish to add, it is impossible to add more than two at a time. Try it. Add $4 + 7 + 5$. Did you do it? Did you add all three at the same time? Think about it. If you added the numbers in the order that they are written, then you first added $4 + 7$ and got 11. Then you added 11 and 5 to get 16. You might have added them in a different order, but you definitely added only two at a time. If you do not feel confident about your ability to add long columns of numerals, then you can take advantage of the binary (two at a time) nature of addition. Here is an example:

```
   45                  45                              45
                     + 57                
   57       First:    102     Then:     102            57
                              + 68
   68                          170      Finally:  170  68
                                                 + 34
   34                          204      So:       34
                                                 204
```

That's right, it really works! Unfortunately, we are getting a little ahead of ourselves. We still have to look at the idea of renaming, or regrouping. Consider the following addition:

$$
\begin{array}{r} 39 \\ +47 \\ \hline \end{array}
$$
First we will mark the places:

T	U
3	9
+4	7

Then add up the units:

T	U
3	9
+4	7
	16

Now, as you may notice, 16 is too large a number to be in the units' place. In fact, 16 is a ten and 6 units. The solution is to rename the ten extra units as one ten:

T	U
1	
3	9
+4	7
	6

Then the tens may be added:

T	U
1	
3	9
+4	7
8	6

So:
$$
\begin{array}{r} 39 \\ +47 \\ \hline 86 \end{array}
$$

Notice that we use the word **rename**, not **carry**. That is because renaming is exactly what we are doing. One ten is another name for ten ones. Either way, they are worth the same amount—ten.

H	T	U
	5	6	4
	+2	8	9

First add the units:

H	T	U
5	6	4
+2	8	9
		13

10 of the 13 units must be renamed as 1 ten:

H	T	U
		1
5	6	4
+2	8	9
		3

Next the tens are added together:

H	T	U
	1	
5	6	4
+2	8	9
	15	3

Now there are 10 extra tens. They must be renamed as 1 hundred:

H	T	U
1	1	
5	6	4
+2	8	9
	5	3

Did you understand that last step? If not, look at it again. Any time there are 10 or more in a single column, 10 must be renamed into the next column to the left. There it becomes one of whatever that new column is. For example, 10 ones become 1 ten; 10 tens become 1 hundred; 10 hundreds become 1 thousand ...

The addition is completed by adding up the digits in the hundreds column:

```
  H | T | U
  1 | 1 |
  5 | 6 | 4
+ 2 | 8 | 9
  8 | 5 | 3
```

Test Yourself

Directions: Try the following problems.

1.
```
  H | T | U
    |   |
  4 | 7 | 5
+ 4 | 9 | 7
```

2.
```
  H | T | U
    |   |
  3 | 8 | 9
+ 5 | 7 | 6
```

3.
```
  H | T | U
    |   |
  6 | 7 | 4
+ 2 | 9 | 7
```

4.
```
  H | T | U
    |   |
  5 | 2 | 8
+ 1 | 8 | 5
```

5.
```
  H | T | U
    |   |
  3 | 4 | 7
+ 2 | 7 | 4
```

6.
```
  H | T | U
    |   |
  2 | 8 | 5
+ 5 | 5 | 8
```

7.
```
  H | T | U
    |   |
  1 | 6 | 7
+ 7 | 7 | 4
```

8.
```
  H | T | U
    |   |
  4 | 7 | 3
+ 2 | 2 | 7
```

Answers

1. 972	**2.** 965	**3.** 971	**4.** 713
5. 621	**6.** 843	**7.** 941	**8.** 700

Now you know almost everything there is to know about addition. There is one remaining situation that we have not yet dealt with. It should not cause you any difficulty, but it is essential that we get it out into the open. Consider this:

57   Let us first

99   add the units:

86   Thirty-one?! Do you know how

<u>49</u>   to deal with that many ones?

You may recall that we just finished dealing with the matter of renaming ten of anything as one in the next column to the left (ten ones = one ten, etc.). The same applies to multiples of ten. Thirty-one is one more than thirty, and thirty is a multiple of ten. Other multiples of ten are twenty, forty, fifty, and so forth. Actually, any numeral ending in a zero represents a multiple of ten. There may never be a multiple of ten written in any single place. If a sum contains a ten in a single place, you know to rename it as 1 in the next column to the left. Well, following the same principle, 20 ones would become 2 tens; 30 ones would be 3 tens; 40 ones, 4 tens, etc. With that in mind:

A)

H	T	U
	3	
	5	7
	9	9
	8	6
	4	9
	29	1

B)

H	T	U
2	3	
	5	7
	9	9
	8	6
	4	9
	9	1

C)

H	T	U
2	3	
	5	7
	9	9
	8	6
	4	9
2	9	1

And that is the complete story of renaming. Now it's time for you to try some.

Test Yourself

Directions: Try the following.

1.

H	T	U
	3	4	6
+ 4	9	6

2.

H	T	U
5	4	3
+ 3	8	6

3.

H	T	U
2	5	9
+ 9	7	7

4.

H	T	U
1	8	7
+ 4	6	6

5.

H	T	U
	7	8
	8	7
	6	5
	8	8

6.

H	T	U
	6	9
	5	6
	8	5
	7	4

7.

H	T	U
3	4	7
	5	8
	9	7
	5	3

8.

H	T	U
	9	6
4	7	7
2	8	5
	9	8

9.

H	T	U
3	5	6
9	6	7
8	9	4
	1	5

10.

H	T	U
7	4	2
8	3	0
2	0	5
6	8	4

11.

H	T	U
4	8	0
6	4	8
2	9	2
1	7	0

12.

H	T	U
7	5	8
5	9	6
	8	7
8	6	3

You may add place-value headings to the following if you wish.

13. 3 6 5 + 2 3 8	**14.** 4 1 7 + 3 8 5	**15.** 5 3 6 + 3 7 5	**16.** 2 9 8 + 1 6 3

17. 4 1 5 + 5 9 4	**18.** 7 3 8 2 7 7 4 6	**19.** 6 8 4 7 9 4 3 5	**20.** 9 2 8 7 7 5 6 7 5 7

21. 7 4 6 5 8 3 8 6 8 8	**22.** 5 8 7 3 6 4 7 4 0 5	**23.** 3 6 4 5 9 5 7 3 7 6 4	**24.** 5 9 3 6 8 6 4 0 8 3 4 5 5 9

Answers

1. 842	**2.** 929	**3.** 1236	**4.** 653
5. 318	**6.** 284	**7.** 555	**8.** 956
9. 2232	**10.** 2461	**11.** 1590	**12.** 2304
13. 603	**14.** 802	**15.** 911	**16.** 461
17. 1009	**18.** 278	**19.** 244	**20.** 378
21. 396	**22.** 1183	**23.** 1760	**24.** 2694

If you still feel shaky in your ability to do addition, refer back to the sections that cover the area(s) of which you are unsure. You may also wish to use the Exercise sections for further reinforcement of your addition skills. It is one thing to understand how something works and quite another to be proficient at exercising skill in applying it. If addition was an area with which you needed to work, we suggest that you not put the subject aside just because you have finished this section. Make up additional exercises for yourself from time to time, and check your work using a calculator or a computer.

DIFFERENCES BETWEEN WHOLE NUMBERS

The key word to recognizing when subtraction is to be performed lies in the title of this section. Subtraction is used to find the difference between two quantities. Observe:

Bill is 19 years old, and Susan is 14. What is the difference in their ages?

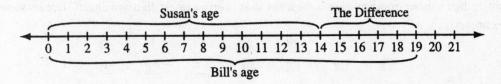

The first brace on the top of the number line shows Susan's age. The brace on the bottom shows Bill's. The second brace on the top indicates the difference in their ages. By counting the spaces between their ages, you will find that the difference is 5. Another way to find the difference in their ages is to subtract 14 from 19: 19 - 14 = 5.

Another key subtraction word is remain, not **remainder**:

Alessandra received a check for $80. She paid a bill for $30. How much of her check remained?

This problem could be solved on the number line, just as the one about Bill's and Susan's ages was, but we would need a very long number line to do it. Since, however, we recognize remained as a word with a similar meaning to difference, indicating that subtraction is called for, we simply subtract: 80 − 30 = 50. Alessandra had $50 left.

Subtraction Facts

We must assume that you already know your addition facts backward and forward, or you would not yet be in this section of the book. Knowing them backward and forward is a good thing, because addition facts backward are subtraction facts. In fact, subtraction is addition backwards. Consider the following "family of facts:"

$$4 + 5 = 9 \qquad 5 + 4 = 9$$
$$9 - 5 = 4 \qquad 9 - 4 = 5$$

If that set of relationships is not quite graphic enough, solve the following "missing addend" relationships:

3 + _____ = 8   5 + _____ = 8   9 + _____ = 17   8 + _____ = 17

The answers are, respectively, 5, 3, 8, and 9. How did you find them? For the first one, did you ask yourself "three plus what equals eight?" Or did you ask yourself "Eight take away three equals what?" Whichever way you did it, you were adding backward. In other words, **you were subtracting**.

Subtracting by Grouping to Ten

Since you are fully conversant with your addition facts through ten, there is no need to go into subtraction facts through ten in any greater detail than has already been done. What makes subtraction even easier than addition is that there is **never** a need to subtract from a number greater than 18. That is because of the way numerals are grouped for subtraction, and because of the fact that subtraction always involves only two numbers at a time. There is no such thing as "column subtraction" the way there is column addition.

Subtraction will always consist of subtracting a single digit number from another, or subtracting a single digit number from a number in the teens so as to give a single digit remainder. Here are some examples*:

$$
\begin{array}{ccccccc}
9 & 7 & 8 & 12 & 15 & 17 & 14 \\
-6 & -4 & -5 & -7 & -8 & -9 & -6
\end{array}
$$

If you are not terribly sure of subtractions from the teens, there is a little trick that you can use. Consider in each case what you would need to take away in order to get to ten. Here is an example:

$$
\begin{array}{ccc}
17 & 17 & \\
-9 & -7 & \text{2 of the 9 remains} \\
& 10 & \text{to be subtracted:}
\end{array}
\qquad
\begin{array}{c}
10 \\
-2 \\
8
\end{array}
\qquad
\text{Therefore:}
\begin{array}{c}
17 \\
-9 \\
8
\end{array}
$$

Here is another example:

$$
\begin{array}{cc}
15 & \text{To get the 15 down to} \\
-8 & \text{10 take away 5:}
\end{array}
\qquad
\begin{array}{c}
15 \\
-5 \\
10
\end{array}
\qquad
\begin{array}{c}
\text{3 of the 8 remains} \\
\text{to be subtracted:}
\end{array}
\qquad
\begin{array}{c}
10 \\
-3 \\
7
\end{array}
$$

$$
\text{Therefore:}
\begin{array}{c}
15 \\
-8 \\
7
\end{array}
$$

If you have difficulty with teen subtractions, try the following Test Yourself quiz. Bear in mind that you must practice this technique until it becomes second nature to you. Remember, when you take your GED examination, speed will count. That is why all the basic techniques must be practiced repeatedly for speed, accuracy, and appropriateness.

---

* The answers to the examples are: 3, 3, 3, 5, 7, 8, and 8.

Test Yourself

Directions: Complete the following.

1.
$$\begin{array}{r} 14 \\ -6 \\ \hline \end{array} \rightarrow \begin{array}{r} 14 \\ -\underline{} \\ \hline 10 \end{array} \rightarrow \begin{array}{r} 10 \\ -\underline{} \\ \hline 8 \end{array} \rightarrow \begin{array}{r} 14 \\ -6 \\ \hline \end{array}$$

2.
$$\begin{array}{r} 13 \\ -7 \\ \hline \end{array} \rightarrow \begin{array}{r} 13 \\ -\underline{} \\ \hline \end{array} \rightarrow \begin{array}{r} 10 \\ -\underline{} \\ \hline \end{array} \rightarrow \begin{array}{r} 13 \\ -7 \\ \hline \end{array}$$

3.
$$\begin{array}{r} 15 \\ -9 \\ \hline \end{array} \rightarrow \begin{array}{r} 15 \\ -\underline{} \\ \hline \end{array} \rightarrow \begin{array}{r} 10 \\ -\underline{} \\ \hline \end{array} \rightarrow \begin{array}{r} 15 \\ -9 \\ \hline \end{array}$$

4.
$$\begin{array}{r} 12 \\ -6 \\ \hline \end{array} \rightarrow \begin{array}{r} 12 \\ -\underline{} \\ \hline \end{array} \rightarrow \begin{array}{r} 10 \\ -\underline{} \\ \hline \end{array} \rightarrow \begin{array}{r} 12 \\ -6 \\ \hline \end{array}$$

5.
$$\begin{array}{r} 16 \\ -9 \\ \hline \end{array} \rightarrow \rightarrow \rightarrow \begin{array}{r} 16 \\ -9 \\ \hline \end{array}$$

6.
$$\begin{array}{r} 18 \\ -9 \\ \hline \end{array} \rightarrow \rightarrow \rightarrow \begin{array}{r} 18 \\ -9 \\ \hline \end{array}$$

Answers

1.
$$\begin{array}{r} 14 \\ -4 \\ \hline 10 \end{array} \rightarrow \begin{array}{r} 10 \\ -2 \\ \hline 8 \end{array} \rightarrow \begin{array}{r} 14 \\ -6 \\ \hline 8 \end{array}$$

2.
$$\begin{array}{r} 13 \\ -3 \\ \hline 10 \end{array} \rightarrow \begin{array}{r} 10 \\ -4 \\ \hline 6 \end{array} \rightarrow \begin{array}{r} 13 \\ -7 \\ \hline 6 \end{array}$$

3.
$$\begin{array}{r} 15 \\ -5 \\ \hline 10 \end{array} \rightarrow \begin{array}{r} 10 \\ -4 \\ \hline 6 \end{array} \rightarrow \begin{array}{r} 15 \\ -9 \\ \hline 6 \end{array}$$

4.
$$\begin{array}{r} 12 \\ -2 \\ \hline 10 \end{array} \rightarrow \begin{array}{r} 10 \\ -4 \\ \hline 6 \end{array} \rightarrow \begin{array}{r} 12 \\ -6 \\ \hline 6 \end{array}$$

5.
$$\begin{array}{r} 16 \\ -6 \\ \hline 10 \end{array} \rightarrow \begin{array}{r} 10 \\ -3 \\ \hline 7 \end{array} \rightarrow \begin{array}{r} 16 \\ -9 \\ \hline 7 \end{array}$$

6.
$$\begin{array}{r} 18 \\ -8 \\ \hline 10 \end{array} \rightarrow \begin{array}{r} 10 \\ -1 \\ \hline 9 \end{array} \rightarrow \begin{array}{r} 18 \\ -9 \\ \hline 9 \end{array}$$

Subtraction without Renaming

The easiest type of subtraction involves no renaming. It is done by moving from the ones place to the left, one place at a time. Here are three examples:

Example 1	345	345	345	345
	-134	-134	-134	-134
		1	11	211
Example 2	435	435	435	435
	-103	-103	-103	-103
		2	32	332
	678	678	678	678
	- 34	- 34	- 34	- 34
		4	44	644

You may have noticed that the main feature of this type of subtraction is that the digit being "taken away" is always smaller than the digit it is being taken away from. That is to say, the top digit in any place is never smaller than the digit below it.

Test Yourself

Directions: Try the subtractions below, just for practice.

1.
   489
   -254

2.
   567
   -165

3.
   794
   -434

4.
   835
   -232

5.
   372
   -140

6.
   758
   -741

7.
   843
   -841

8.
   947
   - 36

9.
   651
   -551

10.
   138
   - 37

Answers

1. 235
2. 402
3. 360
4. 603
5. 232
6. 17
7. 2
8. 911
9. 100
10. 101

Subtraction with Renaming

Look at the following subtraction:
$$\begin{array}{r} 43 \\ -28 \\ \hline \end{array}$$

Try taking the bottom digit in the units' column away from the top digit in the same column, and you will discover that you have a problem. In the realm of natural numbers, there is no way that 8 can be taken away from 3. (In point of fact, 3 - 8 = -5, but we are not yet ready to deal with the realm of negative numbers.)

As the first step in making some sense out of what the solution to this subtraction will be, let us consider the numerals in expanded form:

$$\begin{array}{r} 43 = 40 + 3 \\ -28 = \underline{(20 + 8)} \end{array}$$

Notice that in the expanded form, the top numerals represent a total value of 43—exactly what we started with. The 43, however, is obtained in a configuration that is somewhat different from the original one. In fact, we could configure those numerals in any way that we wished, so long as the total value of the top row remained 43. Let us choose a new configuration that will help us to accomplish what we initially set out to do—to subtract 28 from 43. We will rewrite the expanded numerals as follows:

$$\begin{array}{r} 40 + 3 \;=\; 30 + 13 \\ -(20 + 8) = -(20 + 8) \end{array}$$

Notice that the 40 + 3 has been renamed as 30 + 13. To do so, we simply took one full group of 10 from the 40 and moved it over, combining it with the 3. Notice also that the value of the top row is still 43. Now the subtraction can be accomplished easily:

$$\begin{array}{r} 30 + 13 \\ -(20 + 8) \\ \hline 10 + 5 = 15 \end{array}$$

That means that 43 - 28 = 15.

Did you follow that? If not, look it over again. Now look at the problem that follows:

$$\begin{array}{ccccccccc} 67 & = & 60 + 7 & = & 50 + 17 & = & 50 + 17 & = & 67 \\ -38 & = & -(30 + 8) & = & -(30 + 8) & = & -(30 + 8) & = & -38 \\ & & & & & & \hline 20 + 9 & & \hline 29 \end{array}$$

In keeping with our running commentary, you will notice that the 8 of 38 cannot be subtracted from the 7 of 67. We therefore expanded both the 67 and the 38. Next, the 60 + 7 is renamed by moving a 10 from the 60 over to the 7 and adding. We have actually subtracted 10 from the 60 and added 10 to the 7. By adding and subtracting 10 to the same number, 67, there has been no net change in the value of that number. It still totals 67. Yet, it now has been renamed as 50 + 17. Following that renaming, it is possible to subtract, and we get a difference of 20 + 9, which, when added together makes 29. Hence, 67 - 38 = 29.

NOTE

The minus sign outside the parentheses is there to indicate that both the 20 and the 8 are being subtracted from the 43, even though the 20 and the 8 are related to each other by the plus sign.

Test Yourself

Directions: Here are some subtractions for you to complete.

1. 81 = (80 + 1) = (70 + __)
   -57 = -(50 + 7) = -(50 + 7)
   __ + __ = __

2. 53 = (50 + 3) = (40 + __)
   -26 = -(20 + 6) = -(20 + 6)
   __ + __ = __

3. 74 = (70 + 4) = (__ + __)
   -49 = -(40 + 9) = -(__ + __)
   __ + __ = __

4. 65 = (60 + 5) = (__ + __)
   -27 = -(20 + 7) = -(__ + __)
   __ + __ = __

5. 62 = (__ + __) = (__ + __)
   -33 = -(__ + __) = -(__ + __)
   __ + __ = __

6. 98 = (__ + __) = (__ + __)
   -59 = -(__ + __) = -(__ + __)
   __ + __ = __

7. 31 = (__ + __) = (__ + __)
   -24 = -(__ + __) = -(__ + __)
   __ + __ = __

8. 50 = (__ + __) = (__ + __)
   -26 = -(__ + __) = -(__ + __)
   __ + __ = __

Answers

1. 70 + 11
   -(50 + 7)
   20 + 4 = 24

2. 40 + 13
   -(20 + 6)
   20 + 7 = 27

3. 60 + 14
   -(40 + 9)
   20 + 5 = 25

4. 50 + 15
   -(20 + 7)
   30 + 8 = 38

5. 50 + 12
   -(30 + 3)
   20 + 9 = 29

6. 80 + 18
   -(50 + 9)
   30 + 9 = 39

7. 20 + 11
   -(20 + 4)
   0 + 7 = 7

8. 40 + 10
   -(20 + 6)
   20 + 4 = 24

Subtraction with Two Renamings

Now that you have mastered the technique of renaming to subtract, you should be able to handle any two digit subtraction that may arise. As you may have surmised, however, not all subtractions are two-digit ones. Consider the following example:

$$523$$
$$\underline{-289}$$

If you examine this subtraction, you will notice that as it is now written, both the units' and the tens' places have situations in which the bottom digit is larger than the top one. Let us expand this subtraction, and see what it looks like:

$$523 = \ 500 + 20 + 3$$
$$\underline{-289} = \underline{-(200 + 80 + 9)}$$

(Note once more that the minus sign and parentheses around the bottom, expanded numeral indicate that each part of that numeral is to be subtracted from the quantity above it. The plus signs relate each part of the numeral to the other parts of the same numeral, that is $200 + 80 + 9$ equals 289.)

Now, if we apply the same technique that we used when dealing with two digit numerals, the tens and ones places would be changed as follows:

$$500 \ + \ 20 \ + \ 3 \qquad = \ 500 \ + \ 10 \ + \ 13$$
$$\underline{-(200 \ + \ 80 \ + \ 9)} \qquad = \underline{-(200 \ + \ 80 \ + \ 9)}$$

It is now possible to subtract in the ones place, however the tens place still presents a problem. What should we do about that problem? How about this:

$$500 + 10 + 13 \ = \ \ 400 + 110 + 13$$
$$\underline{-(200 + 80 + \ 9)} = \underline{-(200 + \ 80 + \ 9)}$$

Is that the solution that you thought of? Mind you, it is not the only solution possible, but it is based on the one that we used for two digit numerals. We have done a second renaming, subtracting one of the hundreds from the 500 and adding it back on to the 10, so that the top row now has the same value as it did before. $400 + 110 + 13$ still equals 523. Now it is possible to subtract in all three sections of the expanded notation:

$$400 + 110 + 13$$
$$\underline{-(200 + \ 80 + \ 9)}$$
$$200 + \ 30 + \ 4 = 234$$

Hence: 523
$\underline{-289}$
234

Try this one on your own. Fill in the blanks and then subtract. A detailed solution follows the exercise:

$$647 = (\underline{} + \underline{} + 7) = (\underline{} + \underline{} + 17) = (500 + \underline{} + 17) = 647$$
$$-459 = -(400 + 50 + 9) = -(400 + 50 + 9) = -(400 + 50 + 9) = -459$$
$$\underline{} + \underline{} + \underline{} = \underline{}$$

$$647 = (600 + 40 + 7) = (600 + 30 + 17) = (500 + 130 + 17)$$
$$-459 = -(400 + 50 + 9) = -(400 + 50 + 9) = -(400 + 50 + 9)$$
$$100 + 80 + 8 = 188$$

647 was first expanded to be 600 + 40 + 7. 459 was expanded to 400 + 50 + 9. In the next step, 10 was subtracted from the 40 and added to the 7 to make 600 + 30 + 17. That took care of the problem in the rightmost grouping. Next 100 was subtracted from the 600 and added to the 30 to make 500 + 130 + 17. Check to see that that still adds up to 647. Next, you should have subtracted to get 100 + 80 + 8. Adding that together so as to put it back into standard (place-value) form, you should have found that 647 - 459 = 188.

Test Yourself

Directions: Here are some exercises to try on your own.

1. $345 = (300 + 40 + 5) = (300 + \underline{} + \underline{}) = (\underline{} + \underline{} + 15)$
    $-187 = -(\underline{} + \underline{} + \underline{}) = -(100 + 80 + 7) = -(100 + 80 + 7)$
    $\underline{} + \underline{} + \underline{} = \underline{}$

2. $764 = (\underline{} + \underline{} + \underline{}) = (\underline{} + \underline{} + \underline{}) = (\underline{} + \underline{} + \underline{})$
    $-379 = -(\underline{} + \underline{} + \underline{}) = -(\underline{} + \underline{} + \underline{}) = -(\underline{} + \underline{} + \underline{})$
    $\underline{} + \underline{} + \underline{} = \underline{}$

3. $851 = (\underline{} + \underline{} + \underline{}) = (\underline{} + \underline{} + \underline{}) = (\underline{} + \underline{} + \underline{})$
    $-263 = -(\underline{} + \underline{} + \underline{}) = -(\underline{} + \underline{} + \underline{}) = -(\underline{} + \underline{} + \underline{})$
    $\underline{} + \underline{} + \underline{} = \underline{}$

4. $435 = (\underline{} + \underline{} + \underline{}) = (\underline{} + \underline{} + \underline{}) = (\underline{} + \underline{} + \underline{})$
    $-296 = -(\underline{} + \underline{} + \underline{}) = -(\underline{} + \underline{} + \underline{}) = -(\underline{} + \underline{} + \underline{})$
    $\underline{} + \underline{} + \underline{} = \underline{}$

5. $154 = (\underline{} + \underline{} + \underline{}) = (\underline{} + \underline{} + \underline{}) = (\underline{} + \underline{} + \underline{})$
    $-75 = -(\underline{} + \underline{} + \underline{}) = -(\underline{} + \underline{} + \underline{}) = -(\underline{} + \underline{} + \underline{})$
    $\underline{} + \underline{} + \underline{} = \underline{}$

6. $523 = (\underline{} + \underline{} + \underline{}) = (\underline{} + \underline{} + \underline{}) = (\underline{} + \underline{} + \underline{})$
    $-267 = -(\underline{} + \underline{} + \underline{}) = -(\underline{} + \underline{} + \underline{}) = -(\underline{} + \underline{} + \underline{})$
    $\underline{} + \underline{} + \underline{} = \underline{}$

Answers

1.
$$\begin{array}{r} 200+130+15 \\ -(100+80+7) \\ \hline 100+50+8 = 158 \end{array}$$

2.
$$\begin{array}{r} 600+150+14 \\ -(300+70+9) \\ \hline 300+80+5 = 385 \end{array}$$

3.
$$\begin{array}{r} 700+140+11 \\ -(200+60+3) \\ \hline 500+80+8 = 588 \end{array}$$

4.
$$\begin{array}{r} 300+120+15 \\ -(200+90+6) \\ \hline 100+30+9 = 139 \end{array}$$

5.
$$\begin{array}{r} 0+140+14 \\ -(0+70+5) \\ \hline 0+70+9 = 79 \end{array}$$

6.
$$\begin{array}{r} 400+110+13 \\ -(200+60+7) \\ \hline 200+50+6 = 256 \end{array}$$

Subtraction in Place-Value Form

Everything that you have read and practiced in subtraction up to this point has been preparing you for this section. Expanded form is an excellent way to learn how something works and to gain experience with applying the concept of renaming. Practically speaking, it is too time consuming a process to use for subtraction on a GED examination. You will now see how the technique of renaming can be applied in place-value form, while saving you considerable time and ink (or graphite). Look at the following subtraction:

$$\begin{array}{r} 53 \\ -26 \\ \hline \end{array}$$

Remember, when we rename in addition we often have to exchange ten ones for one ten. When subtracting, the exchange goes the other way. We exchange one ten for ten ones:

(a)
$$\begin{array}{c|c} T & U \\ \hline 5 & 3 \\ -2 & 6 \\ \hline \end{array}$$

(b)
$$\begin{array}{c|c} T & U \\ \hline \overset{4}{\cancel{5}} & 13 \\ -2 & 6 \\ \hline \end{array}$$

(c)
$$\begin{array}{c|c} T & U \\ \hline \overset{4}{\cancel{5}} & 13 \\ -2 & 6 \\ \hline 2 & 7 \end{array}$$

In step (a), Tens and Units headings are placed over each place (as a reminder of the value of the digit beneath). In step (b), one of the 5 tens has been renamed as 10 units and added to the 3 units. That leaves 4 tens in the tens place. Finally, in step (c), we subtract and get a difference of 27.

Try this one.

$$\begin{array}{r} 75 \\ -36 \\ \hline \end{array}$$

$$\begin{array}{c|c} T & U \\ \hline \overset{6}{\cancel{7}} & 5 \\ -3 & 6 \\ \hline \end{array}$$

One ten from the 70 should have been renamed as 10 units and added to the 5 units that are already there. That makes a 15 in the units' column. 15 - 6 = 9. Then, subtracting 3 tens from 6 tens, you should have gotten a remainder of 3 in the tens column. That makes a total remainder of 39.

Test Yourself

> **Directions:** Try the following. (You may add frames and column headings if you wish.)

1. 46	**2.** 67	**3.** 84	**4.** 61	**5.** 93
-27	-38	-49	-26	-58
6. 23	**7.** 54	**8.** 45	**9.** 72	**10.** 88
-18	-16	-37	-27	-49

Answers

1. 19	**2.** 29	**3.** 35	**4.** 35	**5.** 35
6. 5	**7.** 38	**8.** 8	**9.** 45	**10.** 39

Three Places in Place-Value Form

Three-digit numerals in subtraction are handled in place-value form about the same way as two-digit ones. The one difference, of course, is that an extra renaming may be needed. In that case, remember that 10 tens and 1 hundred are interchangeable expressions. Here's an example:

(a)
```
H | T | U
9 | 5 | 4
-6| 8 | 5
```

(b)
```
H |  T  | U
  |  4  |
9 | 5̶  | 14
-6|  8  | 5
```

(c)
```
  8 | 14  |
  H |  T  | U
  9̶ | 5̶  | 14
 -6 |  8  | 5
  2 |  6  | 9
```

Step (a), above, names the subtraction that is to be performed. You will notice the place names (H, T, and U) over each column for reference purposes. Notice that in both the tens' and the units' places, the top digit is less than the bottom digit.

In step (b), one 10 from the five 10s has been renamed as 10 ones. That leaves a 4 in the top line of the "T" column and a 14 in the top line of the "U" column. So far, you will notice that we have done exactly the same thing as when we had only two-digit numerals.

For step (c), we ignore the units' column completely (at least until our renaming is done). To all intents and purposes, we have a two-digit subtraction, but the two digits that we deal with are in the "H" and "T" places. One of the 9 hundreds is renamed as 10 tens. It is then added to the 4 tens that are already there, to make a total of 14 tens. Of course, only 8 hundreds will remain in the "H" place. Finally, we subtract, to get a difference of 269.

Here is another example:

H	T	U
7	3	6
−2	8	9

First we must rename one ten as ten ones:

H	T	U
7	2̸	16
−2	8	9

Next we must rename one hundred as ten tens:

Then subtract:

H	T	U
6̸7	12̸3	16
−2	8	9
4	4	7

You try these:

1.

H	T	U
8	4	6
−4	7	8

2.

H	T	U
5	8	2
−2	9	9

3.

H	T	U
6	1	3
−5	6	4

Solutions:

1.

H	T	U
7̸8	13̸4	16
−4	7	8
3	6	8

2.

H	T	U
4̸5	17̸8	12
−2	9	9
2	8	3

3.

H	T	U
5̸6	10̸1	13
−5	6	4
	4	9

Test Yourself

> **Directions:** The following are provided for your practice. If you wish to add frames and place names, you may. Remember, as you solve each one, not all places necessarily need renaming. Rename only when the lower digit in any place is larger than the top digit in the same place.

1.   5 6 4
    -2 8 5
	2. 3 2 8
	    -1 5 7
		3. 7 2 9
		    -3 8 9
			4. 6 5 3
			    -2 8 8
				5. 8 6 7
				    -1 5 9

6.   4 5 9
    -1 8 9
	7. 8 2 7
	    -3 8 9
		8. 6 2 5
		    -4 6 7
			9. 5 3 2
			    -1 2 2
				10. 2 1 4
				    -1 1 7

Answers

1. 279	2. 171	3. 340	4. 365	5. 708
6. 270	7. 438	8. 158	9. 410	10. 97

Zero in the Top Numeral

One situation that may arise in subtraction remains to be dealt with. That is the situation in which a zero appears in the top numeral. Consider the following three examples:

Example 1
$$650 \\ -286$$

There is no difference in the way you would approach this problem from the way you would approach any place-value subtraction. First, a ten must be renamed as ten ones:

$$\begin{array}{r} {}^{4} \\ 6\ \cancel{5}\ 10 \\ -2\ 8\ 6 \end{array}$$

Then a hundred must be renamed as ten tens:

$$\begin{array}{r} {}^{5}\ {}^{14} \\ \cancel{6}\ \cancel{5}\ 10 \\ -2\ 8\ 6 \end{array}$$

And, finally we subtract:

$$\begin{array}{r} {}^{5}\ {}^{14} \\ \cancel{6}\ \cancel{5}\ 10 \\ -2\ 8\ 6 \\ \hline 3\ 6\ 4 \end{array}$$

Example 2
$$504 \\ -368$$

Here, there is a difference from what we have been doing. That is because, while we need ten ones to add to the 4, there are no tens in the tens place. In order to get some tens in the tens place, we must first rename one hundred as ten tens:

$$\begin{array}{r} {}^{4} \\ \cancel{5}\ 10\ 4 \\ -3\ 6\ 8 \end{array}$$

Now we have ten tens from which to rename one and add it to 4:

$$\begin{array}{r} {}^{4}\ {}^{9} \\ \cancel{5}\ \cancel{10}\ 14 \\ -3\ 6\ 8 \end{array}$$

And, finally we subtract:

$$\begin{array}{r} {}^{4}\ {}^{9} \\ \cancel{5}\ \cancel{10}\ 14 \\ -3\ 6\ 8 \\ \hline 1\ 3\ 6 \end{array}$$

Example 3
$$700 \\ -312$$

The solution to this subtraction is similar to the one used for Example 2. Again, since there are no tens to rename as ones, we must first go to the hundreds and rename a hundred as ten tens:

$$
\begin{array}{r}
{}^{6}\!\!\!\not{7}\ \ 10\ \ 0 \\
-\ 3\ \ 1\ \ 2 \\
\hline
\end{array}
$$

Now there are ten tens from which to rename one as ten ones:

$$
\begin{array}{r}
{}^{6}\!\!\!\not{7}\ \ {}^{9}\!\!\!\not{10}\ \ 10 \\
-\ 3\ \ 1\ \ \ \ 2 \\
\hline
\end{array}
$$

Finally we subtract:

$$
\begin{array}{r}
{}^{6}\!\!\!\not{7}\ \ {}^{9}\!\!\!\not{10}\ \ 10 \\
-\ 3\ \ 1\ \ \ \ 2 \\
\hline
3\ \ 8\ \ \ \ 8 \\
\end{array}
$$

Test Yourself

Directions: Try these for practice.

1.
$$\begin{array}{r}607 \\ -319 \\ \hline\end{array}$$

2.
$$\begin{array}{r}580 \\ -437 \\ \hline\end{array}$$

3.
$$\begin{array}{r}900 \\ -546 \\ \hline\end{array}$$

4.
$$\begin{array}{r}570 \\ -167 \\ \hline\end{array}$$

5.
$$\begin{array}{r}704 \\ -588 \\ \hline\end{array}$$

6.
$$\begin{array}{r}800 \\ -763 \\ \hline\end{array}$$

7.
$$\begin{array}{r}800 \\ -\ 508 \\ \hline\end{array}$$

8.
$$\begin{array}{r}603 \\ -361 \\ \hline\end{array}$$

9.
$$\begin{array}{r}5002 \\ -496 \\ \hline\end{array}$$

10.
$$\begin{array}{r}4050 \\ -1284 \\ \hline\end{array}$$

11.
$$\begin{array}{r}7000 \\ -4173 \\ \hline\end{array}$$

Answers

1. 288   2. 143   3. 354   4. 403   5. 116   6. 37

7. 292   8. 242   9. 4506   10. 2766   11. 2827

MULTIPLICATION OF WHOLE NUMBERS

Multiplication is a combining operation and as such is very close to the other combining operation, addition. In fact, multiplication is shorthand for **repeated addition of the same number**.

$$7 + 7 + 7 + 7 + 7 + 7 = 6 \times 7 = 42$$

Six times seven is another way of writing $7 + 7 + 7 + 7 + 7 + 7$. The first numeral names the number of times the number represented by the second numeral is being added to itself. In this case, 7 is being added to itself six different **times**.

Any problem that can be solved by multiplication can also be solved by addition. Do you think the reverse is true?

Q Martha, Alice, David, Geoffrey, and Erica each have 6 headaches per week. How many headaches do they have altogether in a week?

A Multiplication solution: Addition solution:

$$
\begin{array}{cc}
6 & 6 \\
\underline{\times 5} & 6 \\
30 & 6 \\
 & 6 \\
 & \underline{+6} \\
 & 30
\end{array}
$$

The same number was being added repeatedly. That is why the solution can be obtained by multiplication or addition.

Q Stephanie has $150, Marjorie has $230, and Jonah has $112. How much money do they have altogether?

A Multiplication solution: Addition solution:

$$
\textbf{?}
\qquad
\begin{array}{r}
\$150 \\
230 \\
\underline{+112} \\
\$492
\end{array}
$$

Different numbers are being combined. Multiplication cannot be used to combine different numbers.

Multiplication Facts

The multiplications between 0×0 and 10×10 are known as the multiplication facts. You may have learned them as "times tables." In either case, they are the basis upon which all multiplication is built. You are the judge of how well you know your multiplication facts. To help you to decide how well you know them, the inventory below is provided. Consider it a pretest on multiplication facts. Time yourself while you take it. You should not need more than three minutes to complete the inventory.

1. $1 \times 1 = $ _____ $2 \times 5 = $ _____ $3 \times 3 = $ _____ $4 \times 2 = $ _____ $5 \times 7 = $ _____

2. $5 \times 10 = $ _____ $1 \times 2 = $ _____ $2 \times 6 = $ _____ $3 \times 4 = $ _____ $4 \times 3 = $ _____

3. $4 \times 4 = $ _____ $3 \times 2 = $ _____ $1 \times 3 = $ _____ $2 \times 7 = $ _____ $5 \times 9 = $ _____

4. $3 \times 1 = $ _____ $2 \times 10 = $ _____ $4 \times 10 = $ _____ $1 \times 4 = $ _____ $2 \times 8 = $ _____

5. $2 \times 9 = $ _____ $5 \times 8 = $ _____ $3 \times 5 = $ _____ $4 \times 9 = $ _____ $1 \times 5 = $ _____

6. $1 \times 6 = $ _____ $3 \times 6 = $ _____ $4 \times 5 = $ _____ $5 \times 1 = $ _____ $5 \times 6 = $ _____

7. $3 \times 7 = $ _____ $1 \times 7 = $ _____ $4 \times 1 = $ _____ $5 \times 5 = $ _____ $2 \times 4 = $ _____

8. $5 \times 4 = $ _____ $3 \times 8 = $ _____ $1 \times 8 = $ _____ $2 \times 3 = $ _____ $4 \times 6 = $ _____

9. $4 \times 7 = $ _____ $2 \times 2 = $ _____ $3 \times 9 = $ _____ $1 \times 9 = $ _____ $5 \times 2 = $ _____

10. $2 \times 1 = $ _____ $3 \times 10 = $ _____ $4 \times 8 = $ _____ $5 \times 3 = $ _____ $1 \times 10 = $ _____

11. $9 \times 1 = $ _____ $8 \times 10 = $ _____ $10 \times 5 = $ _____ $7 \times 2 = $ _____ $6 \times 10 = $ _____

12. $8 \times 9 = $ _____ $9 \times 2 = $ _____ $10 \times 6 = $ _____ $6 \times 9 = $ _____ $7 \times 1 = $ _____

13. $7 \times 3 = $ _____ $10 \times 7 = $ _____ $6 \times 8 = $ _____ $9 \times 3 = $ _____ $8 \times 8 = $ _____

14. $9 \times 4 = $ _____ $6 \times 7 = $ _____ $7 \times 4 = $ _____ $8 \times 7 = $ _____ $10 \times 8 = $ _____

15. $6 \times 6 = $ _____ $10 \times 9 = $ _____ $9 \times 5 = $ _____ $7 \times 5 = $ _____ $8 \times 6 = $ _____

16. $10 \times 10 = $ _____ $9 \times 6 = $ _____ $7 \times 6 = $ _____ $8 \times 5 = $ _____ $6 \times 5 = $ _____

17. $7 \times 7 = $ _____ $8 \times 4 = $ _____ $9 \times 7 = $ _____ $6 \times 4 = $ _____ $10 \times 4 = $ _____

18. $8 \times 3 = $ _____ $7 \times 8 = $ _____ $6 \times 3 = $ _____ $10 \times 3 = $ _____ $9 \times 8 = $ _____

19. $9 \times 9 = $ _____ $6 \times 2 = $ _____ $7 \times 9 = $ _____ $8 \times 2 = $ _____ $10 \times 2 = $ _____

20. $6 \times 1 = $ _____ $10 \times 1 = $ _____ $9 \times 10 = $ _____ $7 \times 10 = $ _____ $8 \times 1 = $ _____

Answers

1. 1	10	9	8	35	**2.** 50	2	12	12	12
3. 16	6	3	14	45	**4.** 3	20	40	4	16
5. 18	40	15	36	5	**6.** 6	18	20	5	30
7. 21	7	4	25	8	**8.** 20	24	8	6	24
9. 28	4	27	9	10	**10.** 2	30	32	15	10
11. 9	80	50	14	60	**12.** 72	18	60	54	7
13. 21	70	48	27	64	**14.** 36	42	28	56	80
15. 36	90	45	35	48	**16.** 100	54	42	40	30
17. 49	32	63	24	40	**18.** 24	56	18	30	72
19. 81	12	63	16	20	**20.** 6	10	90	70	8

Traditionalists suggest that it is essential to memorize multiplication facts by rote, so that they are always on the tip of your tongue. While it is not a bad idea to know the tables well, there are alternative methods of learning them—one of which is also quite helpful anytime you happen to forget a multiplication fact and may also prove of assistance in learning to compute more efficiently mentally. We call it two-step multiplication.

Two-Step Multiplication

Remember, multiplication and addition are related. Often, we forget that relationship and fail to take advantage of it. If we know certain multiplication facts, we can put them together to find the ones we do not know. For openers, let us consider the basic—easiest—multiplication tables. From those, we will build the others, using two steps at a time.

The easiest of all multiplication tables is the ones' table. It is a restatement of what mathematicians call the "identity property." Simply stated, the principle says that one times any number is that number. In other words, $1 \times 1 = 1$, $1 \times 2 = 2$, $1 \times 3 = 3$, etc.

Next easiest is the tens' table. To multiply any number by ten, add a zero to the end of the number: $10 \times 3 = 30$, $10 \times 12 = 120$, $10 \times 36 = 360$, for example.

The two times table is commonly known as doubling. If you are not sure of your twos' table, you still probably know how to double already. Simply think of 2x anything as adding the number to itself. $1 + 1$, $2 + 2$, $3 + 3$, $4 + 4$ ... are all examples of doubling, that is, of the twos' times table. To find any member of the twos' table, the alternative method is counting by twos and keeping track of how many twos you have counted. For example, count 2, 4, 6, 8. You've counted by 2 four times, so 8 is 4×2 or 2×4.

The last table you absolutely must know is the fives' table. Once more, counting by fives is as good a way as any to learn and use the fives' table.

IT IS ESSENTIAL THAT YOU KNOW THE 1X, 2X, 5X, AND 10X TABLES!!!

Once you know the 1×, 2×, 5×, and 10× tables, you can develop any other multiplication fact in two steps:

A.
Desired fact:	3 ×8
Rationale:	3 = 2 + 1
Technique:	2 ×8 = 16
	+ (1 ×8) = 8
Therefore:	3 ×8 = 24

B.
Desired fact:	4 ×8
Rationale:	4 = 2 + 2
Technique:	2 ×8 = 16
	+ (2 ×8) = 16
Therefore:	4 ×8 = 32

C.
Desired fact:	6 ×8
Rationale:	6 = 5 + 1
Technique:	5 ×8 = 40
	+ (1 ×8) = 8
Therefore:	6 ×8 = 48

D.
Desired fact:	7 ×8
Rationale:	7 = 5 + 2
Technique:	5 ×8 = 40
	+ (2 ×8) = 16
Therefore:	7 ×8 = 56

Hopefully, you can see from the above that all multiplication facts between 1 ×1 and 7 ×10 can be developed in either a single step or by a simple addition of two numbers. The eight and nine times tables could be developed in the same way—that is, by adding the results of 3 multiplications together; for example, 5x + 2x + 1x makes 8x, or 5x + 2x + 2x makes 9x. However, at the beginning of this section we said you could do any multiplication in two steps; not in two or three steps. Well, why not throw in a little subtraction—which, as we have already seen, is backward addition? Then, we might find that:

E.
Desired fact:	8 ×8
Rationale:	8 = 10 - 2
Technique:	10 ×8 = 80
	- (2 ×8) = 16
Therefore:	8 ×8 = 64

F.
Desired fact:	9 ×8
Rationale:	9 = 10 - 1
Technique:	10 ×8 = 80
	- (1 ×8) = 8
Therefore:	9 ×8 = 72

Two-step multiplication is a means to two ends. It is not an end in itself. The first end that it is a path to is the eventual committing to memory of the multiplication facts. In order for you to succeed with more complex multiplication and with division, it is really essential that you commit the multiplication facts to memory. Otherwise, you will find that the amount of time that you consume in working out each problem requiring multiplication will prevent you from completing the number of problems needed to succeed on the GED examination. This method will permit you to work with multiplication even before you have learned all of your tables. Hopefully, by working with multiplication instead of trying to learn the tables by rote, you will eventually memorize them. The second purpose of two-step multiplication is to better enable you to compute mentally. We will have more to say about that later in this book. The following two-step multiplications are provided for you to practice the skill.

Test Yourself

> **Directions:** Complete the following, using only 1×, 2×, 5×, and 10× tables, or combinations thereof.

1. 4 × 6:   __ × 6 = __
      +(__ × 6) = __
        4 × 6 = __

2. 7 × 7:   __ × 7 = __
      +(__ × 7) = __
        7 × 7 = __

3. 3 × 9:   __ × 9 = __
      + (__ × 9) = __
        3 × 9 = __

4. 8 × 6   __ × 6 = __
      −(__ × 6) = __
        8 × 6 = __

5. 6 × 9:   __ × 9 = __
      +(__ × 9) = __
        6 × 9 = __

6. 9 × 7:   __ × 7 = __
      −(__ × 7) = __
        9 × 7 = __

7. 6 × 8:

8. 4 × 7:

9. 9 × 9:

10. 4 × 9:

11. 8 × 7:

12. 7 × 9:

Answers

1.   2 × 6 = 12
   + (2 × 6) = 12
     4 × 6 = 24

2.   5 × 7 = 35
   + (2 × 7) = 14
     7 × 7 = 49

3.   2 × 9 = 18
   + (1 × 9) = 9
     3 × 9 = 27

4.   10 × 6 = 60
   − (2 × 6) = 12
     8 × 6 = 48

5.   5 × 9 = 45
   + (1 × 9) = 9
     6 × 9 = 54

6.   10 × 7 = 70
   − (1 × 7) = 7
     9 × 7 = 63

7.   5 × 8 = 40
   + (1 × 8) = 8
     6 × 8 = 48

8.   2 × 7 = 14
   + (2 × 7) = 14
     4 × 7 = 28

9.   10 × 9 = 90
   − (1 × 9) = 9
     9 × 9 = 81

10.   2 × 9 = 18
   + (2 × 9) = 18
     4 × 9 = 36

11.   10 × 7 = 70
   − (2 × 7) = 14
     8 × 7 = 56

12.   5 × 9 = 45
   + (2 × 9) = 18
     7 × 9 = 63

Multiplying Decades

Multiples of ten often are called decades. 10, 20, and 30 are examples of decades. You may recall that any number may be multiplied by ten simply by placing a zero at the end of it, for example: $10 \times 4 = 40$ and $10 \times 23 = 230$. Now consider the case of multiplying any decade by a single digit number:

Q $7 \times 30 = ?$

A We know that $30 = 3 \times 10$. Therefore, we may rewrite this multiplication to look like this:

$7 \times (3 \times 10)$

Now, when two or more numbers are being multiplied together, the way they are grouped for multiplication does not affect the result.* We may, therefore, regroup the numerals as follows:

$(7 \times 3) \times 10$

Since $7 \times 3 = 21$, we now have:

21×10

By placing a zero after the 21, we complete the multiplication:

$21 \times 10 = 210$

Therefore, $7 \times 30 = 210$.

Do you see a shortcut that could have been used to solve this problem? Look at the two numerals, 7 and 30. To multiply them, all that had to be done was to multiply the tens digit, 3, by 7, and then put a zero at the end. $7 \times 3 = 21$. Put a zero after the 21 and you have 210.

Q $6 \times 40 = ?$

A Multiply 6×4:

$6 \times 4 = 24$

Then position a zero at the end:

$6 \times 40 = 240$

*This is known technically as the Associative Property for Multiplication. As in addition, or any other arithmetic operation, only two numbers at a time can actually be multiplied. Try multiplying $2 \times 3 \times 2$ in your head, and you will discover that you actually perform two separate multiplications. You will also see that it makes no difference which two of the three numbers you multiply together first.

Test Yourself

Directions: Complete the following.

1. 5 × 80 = _____

2. 4 × 70 = _____

3. 9 × 50 = _____

4. 60
   × 3

5. 40
   × 4

6. 20
   × 5

7. 30
   × 6

8. 70
   × 7

9. 90
   × 8

10. 60
    × 9

Directions: Multiplying by hundreds works the same, but you place two zeros at the end. 7 × 300 = 2100. How do you think you multiply thousands? Complete the following.

11. 300
    × 6

12. 400
    × 8

13. 500
    × 4

14. 600
    × 7

15. 700
    × 3

16. 800
    × 5

17. 900
    × 6

18. 3000
    × 4

19. 4000
    × 9

20. 7000
    × 6

21. 8000
    × 3

22. 6000
    × 7

23. 5000
    × 8

Answers

1. 400
2. 280
3. 450
4. 180
5. 160
6. 100
7. 180
8. 490
9. 720
10. 540
11. 1800
12. 3200
13. 2000
14. 4200
15. 2100
16. 4000
17. 5400
18. 12,000
19. 36,000
20. 42,000
21. 24,000
22. 42,000
23. 40,000

Multiplying Two Digits by One

There are several alternative methods of multiplication that will lead to the desired product (answer). If you now are able to multiply two or more digits by two or more digits, stick with the method that works for you. If your current method of multiplication is not working well for you, then perhaps one of the alternatives examined below will work better.

Follow the model multiplication below:

36
×7   This multiplication means that you must find the total of seven 36s.

Begin by finding seven 6s (i.e., multiply 7 ×6).

36
×7   42 is the result of multiplying 7 ×6. There is still a 30 (from the 36) to be multiplied.
42

Remember, multiplying a decade requires multiplying the tens digit and then placing a zero at the end.

36
×7   This time we will place the zero first.
42
 0

36
×7   Then we multiply 7 ×3 and get 21.
42
210

Finally, we add the two partial products together:

$$
\begin{array}{r}
36 \\
\times 7 \\
\hline
\end{array}
$$

Partial products: 42

210

Final product: 252

Test Yourself

Directions: Complete the following.

1.
```
    5 8
  ×   6
    4 8
  _ _ 0
  _ 4 8
```

2.
```
    3 9
  ×   4
  _   6
  _ 2 _
  _ 5 _
```

3.
```
    6 4
  ×   9
    3 _
  _ _ _
  5 _ _
```

4.
```
    8 7
  ×   3
  _ _
  _ _ _
  _ 6 _
```

5.
```
    7 6
  ×   5
  _ _
  _ _ _
  _ _ _
```

6.
```
    8 3
  ×   7
  _ _
  _ _ _
  _ _ _
```

7.
```
    5 9
  ×   8
  _ _
  _ _ _
  _ _ _
```

8.
```
    2 5
  ×   4
  _ _
  _ _ _
  _ _ _
```

9.
```
    4 4
  ×   6
  _ _
  _ _ _
  _ _ _
```

10.
```
    6 8
  ×   3
  _ _
  _ _ _
  _ _ _
```

11.
```
    9 5
  ×   8
  _ _
  _ _ _
  _ _ _
```

12.
```
    4 8
  ×   7
  _ _
  _ _ _
  _ _ _
```

Answers

1.
```
   48
  300
  348
```

2.
```
   36
  120
  156
```

3.
```
   36
  540
  576
```

4.
```
   21
  240
  261
```

5.
```
   30
  350
  380
```

6.
```
   21
  560
  581
```

7.
```
   72
  400
  472
```

8.
```
   20
   80
  100
```

9.
```
   24
  240
  264
```

10.
```
   24
  180
  204
```

11.
```
   40
  720
  760
```

12.
```
   56
  280
  336
```

Two Digits Times Two Digits

Multiplying by a two-digit numeral requires one step more than multiplying by a one-digit numeral does. The same scheme followed when multiplying two digits by one may be followed. In addition, however, multiplication by the tens' digit must be done. Observe the model:

```
  35
× 28
```

1. First do 8 ×5:
```
  35
× 28
  40
```

2. Then 8 ×30:
```
  35
× 28
  40
 240
```

3. Next comes 20 ×5:
```
  35
× 28
  40
 240
 100
```

4. Last, 20 ×30:
```
  35
× 28
  40
 240
 100
 600
```

5. Finally, add:
```
  35
× 28
  40
 240
 100
 600
 980
```

Test Yourself

Directions: When you feel comfortable with the model, try the following exercises. The notes on the side are there for your convenience.

1.

```
     4  6
  ×  3  4
    2  4 = 4 × 6
  _  _  0 = 4 × 40
  _  _  _ = 30 × 6
_  _  0  0 = 30 × 40
_  _  _  _
```

2.

```
     5  9
  ×  2  5
  _  _   = 5 × 9
  _  _  0 = 5 × 50
  _  _  _ = 20 × 9
_  _  0  0 = 20 × 50
_  _  _  _
```

3.

```
     6  3
  ×  2  9
  _  _   = 9 × 3
  _  _  0 = 9 × 60
  _  _  _ = 20 × 3
_  _  0  0 = 20 × 60
_  _  _  _
```

4.

```
     7  4
  ×  3  7
  _  _   = 7 × 4
  _  _   = 7 × 70
  _  _   = 30 × 4
_  _  _  _ = 30 × 70
_  _  _  _
```

5.

```
     8  5
  ×  1  9
  _  _   = 9 × 5
  _  _  _ = 9 × 80
  _  _  _ = 10 × 5
_  _  _  _ = 10 × 80
_  _  _
```

6.

```
     9  7
  ×  6  3
  _  _   = 3 × 7
  _  _  _ = 3 × 90
  _  _  _ = 60 × 7
_  _  _  _ = 60 × 90
_  _  _  _
```

7.

```
     3  9
  ×  2  6
  _  _   = 6 × _
  _  _  _ = 6 × _
  _  _  _ = _ × _
_  _  _  _ = _ × _
_  _  _  _
```

8.

```
     5  4
  ×  3  7
  _  _   = 7 × _
  _  _  _ = 7 × _
  _  _  _ = _ × _
_  _  _  _ = _ × _
_  _  _  _
```

9.

```
     8  2
  ×  5  8
  _  _   = 8 × _
  _  _  _ = 8 × _
  _  _  _ = _ × _
_  _  _  _ = _ × _
_  _  _  _
```

Answers

1.
```
  24
 160
 180
1200
1564
```

2.
```
  45
 250
 180
1000
1475
```

3.
```
  27
 540
  60
1200
1827
```

4.
```
  28
 490
 120
2100
2738
```

5.
```
  45
 720
  50
 800
1615
```

6.
```
  21
 270
 420
5400
6111
```

7.
```
  54
 180
 180
 600
1014
```

8.
```
  28
 350
 120
1500
1998
```

9.
```
  16
 640
 100
4000
4756
```

Streamlining Two-Digit Multiplication

The exercises just done were provided to help you get a better understanding of how two-digit multiplication works. In practice, shortcuts are available and can be easily used. The main shortcut favored in two-digit multiplication consists of performing multiplication by each digit in a single line.

Observe:

$$\begin{array}{r} 35 \\ \times 28 \end{array}$$

1. Multiply 8×5, but do not write the tens' digit in the answer:

$$\begin{array}{r} 4 \\ 35 \\ \times 28 \\ \hline 0 \end{array}$$

Note where the tens' digit from $8 \times 5 = 40$ is placed.

2. Multiply 8×3, and add the regrouped 4 to the product:

$$\begin{array}{r} 4 \\ 35 \\ \times 28 \\ \hline 280 \end{array}$$

Note: $8 \times 3 = 24$. Add 4, and get 28.

3. Place a zero in the ones' place before multiplying by 2 (tens):

$$\begin{array}{r} 35 \\ \times 28 \\ \hline 280 \\ 0 \end{array}$$

Since we next multiply by 2 tens, there will be no ones in the answer.

4. Multiply 2×5: Notice where the ones' and tens' digits are placed.

$$\begin{array}{r} 1 \\ 35 \\ \times 28 \\ \hline 280 \\ 00 \end{array}$$

$2 \times 5 = 10$

5. Multiply 2×3, and add the regrouped 1:

$$\begin{array}{r} 1 \\ 35 \\ \times 28 \\ \hline 280 \\ 700 \end{array}$$

$2 \times 3 = 6; 6 + 1 = 7$

6. Add the partial products:

$$\begin{array}{r} 35 \\ \times 28 \\ \hline 280 \\ 700 \\ \hline 980 \end{array}$$

Add

See whether you can follow this one.

$$
\begin{array}{r}
5\ 8 \\
\times\ 4\ 6 \\
\end{array}
\qquad
\begin{array}{r}
^{4}\ \ \\
5\ \ 8 \\
\times\ 4\ \ 6 \\
\hline
8 \\
\end{array}
$$

Note that the 4 "carried" from 6×8 is added to the product of 6×5 (30).

Before multiplying by 4 tens, place the "0."

$$
\begin{array}{r}
58 \\
\times 46 \\
\hline
348 \\
0 \\
\end{array}
\qquad
\begin{array}{r}
^{4}\ \\
5\ \ 8 \\
\times\ 4\ \ 6 \\
\hline
3\ 4\ 8 \\
\end{array}
\qquad
\begin{array}{r}
^{3}\ \\
5\ \ 8 \\
\times\ 4\ \ 6 \\
\hline
3\ 4\ 8 \\
2\ 0\ \ \\
\end{array}
\qquad
\begin{array}{r}
^{3}\ \\
5\ \ 8 \\
\times\ 4\ \ 6 \\
\hline
3\ 4\ 8 \\
2\ 3\ 2\ 0 \\
\hline
2\ 6\ 6\ 8 \\
\end{array}
$$

Notice that each time you multiply (be it by the ones or the tens digit) the top numeral is being treated as a whole entity.

Test Yourself

Directions: Try completing the multiplications below.

1.
$$
\begin{array}{r}
5\ 7 \\
\times\ 3\ 4 \\
\hline
2\ 2\ 8 \\
\_\ \_\ \_\ 0 \\
\hline
1\ \_\ 3\ 8 \\
\end{array}
$$

2.
$$
\begin{array}{r}
6\ 4 \\
\times\ 4\ 6 \\
\hline
3\ \_\ 4 \\
\_\ 5\ \_\ 0 \\
\hline
2\ 9\ \_\ 4 \\
\end{array}
$$

3.
$$
\begin{array}{r}
9\ 8 \\
\times\ 2\ 7 \\
\hline
\_\ 8\ \_ \\
\_\ 9\ \_\ 0 \\
\hline
\_\ 6\ \_\ \_ \\
\end{array}
$$

4.
$$
\begin{array}{r}
8\ 3 \\
\times\ 4\ 9 \\
\hline
7\ \_\ 7 \\
\_\ \_\ 2\ \_ \\
\hline
4\ \_\ \_\ \_ \\
\end{array}
$$

Answers

1.
$$
\begin{array}{r}
57 \\
\times 34 \\
\hline
228 \\
\underline{1710} \\
1938 \\
\end{array}
$$

2.
$$
\begin{array}{r}
64 \\
\times 46 \\
\hline
384 \\
\underline{2560} \\
2944 \\
\end{array}
$$

3.
$$
\begin{array}{r}
98 \\
\times 27 \\
\hline
686 \\
\underline{1960} \\
2646 \\
\end{array}
$$

4.
$$
\begin{array}{r}
83 \\
\times 49 \\
\hline
747 \\
\underline{3320} \\
4067 \\
\end{array}
$$

Test Yourself

Directions: Complete the following.

1. 86 ×18	2. 69 ×18	3. 65 ×53	4. 97 ×90	5. 86 ×27
6. 25 ×67	7. 90 ×37	8. 72 ×15	9. 91 ×26	10. 33 ×70
11. 82 ×73	12. 86 ×69	13. 32 ×23	14. 40 ×29	15. 84 ×66
16. 75 ×24	17. 44 ×26	18. 94 ×74	19. 91 ×41	20. 40 ×10
21. 70 ×41	22. 31 ×19	23. 99 ×39	24. 73 ×17	25. 83 ×48
26. 48 ×33	27. 59 ×98	28. 94 ×81	29. 61 ×34	30. 66 ×23
31. 53 ×14	32. 50 ×97	33. 62 ×34	34. 85 ×50	35. 41 ×23

Answers

1. 1548	2. 1242	3. 3445	4. 8730	5. 2322
6. 1675	7. 3330	8. 1080	9. 2366	10. 2310
11. 5986	12. 5934	13. 736	14. 1160	15. 5544
16. 1800	17. 1144	18. 6956	19. 3731	20. 400
21. 2870	22. 589	23. 3861	24. 1241	25. 3984
26. 1584	27. 5782	28. 7614	29. 2074	30. 1518
31. 742	32. 4850	33. 2108	34. 4250	35. 943

Multiplying Larger Numbers

Three-digit multiplication works in almost exactly the same way as two-digit multiplication. Since there are three digits in the multiplier (the bottom numeral), there will be three lines of partial products before the final addition. Also, as the second line in a two-digit multiplication begins with the placing of the zero in the ones' place, the third line begins with the placing of two zeros. Since the third line is multiplication by hundreds, there will be zero ones and zero tens:

Example

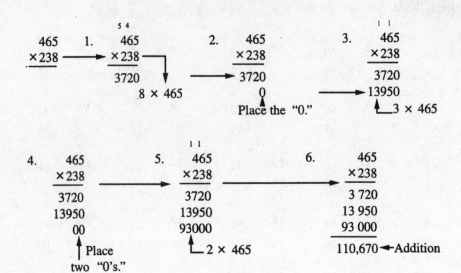

Test Yourself

Directions: Try these. Refer to the example above if necessary.

1.	447 ×265	**2.**	970 ×206	**3.**	834 ×294	**4.**	207 ×118
5.	317 ×186	**6.**	785 ×583	**7.**	380 ×286	**8.**	855 ×703
9.	822 ×480	**10.**	757 ×485	**11.**	576 ×167	**12.**	610 ×112
13.	438 ×254	**14.**	485 ×159	**15.**	279 ×265	**16.**	489 ×296

Answers

1.	118,455	**2.**	199,820	**3.**	245,196	**4.**	24,426
5.	58,962	**6.**	457,655	**7.**	108,680	**8.**	601,065
9.	394,560	**10.**	367,145	**11.**	96,192	**12.**	68,320
13.	111,252	**14.**	77,115	**15.**	73,935	**16.**	144,744

DIVIDING WHOLE NUMBERS

Addition and multiplication are combining operations. Both combine smaller quantities to make larger quantities. Add two dollars to five dollars, and you get seven dollars, a quantity larger than either two or five. If you have four seven-dollar checks, to find out the total amount of money that you have, you may multiply four by seven. Four times seven is twenty-eight, so you have a total of twenty-eight dollars—an amount larger than either the four or the seven that you began with.

Since mathematics is nothing if not logical, it makes sense that there should be two operations to undo the combining made possible by addition and multiplication. Those "taking apart" operations are subtraction and division. With both subtraction and division of whole numbers, you start out with a quantity, and you end up with a smaller one. The operation of subtraction was already explored earlier in this chapter. Let us now take a look at division.

Division is the operation with which most students of whole number arithmetic have the most difficulty. That difficulty occurs most often from a lack of understanding of division. In fact, division is related to two other operations, subtraction and multiplication. Division may be defined as the undoing of multiplication. For example, if 3 times 5 makes 15, then 15 divided by 5 equals 3. The sign, $\div$, is mathematical shorthand for "divided by."

Below you will find a series of exercises designed to emphasize and demonstrate the relationship between multiplication and division. Study the model example first, to make sure that you understand what to do, then try the exercises.

Example

$$3 \times 4 = 12 \qquad\qquad 4 \times 3 = 12$$
$$12 \div 4 = 3 \qquad\qquad 12 \div 3 = 4$$

Test Yourself

Directions: Complete the following.

1. $5 \times 6 = 30 \qquad 6 \times 5 = 30$
   $30 \div 6 = \underline{\hspace{0.6cm}} \qquad 30 \div 5 = \underline{\hspace{0.6cm}}$

2. $7 \times 9 = 63 \qquad 9 \times 7 = 63$
   $63 \div 9 = \underline{\hspace{0.6cm}} \qquad 63 \div 7 = \underline{\hspace{0.6cm}}$

3. $8 \times 7 = 56 \qquad 7 \times 8 = 56$
   $56 \div 7 = \underline{\hspace{0.6cm}} \qquad 56 \div 8 = \underline{\hspace{0.6cm}}$

4. $4 \times 6 = 24 \qquad 6 \times 4 = 24$
   $24 \div 4 = \underline{\hspace{0.6cm}} \qquad 24 \div 6 = \underline{\hspace{0.6cm}}$

5. $3 \times 9 = \underline{\hspace{0.6cm}} \qquad 9 \times 3 = \underline{\hspace{0.6cm}}$
   $27 \div 9 = \underline{\hspace{0.6cm}} \qquad 27 \div 3 = \underline{\hspace{0.6cm}}$

6. $2 \times 8 = \underline{\hspace{0.6cm}} \qquad 8 \times 2 = \underline{\hspace{0.6cm}}$
   $16 \div 8 = \underline{\hspace{0.6cm}} \qquad 16 \div 2 = \underline{\hspace{0.6cm}}$

7. $6 \times 7 = \underline{\hspace{0.6cm}} \qquad 7 \times 6 = \underline{\hspace{0.6cm}}$
   $42 \div 7 = \underline{\hspace{0.6cm}} \qquad 42 \div 6 = \underline{\hspace{0.6cm}}$

8. $5 \times 9 = \underline{\hspace{0.6cm}} \qquad 9 \times 5 = \underline{\hspace{0.6cm}}$
   $45 \div 9 = \underline{\hspace{0.6cm}} \qquad 45 \div 5 = \underline{\hspace{0.6cm}}$

9. $3 \times 7 = \underline{\hspace{0.6cm}} \qquad 7 \times 3 = \underline{\hspace{0.6cm}}$
   $21 \div 7 = \underline{\hspace{0.6cm}} \qquad 21 \div 3 = \underline{\hspace{0.6cm}}$

10. $6 \times 8 = \underline{\hspace{0.6cm}} \qquad 8 \times 6 = \underline{\hspace{0.6cm}}$
    $48 \div 8 = \underline{\hspace{0.6cm}} \qquad 48 \div 6 = \underline{\hspace{0.6cm}}$

Answers

1. 5, 6	**2.** 7, 9	**3.** 8, 7	**4.** 6, 4	**5.** 27, 27, 3, 9

6. 16, 16, 2, 8   **7.** 42, 42, 6, 7   **8.** 45, 45, 5, 9   **9.** 21, 21, 3, 7   **10.** 48, 48, 6, 8

Division also bears the same relationship to subtraction as multiplication does to addition. Multiplication, you may recall, is an operation for performing repeated addition of the same number. Division is an operation for accomplishing repeated subtraction of the same number. The problem $42 \div 6$ may be thought of as asking "How many times can 6 be subtracted from 42?"

$$
\begin{array}{r}
42 \div 6 = \quad 42 \\
\underline{-6} \\
36 \\
\underline{-6} \\
30 \\
\underline{-6} \\
24 \\
\underline{-6} \\
18 \\
\underline{-6} \\
12 \\
\underline{-6} \\
6 \\
\underline{-6} \\
0
\end{array}
$$

Count the number of subtracted 6s.

Count the number of 6s that were subtracted, and you will find that there are seven of them. Therefore, $42 \div 6 = 7$.

Division may be thought of as separating an amount of things into one or more equal groups. So, for example, suppose a woman has 45 poker chips and she wishes to place five equal bets. To find out how many chips must be bet each time, we would divide 45 by 5. Since $5 \times 9 = 45$, $45 \div 5 = 9$.

There are many different techniques for performing division. If you are already familiar with one that works for you, continue to use it. The exercises will give you the chance to polish your skills with the operation. If, however, you are not comfortable with division of larger numbers, you may find a method here that will work better for you than the method you once learned.

Before launching into the different forms of division, we should say a word or two about remainders. A remainder is what you get when one number does not divide into another an exact number of times.* For example, when 7 is divided by 2, one rapidly discovers that three 2s fit into 7, and 1 will be left over. That 1 is called the remainder. Representing a remainder as a fraction was taught as the "grown-up way" to treat remainders, and from about 4th grade on, remainders were always expressed as fractions. A fractional remainder may be formed by placing the remainder above the number being divided by (otherwise known as the divisor). Observe:

$$\begin{array}{cc} 3\ \text{R1}^{**} & 3\frac{1}{2} \\ 2\overline{)7} \quad \text{or} \quad & 2\overline{)7} \end{array}$$

The so-called "grown-up" notion of expressing remainders exclusively as fractions is no longer favored. Rather, the decision to represent a remainder as a fraction or to leave it in the "R" form must depend upon the wording of the problem that is being solved. Here are two problems with identical numbers. Read and solve each:

1. An electrician wishes to cut a 38 foot length of wire into four equal parts. How long must each part be?

2. 38 pupils are to be seated equally at 4 tables. How many pupils will sit at each table?

---

* Just to keep the record straight, you may remember "remainders" as the answers in subtraction problems. That meaning has in no way changed. The remainder in a division has not been divided, but was subtracted from the divisible part. This should become clearer as we go on.

** The word "remainder" is usually abbreviated by an R, hence this answer would be read: "Three, remainder 1."

In problem 1, it is stated that the wire must be cut into four equal lengths. Now 38 divided by 4 gives a result of 9, Remainder 2. We cannot, however, have four 9 foot lengths of wire and a 2 foot length and still satisfy the conditions of the problem. It is therefore necessary to express the remainder as a fraction, so that each length will be $9\dfrac{2}{4}$, or $9\dfrac{1}{2}$ feet long:

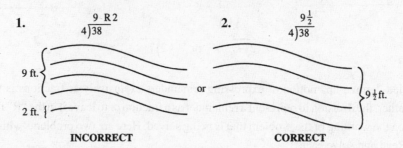

The figures in problem 2, as previously stated, are identical to those in problem 1. But, there is a difference. One cannot cut pupils in half in order to get an equal number at each table. 9, R2 is then the required solution, meaning that 9 students will sit at each table, and 2 will remain standing.

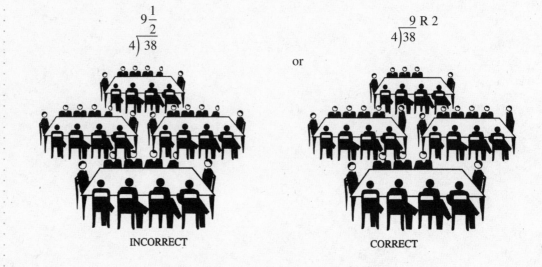

Dividing Two Digits by One

Any form of division, except for the basic division facts (which are, of course, the multiplication facts backward) consists of repeating three steps:

❶ Divide

❷ Multiply

❸ Subtract

Sometimes, the division and multiplication steps are so closely combined that it is difficult to recognize that two separate steps have taken place, but they always do. Consider the following division:

$4\overline{)26}$ Which may be read "26 divided by 4 equals what?"

or

"How many 4s are there in 26?"

In order to find the solution, the following three steps must be done:

❶ **Divide** 26 by 4: Ask yourself how many 4s there are in 26, or "What is the multiple of 4 nearest to 26 without going over 26 (i.e., 4, 8, 12, 16, 20, 24, 28, 32, 36, 40)?" Obviously, it is 24. You will discover that 6 4s fit into 26. Write the 6 above the bracket (in the ones' place). This is called the quotient.

$$\begin{array}{r} 6 \\ 4\overline{)26} \end{array}$$

❷ **Multiply** the number in the quotient (the 6) times the divisor

$$\begin{array}{r} 6 \\ 4\overline{)26} \\ \underline{24} \leftarrow 6 \times 4 = 24 \end{array}$$

❸ **Subtract** to find the remainder (if any).

$$\begin{array}{r} 6 \\ 4\overline{)26} \\ \underline{-24} \\ 2 \end{array}$$

You should write the remainder as shown:

$$\begin{array}{r} 6 \ \ R2 \\ 4\overline{)26} \\ \underline{24} \\ 2 \end{array}$$

The multiplication and division that took place in steps one and two are actually so intermingled that it is at times difficult to separate one from the other. Nevertheless, both are occurring. The subtraction, on the other hand, is easy to see.

Follow the steps in the example below:

$$8 \overline{)35} \longrightarrow \underset{8 \times \underline{\ } = 35?}{8 \overline{)35}} \longrightarrow \underset{8 \times 4 = 32}{8 \overline{)35}} \longrightarrow \overset{4}{8 \overline{)35}} \longrightarrow \overset{4}{\underset{= 32}{8 \overline{)35}}} \longrightarrow \overset{4}{\underset{-32}{8 \overline{)35}}} \longrightarrow \overset{4\,R3}{\underset{\;3}{8 \overline{)35}}}$$

Again, notice the steps: **Divide, Multiply, Subtract.**

Test Yourself

Directions: Domplete the following.

1.	$2\overline{)11}$	**2.**	$3\overline{)14}$	**3.**	$5\overline{)29}$	**4.**	$4\overline{)27}$	**5.**	$7\overline{)56}$
6.	$6\overline{)51}$	**7.**	$9\overline{)43}$	**8.**	$5\overline{)38}$	**9.**	$6\overline{)54}$	**10.**	$8\overline{)71}$
11.	$6\overline{)49}$	**12.**	$5\overline{)46}$	**13.**	$4\overline{)35}$	**14.**	$7\overline{)53}$	**15.**	$6\overline{)19}$
16.	$8\overline{)45}$	**17.**	$3\overline{)22}$	**18.**	$9\overline{)84}$	**19.**	$4\overline{)39}$	**20.**	$8\overline{)74}$

Answers

1.	5, R1	**2.**	4, R2	**3.**	5, R4	**4.**	6, R3	**5.**	8
6.	8, R3	**7.**	4, R7	**8.**	7, R3	**9.**	9	**10.**	8, R7
11.	8, R1	**12.**	9, R1	**13.**	8, R3	**14.**	7, R4	**15.**	3, R1
16.	5, R5	**17.**	7, R1	**18.**	9, R3	**19.**	9, R3	**20.**	9, R2

Adding a Fourth Step to the Division Process

As you have probably suspected, not all divisions are going to work out quite as conveniently as the ones above. However, no matter what the division, the three steps of divide, multiply, and subtract will still apply. Sometimes, it will often be necessary to add a fourth step. That fourth step is known as "bringing down the next digit." Bringing down is necessitated in divisions such as in the model example that follows:

Example

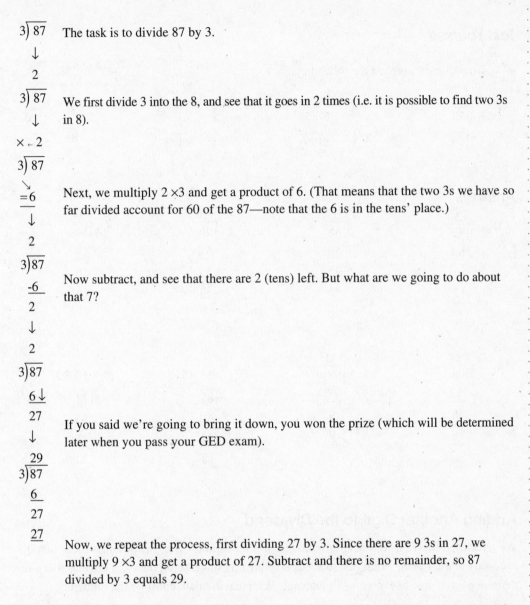

The task is to divide 87 by 3.

We first divide 3 into the 8, and see that it goes in 2 times (i.e. it is possible to find two 3s in 8).

Next, we multiply 2 ×3 and get a product of 6. (That means that the two 3s we have so far divided account for 60 of the 87—note that the 6 is in the tens' place.)

Now subtract, and see that there are 2 (tens) left. But what are we going to do about that 7?

If you said we're going to bring it down, you won the prize (which will be determined later when you pass your GED exam).

Now, we repeat the process, first dividing 27 by 3. Since there are 9 3s in 27, we multiply 9 ×3 and get a product of 27. Subtract and there is no remainder, so 87 divided by 3 equals 29.

Note that when we first divided by 3 and said that there are two 3s in eight, we were perpetrating a fiction. It was not really 8, but, rather, 80 that we were dividing by 30. When we multiplied, we placed the 6 beneath the 8, in the tens column, so that 6 was really worth 6 tens, or 60. This particular form of division uses many fictions but has been popular for a number of years. If it has given you difficulty before, or if you never really knew how to divide before, we recommend that you skip this form completely and go directly to the "Ladder Division" section later in this chapter. Ladder division plays fewer games with place-value concepts and is, in the author's opinion, easier to learn, while every bit as efficient as the form being dealt with in this section.

If you are still with us at this point and have not yet turned to "Ladder Division," try the exercises below.

Test Yourself

Directions: Complete the following.

1. $3\overline{)72}$ 2. $5\overline{)85}$ 3. $6\overline{)96}$ 4. $4\overline{)96}$
5. $7\overline{)98}$ 6. $9\overline{)99}$ 7. $6\overline{)83}$ 8. $5\overline{)72}$

Answers

1.
$$
\begin{array}{r}
24 \\
3\overline{)72} \\
6\downarrow \\
\overline{12} \\
\underline{12}
\end{array}
$$

2.
$$
\begin{array}{r}
17 \\
5\overline{)85} \\
5\downarrow \\
\overline{35} \\
\underline{35}
\end{array}
$$

3.
$$
\begin{array}{r}
16 \\
6\overline{)96} \\
6\downarrow \\
\overline{36} \\
\underline{36}
\end{array}
$$

4.
$$
\begin{array}{r}
24 \\
4\overline{)96} \\
8\downarrow \\
\overline{16} \\
\underline{16}
\end{array}
$$

5.
$$
\begin{array}{r}
14 \\
7\overline{)98} \\
7\downarrow \\
\overline{28} \\
\underline{28}
\end{array}
$$

6.
$$
\begin{array}{r}
11 \\
9\overline{)99} \\
9\downarrow \\
\overline{9} \\
\underline{9}
\end{array}
$$

7.
$$
\begin{array}{r}
13\ R5 \\
6\overline{)83} \\
6\downarrow \\
\overline{23} \\
\underline{18} \\
5
\end{array}
$$

8.
$$
\begin{array}{r}
14\ R2 \\
5\overline{)72} \\
5\downarrow \\
\overline{22} \\
\underline{20} \\
2
\end{array}
$$

Adding Another Digit to the Dividend

We have tried to avoid technical terminology in this volume. When it becomes clear that it is necessary to use fifteen words just to avoid defining one single term, then it is obvious that a few definitions would be to everyone's advantage. With that in mind, consider the following:

$$\text{Divisor}\overline{)\text{Dividend}}^{\text{Quotient}}$$

As an example of the above in division:

$$5\overline{)35}^{\,7}$$

The divisor is 5.

The dividend is 35.

The quotient is 7.

We have been working until now with two-digit dividends and one-digit divisors. There are two possibilities that arise when a third digit is added to the dividend. The following model examples explore both of those possibilities.

Example 1

$5\overline{)723}$ Since 7 is divisible by 5
$1\downarrow$ (can be divided by 5)
$5\overline{)723}$ everything proceeds
$\underline{-5}\downarrow$ as it did before . . .
22
$\downarrow$
14
$5\overline{)723}$
$\underline{5}$
22
$\underline{-20}$
2
$\downarrow$
144
$5\overline{)723}$ except that after the
$\underline{5}\;|$ second subtraction, a
$22\;|$ second "bringing down"
$\underline{20}\downarrow$ is needed.
23
$\underline{-20}$
3
$\downarrow$
$144\;$ R3

Example 2

$9\overline{)345}$ Since 3 is not divisible by 9,
$\downarrow$ the 3 is put together with the digit
3 to its right, and 34 is divided by 9.
$9\overline{)345}$ Note that the 3 in the quotient is
$\underline{27}$ written over the 4—not over the 3
7 in the dividend.
$\downarrow$
3
$9\overline{)345}$
$\underline{27}$
75
$|$
$\downarrow$
38
$9\overline{)345}$
$\underline{27}$ Everything else then proceeds
75 as before.
$\underline{-72}$
3
$|$
$|$
$\downarrow$
$38\;$ R3

Test Yourself

Directions: Complete the following.

1. $4\overline{)273}$
2. $6\overline{)985}$
3. $7\overline{)346}$
4. $8\overline{)957}$

5. $5\overline{)738}$
6. $9\overline{)217}$
7. $3\overline{)849}$
8. $8\overline{)361}$

9. $6\overline{)954}$
10. $5\overline{)875}$
11. $9\overline{)311}$
12. $7\overline{)862}$

13. $4\overline{)935}$
14. $3\overline{)972}$
15. $7\overline{)581}$
16. $9\overline{)634}$

Answers

1. $68\frac{1}{4}$

$$4\overline{)273}$$
$$\underline{24}\downarrow$$
$$33$$
$$\underline{32}$$
$$1$$

2. $164\frac{1}{6}$

$$6\overline{)985}$$
$$\underline{6}\downarrow$$
$$38\,|$$
$$\underline{36}\downarrow$$
$$25$$
$$\underline{24}$$
$$1$$

3. $49\frac{3}{7}$

$$7\overline{)346}$$
$$\underline{28}\downarrow$$
$$66$$
$$\underline{63}$$
$$3$$

4. $119\frac{5}{8}$

$$8\overline{)957}$$
$$\underline{8}\downarrow$$
$$15\,|$$
$$\underline{8}\downarrow$$
$$77$$
$$\underline{72}$$
$$5$$

5. $147\frac{3}{5}$

$$5\overline{)738}$$
$$\underline{5}\downarrow$$
$$23\,|$$
$$\underline{20}\downarrow$$
$$38$$
$$\underline{35}$$
$$3$$

6. $24\frac{1}{9}$

$$9\overline{)217}$$
$$\underline{18}\downarrow$$
$$37$$
$$\underline{36}$$
$$1$$

7. 283

$$3\overline{)849}$$
$$\underline{6}\downarrow$$
$$24\,|$$
$$\underline{24}\downarrow$$
$$9$$
$$\underline{9}$$

8. $45\frac{1}{8}$

$$8\overline{)361}$$
$$\underline{32}\downarrow$$
$$41$$
$$\underline{40}$$
$$1$$

9. 159

$$6\overline{)954}$$
$$\underline{6}\downarrow$$
$$35\,|$$
$$\underline{30}\downarrow$$
$$54$$
$$\underline{54}$$

10. 175

$$5\overline{)875}$$
$$\underline{5}\downarrow$$
$$37\,|$$
$$\underline{35}\downarrow$$
$$25$$
$$\underline{25}$$

11. $34\frac{5}{9}$

$$9\overline{)311}$$
$$\underline{27}\downarrow$$
$$41$$
$$\underline{36}$$
$$5$$

12. $123\frac{1}{7}$

$$7\overline{)862}$$
$$\underline{7}\downarrow$$
$$16\,|$$
$$\underline{14}\downarrow$$
$$22$$
$$\underline{21}$$
$$1$$

13. $233\frac{3}{4}$

$$4\overline{)935}$$
$$\underline{8}\downarrow$$
$$13\,|$$
$$\underline{12}\downarrow$$
$$15$$
$$\underline{12}$$
$$3$$

14. 324

$$3\overline{)972}$$
$$\underline{9}\downarrow$$
$$7\,|$$
$$\underline{6}\downarrow$$
$$12$$
$$\underline{12}$$

15. 83

$$7\overline{)581}$$
$$\underline{56}\downarrow$$
$$21$$
$$\underline{21}$$

16. $70\frac{4}{9}$

$$9\overline{)634}$$
$$\underline{63}\downarrow$$
$$4$$
$$\underline{0}$$
$$4$$

You may have noticed that the answers to all the previous problems were expressed with fractional remainders. That was done as a reminder that there is, indeed, more than one way in which to express the remainder. Your answer to problem 1, for example, would be correct whether you wrote it as $68\frac{1}{4}$ or 68 R 1.

Did problem 16 throw you, or did you figure it out for yourself? If you had trouble with 16, take a look at it now. After dividing 9 into 63, and multiplying 7 times 9, you subtract to get a remainder (difference) of 0. Bringing down a 4 (from the 634), it becomes necessary to divide 4 by 9. Thinking about that momentarily, it should be obvious that there are no 9's in 4. That means that a 0 must be placed in the quotient, above the 4. Multiplying, we find that 0 times 9 equals 0. Subtract, and you get a remainder of 4.

Two-Digit Divisors

Up to this point, our study of division has dealt exclusively with single-digit divisors. In the real world, and in the GED examination, single-digit divisors do not appear nearly as frequently as two-digit ones. The most important thing to remember when approaching division by two digits is that the four steps that were used with one-digit divisors remain the same: divide, multiply, subtract, bring down. Those four steps are repeated as many times as necessary until all the digits of the dividend have been brought down and divided.

Follow the solution of the example below, and you will see that there is nothing really new in the process.

Example

$24\overline{)385}$ As before, the dividend is approached one digit at a time, moving from left to right.

↓

$$\begin{array}{r} 1 \\ 24\overline{)385} \\ 24 \end{array}$$

How many 24s are there in 3? None? Then put the 3 together with the 8. How many 24s are there in 38? There is one, so a 1 is written in the quotient above the 8. Then, multiplying, we find that 1 times 24 equals 24, which we write below the "38."*

↓

$$\begin{array}{r} 1 \\ 24\overline{)385} \\ \underline{-24} \\ 14 \end{array}$$

Subtract, and get a difference of 14.

↓

$$\begin{array}{r} 1 \\ 24\overline{)385} \\ \underline{24} \\ 145 \end{array}$$

Bring down the next digit from the dividend.

↓

$$\begin{array}{r} 16 \\ 24\overline{)385} \\ \underline{24} \\ 145 \\ \underline{-144} \\ 1 \end{array}$$

Divide 145 by 24 (the hard part, because you must first estimate how many 24s are in 145) and find 6 of them. Multiply 6 ×24 and get 144. Subtract.

↓ Since there are no more digits in the dividend to bring down, 1 is the remainder.

16 R1 or

$16\frac{1}{24}$

*Since the 1 in the quotient is in the ten's place, we are really multiplying 10 times 24 and getting 240, which we subtract from 385 to get 145. In practice, you need not worry about that, but that is why it works.

The major difference between single-digit division and dividing by two-digit divisors is that a familiarity with the multiplication tables makes dividing by one digit relatively simple. It is not too difficult, for example, to recognize that 4 divides into 25 6 times. How many 24s there are in 145, however, is a somewhat more difficult question. After all, who is familiar with the 24s' table? The solution lies in estimating what the quotient will be (that is, approximating how many 24s there are in 145).

To estimate a quotient, first look at the numbers with which the division is to be performed:

24   145

Then remove the last digit of each:

24   145

That leaves us with:

2   14

How many 2s are there in 14?

Since there are 7 2s in 14, we can estimate that there will be 7 24s in 145. Now remember, the answer that we get from estimating is not necessarily accurate, but it is a ballpark figure. That is to say, it will be in the neighborhood of the number we are looking for. To find out how close we are, let's take our estimated 7 and multiply it by 24, the actual divisor, to see how close we come to the actual dividend, 145:

$$\begin{array}{r} 24 \\ \times\,7 \\ \hline 168 \end{array}$$

168 is larger than the 145 that we are looking for. That tells us that our estimated quotient, 7, is too big. Since 7 24s are too big, try 6 24s:

$$\begin{array}{r} 24 \\ \times\,6 \\ \hline 144 \end{array}$$

There we have it. Six 24s is just the number we were looking for. It is almost 145 and does not exceed 145.

In the examples below, you will see two estimated quotients and then the way in which those estimations are refined.

Example

1. $2\,3\,)\,\overline{1\,8\,7}$

2. $2\,)\,\overline{18}$ (9)

3. Estimate: 9 23s in 187.

4. Check: $\begin{array}{r} 23 \\ \times\,9 \\ \hline 207 \end{array}$ ← Too big.

5. Try 8 23s.

6. Check: $\begin{array}{r} 23 \\ \times\,8 \\ \hline 184 \end{array}$ ← Perfect.

7. $\begin{array}{r} 8\ \text{R}3 \\ 23\,)\,\overline{187} \\ \underline{184} \\ 3 \end{array}$

Example

1. $2\,\overline{9)16\,5}$

2. $2\,\overline{)16}$ → 8

3. Estimate: 8 29s in 165.

4. Check: $\begin{array}{r} 29 \\ \times\ 8 \\ \hline 232 \end{array}$ ← Way too big.

5. Try 6 29s.

6. Check: $\begin{array}{r} 29 \\ \times\ 6 \\ \hline 174 \end{array}$ ← Slightly too big. 5 29's should do it.

7. $\begin{array}{r} 5\ R20 \\ 29\overline{)165} \\ \underline{145} \\ 20 \end{array}$

Once more, note that the estimate does not give an exact answer each time, but it can speed the process of finding the correct quotient by leading you to it.

Test Yourself

Directions: Solve the following divisions. First estimate the quotients, then divide.

1. $19\overline{)178}$ **2.** $23\overline{)217}$ **3.** $35\overline{)295}$ **4.** $46\overline{)398}$

5. $52\overline{)417}$ **6.** $61\overline{)579}$ **7.** $37\overline{)231}$ **8.** $84\overline{)719}$

9. $34\overline{)458}$ **10.** $43\overline{)692}$ **11.** $28\overline{)736}$ **12.** $31\overline{)945}$

Answers

1. $\begin{array}{r} 9\ R7 \\ 19\overline{)178} \\ \underline{171} \\ 7 \end{array}$ **2.** $\begin{array}{r} 9\frac{10}{23} \\ 23\overline{)217} \\ \underline{207} \\ 10 \end{array}$ **3.** $\begin{array}{r} 8\ R15 \\ 35\overline{)295} \\ \underline{280} \\ 15 \end{array}$ **4.** $\begin{array}{r} 8\ R30 \\ 46\overline{)398} \\ \underline{368} \\ 30 \end{array}$

5. $\begin{array}{r} 8\frac{1}{52} \\ 52\overline{)417} \\ \underline{416} \\ 1 \end{array}$ **6.** $\begin{array}{r} 9\frac{30}{61} \\ 61\overline{)579} \\ \underline{549} \\ 30 \end{array}$ **7.** $\begin{array}{r} 6\ R9 \\ 37\overline{)231} \\ \underline{222} \\ 9 \end{array}$ **8.** $\begin{array}{r} 8\frac{47}{84} \\ 84\overline{)719} \\ \underline{672} \\ 47 \end{array}$

9. $\begin{array}{r} 13\ R16 \\ 34\overline{)458} \\ \underline{34} \\ 118 \\ \underline{102} \\ 16 \end{array}$ **10.** $\begin{array}{r} 16\frac{4}{43} \\ 43\overline{)692} \\ \underline{43} \\ 262 \\ \underline{258} \\ 4 \end{array}$ **11.** $\begin{array}{r} 26\ R8 \\ 28\overline{)736} \\ \underline{56} \\ 176 \\ \underline{168} \\ 8 \end{array}$ **12.** $\begin{array}{r} 30\ R15 \\ 31\overline{)945} \\ \underline{93} \\ 15 \end{array}$

More on Estimating Quotients

You may have noticed that some of the estimates for the quotients in the last set of exercises were far away from the actual quotient you were looking for. That was especially true in model example 2 in the previous section. The reason for that is in the nature of the estimating that you were doing. If the divisor in a given division example were 20, you would estimate by using the 2. If the divisor in another division were 29, you would still estimate by using the 2. Now there is a considerable gap between 20 and 29—large enough, in fact, to make using the 2 to estimate with both times somewhat less than wholly satisfactory.

The form of estimating that you have been doing is known as "estimating by rounding down." That is because the digit that you used to make the estimation was always the name of the tens' digit lower than the two-digit number you were dealing with. In other words, if you were dealing with 37, you rounded down to 30 and used the 3. If the divisor were 43, you rounded down to 40 and used the 4, and so on. Now it should be clear that while 43 is certainly closer to 40 than it is to any other ten, 37 is closer to 40 than it is to 30. We would, therefore, be much more likely to get an accurate estimate of the quotient if we used 4 rather than 3 as our estimating divisor. Look at the following two examples.

Estimating by Rounding Down	*Estimating by Rounding Up*
1.	1. 87)2̲4̲3
2. 8) 24	2. Close to 90
	3. 9)2̲4
3. Check: 87	4. Check: 87
⟨×3⟩	×2
261 ← Too big!	174
4. Try 2 87's.	
5. Check: 87	5. 2 R69
×2	87)243
174	174
	69
6. 2 R69	
87)243	
174	
69	

Clearly, in this particular case, estimating by rounding up to the next decade (90) was a more efficient method than rounding down. Does that mean that you should always estimate by rounding up? Hopefully, you have hit upon the answer for yourself. Sometimes it is better to round down, and sometimes it is better to round up. When the ones' digit of the divisor is less than 5 (i.e., 0, 1, 2, 3, 4), round down as we did in the last section. When the ones' digit of the divisor is 5 or greater, round up.

The rounded tens' digit that we use in order to estimate the quotient is known as **a trial divisor.**

Test Yourself

Directions: Tell what the trial divisor would be for each of the following divisors. The first two have been done for you.

1. 32 is closer to _____, so the trial divisor would be _____.

2. 46 is closer to _____, so the trial divisor would be _____.

3. 29 is closer to _____, so the trial divisor would be _____.

4. 82 is closer to _____, so the trial divisor would be _____.

5. 73 is closer to _____, so the trial divisor would be _____.

6. 54 is closer to _____, so the trial divisor would be _____.

7. 68 is closer to _____, so the trial divisor would be _____.

8. 25 is closer to _____, so the trial divisor would be _____.

9. 66 is closer to _____, so the trial divisor would be _____.

10. 15 is closer to _____, so the trial divisor would be _____.

Answers

1.	30, 3	2.	50, 5	3.	30, 3	4.	80, 8	5.	70, 7
6.	50, 5	7.	70, 7	8.	30, 3	9.	70, 7	10.	20, 2

Now, try applying the two different forms of estimation to some slightly more complex division exercises. Remember, once you have determined your trial divisor for an exercise, that trial divisor will remain the same, no matter how many times you need to divide to complete that division.

Test Yourself

> **Directions:** Estimate by rounding up or rounding down. Record your trial divisor, then divide.

1. $38\overline{)5943}$

T.D. = _____

2. $51\overline{)6782}$

T.D. = _____

3. $47\overline{)3895}$

T.D. = _____

4. $62\overline{)9437}$

T.D. = _____

5. $55\overline{)7482}$

T.D. = _____

6. $38\overline{)4729}$

T.D. = _____

7. $42\overline{)3816}$

T.D. = _____

8. $66\overline{)9583}$

T.D. = _____

9. $79\overline{)8243}$

T.D. = _____

10. $24\overline{)6357}$

T.D. = _____

11. $35\overline{)8124}$

T.D. = _____

12. $56\overline{)4982}$

T.D. = _____

Answers

1. 4; 156 R15

2. 5; 132 R50

3. 5; 82 R41

4. 6; 152 R13

5. 6; 136 R2

6. 4; 124 R17

7. 4; 90 R36

8. 7; 145 R13

9. 8; 104 R27

10. 2; 264 R21

11. 4; 232 R4

12. 6; 88 R54

Ladder Division

Ladder division is an alternate method of dividing based upon the relationship between division and subtraction and taking advantage of the students' knowledge of place-value. Many people find it far more logical than the traditional approach to division—especially when dealing with larger numbers. Ladder division may be done with much less mastery of multiplication than is required to use the traditional form discussed in the preceding pages. If you are shaky in traditional division and/or traditional multiplication, then this may be just what you have been looking for. If division is not a serious problem for you, skip this section. Nobody needs to know how to divide two different ways.

Ladder division is built upon the idea with which we opened the discussion of division, namely that the operation can be thought of as repeated subtraction of the same number. In the following example, it is stated that 42 divided by 6 could be thought of as asking: "How many 6s can be subtracted from 42?" That question can be answered by subtracting 6 from 42 over and over again, until there is nothing left:

```
6)42
 - 6
 ──
  36
 - 6
 ──
  30
 - 6
 ──
  24
 - 6
 ──
  18
 - 6
 ──
  12
 - 6
 ──
   6
 - 6
 ──
   0
```

By counting the number of times 6 has been subtracted from 42, you may conclude that there are seven 6s in 42, or, put another way, 42 divided by 6 is 7.

Ladder division takes this very simple concept and builds a completely different method of division—one which is really not difficult to master (with practice).

While the illustration above shows single 6s being subtracted from the 42, there is no reason why larger groupings of 6 could not have been subtracted. In the examples below, the same division is worked in different ways. The number to the right of each vertical line keeps track of how many 6s were removed:

```
           6 ) 42         6 ) 42
   6 ) 42   - 18 | 3      - 12 | 2
   - 30 | 5   24          30
1.   12    - 18 | 3    3. - 24 | 4
   - 12 | 2    6           6
2.    0 | 7  - 6 | 1     - 6 | 1
               0 | 7       0 | 7
```

When all the 6s that can be removed have been removed, the numbers to the right of the vertical line are added up and the quotient is found: 7. Notice how the number of 6s removed are tracked on the right side of the vertical line, while the total that has been removed from the dividend (e.g., five 6s = 30) are tracked on the left. This gives the impression of rungs on a ladder, hence the name, ladder division. The beauty of ladder division is that there is no prescription as to how many divisors you must remove from the dividend at each step. Take out as many as you can find, or as many as you feel comfortable with. That is why in examples 1, 2, and 3, above, there were different numbers of 6's removed at each step in each division, yet the quotient still ended up as 7 each time.

When solving a division by the ladder method, it is helpful to estimate the total quotient before beginning. Consider, for instance:

$8\overline{)322}$

It is essential to know about how large the quotient is going to be. Estimating takes place in multiples of 10. So, for example, we begin by asking ourselves whether there are 10 groups of 8 in 322. Since 10 8s make 80, and 80 is less than 322, we can conclude that there are. Next, ask whether there are 100 groups of 8 in 322. Since 100 8s make 800, and 800 is larger than 322, there are not. We have now estimated that our quotient will be greater than 10, but less than 100. We can now solve the problem. Here is a solution that requires the least amount of thinking, but which is somewhat lengthy:

$$8\overline{)322}$$

80	10
242	
80	10
162	
80	10
82	
80	10
2	40 R2

Since we already know that there are 10 groups of 8, let's remove 10 8s from 322 and see what's left. Then let's do it again ...

... and again ..

... and again.

Since there is not another 8 remaining, we cannot subtract again. Add up the column to the right of the line and get 40 R 2.

Now, if you would rather get to the quotient more quickly, we can sophisticate our approach:

$$8\overline{)322}$$

If 10 8s = 80, then twice 10 8s, or 20 8s =

$$\begin{array}{r} 80 \\ + 80 \\ \hline 160 \end{array}$$ (20 8s)

But 20 8s and another 20 8s =

$$\begin{array}{r} 160 \\ +160 \\ \hline 320 \end{array}$$ (40 8s)

$$8\overline{)322}$$

320	40
2	40 R2

That's pretty close. Let's write it.

And that's that!

If you understand the principle involved—removing as much at one time from the dividend as you are comfortable removing—then you have the whole story. Just remember to keep track (on the right of the line) of the number of divisors that you have removed at each step.

Test Yourself

Directions: Complete the following.

1. $7\overline{)532}$

2. $5\overline{)168}$

3. $4\overline{)234}$

4. $8\overline{)567}$

5. $7\overline{)384}$

6. $9\overline{)638}$

7. $7\overline{)461}$

8. $5\overline{)316}$

9. $6\overline{)225}$

NOTE

In these problems, your solution might well be quite different from the one shown. Your answer, however, should be the same as the answer shown.

Answers

1.

$$7\overline{)532}$$
$$\underline{490}\quad 70$$
$$42$$
$$\underline{42}\quad\underline{6}$$
$$\quad 76$$

2.

$$5\overline{)168}$$
$$\underline{150}\quad 30$$
$$18$$
$$\underline{15}\quad\underline{3}$$
$$\quad 33\quad R3$$

3.

$$4\overline{)234}$$
$$\underline{200}\quad 50$$
$$34$$
$$\underline{32}\quad\underline{8}$$
$$2\quad 58\quad R2$$

4.

```
8)567
  560 | 70
    7 | 70   R7
```

5.

```
7)384
  350 | 50
   34 |
   28 | 4
    6 | 54   R6
```

6.

```
9)638
  630 | 70
    8 | 70   R8
```

7.

```
7)461
  420 | 60
   41 |
   35 | 5
    6 | 65   R6
```

8.

```
5)316
  300 | 60
   16 |
   15 | 3
    1 | 63   R1
```

9.

```
6)225
  180 | 30
   45 |
   42 | 7
    3 | 37   R3
```

Ladder Division with Two-Digit Divisors

There is essentially no difference between ladder division with single-digit divisors and ladder division with multiple-digit divisors. Exactly the same rules apply. Indeed, the only significant change is in the magnitude of the numbers being worked with.

Examine the example below.

Example

```
32)6843
  6400 | 200
   443 |
   320 | 10
   123 |
    96 | 3
    27 | 213   R27
```

First, the quotient must be estimated. Since 100 32s make 3200, and 1000 make 32,000, it is obvious that the answer must be in the hundreds.

200 32s make 6400. That fact is taken care of on the first step of the ladder. 10 more 32s are removed on the second step, and finally the last 3 are taken care of.

As before, there is no single correct way to perform a ladder division, and so the model example shows just one possible solution. There are many other possible solutions—most a bit longer than the one shown. Try your hand at the divisions below.

Test Yourself

Directions: Divide using the ladder method.

1. $34\overline{)688}$

2. $53\overline{)792}$

3. $46\overline{)983}$

4. $22\overline{)517}$

5. $18\overline{)376}$

6. $57\overline{)697}$

7. $38\overline{)6431}$

8. $29\overline{)8537}$

9. $53\overline{)6947}$

10. $82\overline{)7386}$

11. $47\overline{)8526}$

12. $34\overline{)5264}$

13. $23\overline{)9537}$

14. $38\overline{)9438}$

15. $42\overline{)7159}$

16. $63\overline{)27,584}$

17. $48\overline{)35,829}$

18. $26\overline{)86,635}$

Answers

1.
$$
\begin{array}{r}
34\overline{)688} \\
\underline{680} \quad 20 \\
8 \quad 20 \quad \text{R8}
\end{array}
$$

2.
$$
\begin{array}{r}
53\overline{)792} \\
\underline{530} \quad 10 \\
262 \\
\underline{212} \quad 4 \\
50 \quad 14 \quad \text{R50}
\end{array}
$$

3.
$$
\begin{array}{r}
46\overline{)983} \\
\underline{920} \quad 20 \\
63 \\
\underline{46} \quad 1 \\
17 \quad 21 \quad \text{R17}
\end{array}
$$

4.
$$
\begin{array}{r}
22\overline{)517} \\
\underline{440} \quad 20 \\
77 \\
\underline{66} \quad 3 \\
11 \quad 23 \quad \text{R11}
\end{array}
$$

5.
$$
\begin{array}{r}
18\overline{)376} \\
\underline{360} \quad 20 \\
16 \quad 20 \quad \text{R16}
\end{array}
$$

6.
$$
\begin{array}{r}
57\overline{)697} \\
\underline{570} \quad 10 \\
127 \\
\underline{114} \quad 2 \\
13 \quad 12 \quad \text{R13}
\end{array}
$$

7.
$$
\begin{array}{r}
38\overline{)6431} \\
\underline{3800} \quad 100 \\
2631 \\
\underline{1900} \quad 50 \\
731 \\
\underline{380} \quad 10 \\
351 \\
\underline{342} \quad 9 \\
9 \quad 169 \quad \text{R9}
\end{array}
$$

8.
$$
\begin{array}{r}
29\overline{)8537} \\
\underline{5800} \quad 200 \\
2737 \\
\underline{2610} \quad 90 \\
127 \\
\underline{116} \quad 4 \\
11 \quad 294 \quad \text{R11}
\end{array}
$$

9.
$$
\begin{array}{r}
53\overline{)6947} \\
\underline{5300} \quad 100 \\
1647 \\
\underline{1590} \quad 30 \\
57 \\
\underline{53} \quad 1 \\
4 \quad 131 \quad \text{R4}
\end{array}
$$

10.
$$
\begin{array}{r}
82\overline{)7386} \\
\underline{7380} \quad 90 \\
6 \quad 90 \quad \text{R6}
\end{array}
$$

11.
$$
\begin{array}{r}
47\overline{)8526} \\
\underline{4700} \quad 100 \\
3826 \\
\underline{3760} \quad 80 \\
66 \\
\underline{47} \quad 1 \\
19 \quad 181 \quad \text{R19}
\end{array}
$$

12.
$$
\begin{array}{r}
34\overline{)5264} \\
\underline{3400} \quad 100 \\
1864 \\
\underline{1700} \quad 50 \\
164 \\
\underline{136} \quad 4 \\
28 \quad 154 \quad \text{R28}
\end{array}
$$

13.

```
23)9537
   9200  400
    337
    230   10
    107
     92    4
     15  414   R15
```

14.

```
38)9438
   7600  200
   1838
   1520   40
    318
    304    8
     14  248   R14
```

15.

```
26)7159
   5200  200
   1959
   1820   70
    139
    130    5
      9  275   R9
```

16.

```
63)27,584
   25,200  400
    2384
    1890   30
     494
     441    7
      53  437   R53
```

17.

```
48)35,829
   33,600  700
    2229
    1920   40
     309
     288    6
      21  746   R21
```

18.

```
26)86,635
   78,000  3000
    8635
    7800   300
     835
     780    30
      55
      52     2
       3  3332   R3
```

EXERCISES: USING ZERO AS A PLACEHOLDER

Directions: Write each number as a place-value numeral.

1. Four thousand twenty ....

2. Three thousand, eight hundred ....

3. Fifty thousand ....

4. Nine million ....

5. Seven million, six thousand ....

6. Fifty-four thousand ....

7. Two hundred five thousand, eighty ....

8. Four million, four hundred ....

9. Seven million, six ....

ANSWERS

1. 4020	**2.** 3800	**3.** 50,000	**4.** 9,000,000	**5.** 7,006,000
6. 54,000	**7.** 205,080	**8.** 4,000,400	**9.** 7,000,006	

EXERCISES: COLUMN ADDITION

1.
```
H T U
3 4 5
+ 2 7 8
```

2.
```
H T U
6 7 9
+ 1 9 6
```

3.
```
H T U
2 4 7
  8 5
+ 4 0 6
```

4.
```
H T U
  8 9
  6 4
  5 3
  8 6
+ 5 8
```

5.
```
H T  U
4 3  7
  9  5
  4  3
  9  4
+ 5 3 6
```

6.
```
 435
+268
```

7.
```
 618
 278
+ 54
```

8.
```
 458
   9
+867
```

9.
```
 807
  52
  47
+633
```

10.
```
 4315
  914
   85
  579
+6802
```

ANSWERS

1.	623	**2.**	875	**3.**	738	**4.**	350	**5.**	1205
6.	703	**7.**	950	**8.**	1334	**9.**	1539	**10.**	12,695

EXERCISES: SUBTRACTION WITH RENAMING

1.
$$53 = 50 + 3 = 40 + \underline{\quad}$$
$$-28 = -(20 + 8) = -(20 + 8)$$
$$\underline{\quad} + \underline{\quad} = \underline{\quad}$$

2.
$$64 = \underline{\quad} + \underline{\quad} = \underline{\quad} + \underline{\quad}$$
$$-37 = -(\underline{\quad} + \underline{\quad}) = -(\underline{\quad} + \underline{\quad})$$
$$\underline{\quad} + \underline{\quad} = \underline{\quad}$$

3.
$$85 = \underline{\quad} + \underline{\quad} = \underline{\quad} + \underline{\quad}$$
$$-58 = -(\underline{\quad} + \underline{\quad}) = -(\underline{\quad} + \underline{\quad})$$
$$\underline{\quad} + \underline{\quad} = \underline{\quad}$$

4.
$$41 = \underline{\quad} + \underline{\quad} = \underline{\quad} + \underline{\quad}$$
$$-18 = -(\underline{\quad} + \underline{\quad}) = -(\underline{\quad} + \underline{\quad})$$
$$\underline{\quad} + \underline{\quad} = \underline{\quad}$$

5.
$$72 = \underline{\quad} + \underline{\quad} = \underline{\quad} + \underline{\quad}$$
$$-27 = -(\underline{\quad} + \underline{\quad}) = -(\underline{\quad} + \underline{\quad})$$
$$\underline{\quad} + \underline{\quad} = \underline{\quad}$$

6.
$$96 = \underline{\quad} + \underline{\quad} = \underline{\quad} + \underline{\quad}$$
$$-49 = -(\underline{\quad} + \underline{\quad}) = -(\underline{\quad} + \underline{\quad})$$
$$\underline{\quad} + \underline{\quad} = \underline{\quad}$$

7.
$$561 = 400 + 150 + 11$$
$$-285 = -(200 + 80 + 5)$$
$$\underline{\quad} + \underline{\quad} + \underline{\quad} = \underline{\quad}$$

8.
$$485 = \underline{\quad} + 170 + \underline{\quad}$$
$$-197 = -(\underline{\quad} + \underline{\quad} + \underline{\quad})$$
$$\underline{\quad} + \underline{\quad} + \underline{\quad} = \underline{\quad}$$

exercises

9.   612 = ___ + ___ + ___

         -437 = -(___ + ___ + ___)

                ___ + ___ + ___ = _____

10.   736 = ___ + ___ + ___

         -359 = -(___ + ___ + ___)

                ___ + ___ + ___ = _____

11.   378 = ___ + ___ + ___

         -199 = -(___ + ___ + ___)

                ___ + ___ + ___ = _____

12.   943 = ___ + ___ + ___

         -576 = -(___ + ___ + ___)

                ___ + ___ + ___ = _____

13.   834 = ___ + ___ + ___

         -257 = -(___ + ___ + ___)

                ___ + ___ + ___ = _____

14.   530 = ___ + ___ + ___

         -163 = -(___ + ___ + ___)

                ___ + ___ + ___ = _____

ANSWERS

1.
$$\begin{array}{r} 40 + 13 \\ -(20 + 8) \\ \hline 20 + 5 = 25 \end{array}$$

2.
$$\begin{array}{r} 50 + 14 \\ -(30 + 7) \\ \hline 20 + 7 = 27 \end{array}$$

3.
$$\begin{array}{r} 70 + 15 \\ -(50 + 8) \\ \hline 20 + 7 = 27 \end{array}$$

4.
$$\begin{array}{r} 30 + 11 \\ -(10 + 8) \\ \hline 20 + 3 = 23 \end{array}$$

5.
$$\begin{array}{r} 60 + 12 \\ -(20 + 7) \\ \hline 40 + 5 = 45 \end{array}$$

6.
$$\begin{array}{r} 80 + 16 \\ -(40 + 9) \\ \hline 40 + 7 = 47 \end{array}$$

7. $200 + 70 + 6 = 276$
8. $200 + 80 + 8 = 288$
9. $100 + 70 + 5 = 175$

10. $300 + 70 + 7 = 377$
11. $100 + 70 + 9 = 179$
12. $300 + 60 + 7 = 367$

13. $500 + 70 + 7 = 577$
14. $300 + 60 + 7 = 367$

EXERCISES: MULTIPLYING TWO DIGITS

1.
$$\begin{array}{r} 75 \\ \times\, 8 \\ \hline \end{array}$$

2.
$$\begin{array}{r} 84 \\ \times\, 6 \\ \hline \end{array}$$

3.
$$\begin{array}{r} 64 \\ \times\, 9 \\ \hline \end{array}$$

4.
$$\begin{array}{r} 38 \\ \times\, 7 \\ \hline \end{array}$$

5.
$$\begin{array}{r} 29 \\ \times\, 4 \\ \hline \end{array}$$

6.
$$\begin{array}{r} 97 \\ \times\, 5 \\ \hline \end{array}$$

7.
$$\begin{array}{r} 42 \\ \times\, 3 \\ \hline \end{array}$$

8.
$$\begin{array}{r} 56 \\ \times\, 4 \\ \hline \end{array}$$

9.
$$\begin{array}{r} 67 \\ \times\, 5 \\ \hline \end{array}$$

10.
$$\begin{array}{r} 78 \\ \times\, 6 \\ \hline \end{array}$$

11.
$$\begin{array}{r} 83 \\ \times\, 7 \\ \hline \end{array}$$

12.
$$\begin{array}{r} 94 \\ \times\, 9 \\ \hline \end{array}$$

13.
$$\begin{array}{r} 82 \\ \times 16 \\ \hline \end{array}$$

14.
$$\begin{array}{r} 57 \\ \times 12 \\ \hline \end{array}$$

15.
$$\begin{array}{r} 44 \\ \times 11 \\ \hline \end{array}$$

16.
$$\begin{array}{r} 73 \\ \times 72 \\ \hline \end{array}$$

17.
$$\begin{array}{r} 68 \\ \times 12 \\ \hline \end{array}$$

18.
$$\begin{array}{r} 30 \\ \times 20 \\ \hline \end{array}$$

19.
$$\begin{array}{r} 70 \\ \times 42 \\ \hline \end{array}$$

20.
$$\begin{array}{r} 62 \\ \times 55 \\ \hline \end{array}$$

21.
$$\begin{array}{r} 37 \\ \times 26 \\ \hline \end{array}$$

22.
$$\begin{array}{r} 68 \\ \times 69 \\ \hline \end{array}$$

23.
$$\begin{array}{r} 88 \\ \times 14 \\ \hline \end{array}$$

24.
$$\begin{array}{r} 89 \\ \times 30 \\ \hline \end{array}$$

25.
$$\begin{array}{r} 44 \\ \times 88 \\ \hline \end{array}$$

26.
$$\begin{array}{r} 27 \\ \times 26 \\ \hline \end{array}$$

27.
$$\begin{array}{r} 45 \\ \times 21 \\ \hline \end{array}$$

28.
$$\begin{array}{r} 97 \\ \times 83 \\ \hline \end{array}$$

29.
$$\begin{array}{r} 30 \\ \times 21 \\ \hline \end{array}$$

30.
$$\begin{array}{r} 32 \\ \times 12 \\ \hline \end{array}$$

ANSWERS

1. 600	**2.** 504	**3.** 576	**4.** 266	**5.** 116
6. 485	**7.** 126	**8.** 224	**9.** 335	**10.** 468
11. 581	**12.** 846	**13.** 1312	**14.** 684	**15.** 484
16. 5256	**17.** 816	**18.** 600	**19.** 2940	**20.** 3410
21. 962	**22.** 4692	**23.** 1232	**24.** 2670	**25.** 3872
26. 702	**27.** 945	**28.** 8051	**29.** 630	**30.** 384

EXERCISES: DIVIDING BY ONE DIGIT

1. $2\overline{)15}$	**2.** $3\overline{)24}$	**3.** $5\overline{)19}$	**4.** $6\overline{)39}$	**5.** $7\overline{)61}$
6. $8\overline{)54}$	**7.** $4\overline{)23}$	**8.** $9\overline{)54}$	**9.** $5\overline{)49}$	**10.** $7\overline{)51}$
11. $6\overline{)31}$	**12.** $8\overline{)17}$	**13.** $3\overline{)598}$	**14.** $6\overline{)379}$	**15.** $5\overline{)842}$
16. $7\overline{)536}$	**17.** $7\overline{)638}$	**18.** $4\overline{)742}$	**19.** $8\overline{)539}$	**20.** $3\overline{)948}$

ANSWERS

1.	7, R1	**2.**	8	**3.**	3, R4	**4.**	6, R3
5.	8, R5	**6.**	6, R6	**7.**	5, R3	**8.**	6
9.	9, R4	**10.**	7, R2	**11.**	5, R1	**12.**	2, R1

13.	$199\dfrac{1}{3}$	**14.**	63 R1	**15.**	$168\dfrac{2}{5}$	**16.**	76 R4

17.	$91\dfrac{1}{7}$	**18.**	185 R2	**19.**	$67\dfrac{3}{8}$	**20.**	316

EXERCISES: DIVIDING BY TWO DIGITS

1.	$34\overline{)586}$	**2.**	$27\overline{)318}$	**3.**	$45\overline{)269}$	**4.**	$54\overline{)498}$
5.	$46\overline{)732}$	**6.**	$39\overline{)267}$	**7.**	$34\overline{)700}$	**8.**	$63\overline{)470}$
9.	$53\overline{)389}$	**10.**	$78\overline{)912}$	**11.**	$61\overline{)874}$	**12.**	$92\overline{)653}$

exercises

ANSWERS

1.
$$\begin{array}{r} 17\ \text{R8} \\ 34\overline{)586} \\ \underline{34}\downarrow \\ 246 \\ \underline{238} \\ 8 \end{array}$$

2.
$$\begin{array}{r} 11\ \text{R21} \\ 27\overline{)318} \\ \underline{27}\downarrow \\ 48 \\ \underline{27} \\ 21 \end{array}$$

3.
$$\begin{array}{r} 5\frac{44}{45} \\ 45\overline{)269} \\ \underline{225} \\ 44 \end{array}$$

4.
$$\begin{array}{r} 9\ \text{R12} \\ 54\overline{)498} \\ \underline{486} \\ 12 \end{array}$$

5.
$$\begin{array}{r} 15\ \text{R42} \\ 46\overline{)732} \\ \underline{46}\downarrow \\ 272 \\ \underline{230} \\ 42 \end{array}$$

6.
$$\begin{array}{r} 6\ \text{R33} \\ 39\overline{)267} \\ \underline{234} \\ 33 \end{array}$$

7.
$$\begin{array}{r} 20\ \text{R20} \\ 34\overline{)700} \\ \underline{68}\downarrow \\ 20 \end{array}$$

8.
$$\begin{array}{r} 7\frac{29}{63} \\ 63\overline{)470} \\ \underline{441} \\ 29 \end{array}$$

9.
$$\begin{array}{r} 7\frac{18}{53} \\ 53\overline{)389} \\ \underline{371} \\ 18 \end{array}$$

10.
$$\begin{array}{r} 11\ \text{R54} \\ 78\overline{)912} \\ \underline{78}\downarrow \\ 132 \\ \underline{78} \\ 54 \end{array}$$

11.
$$\begin{array}{r} 14\frac{20}{61} \\ 61\overline{)874} \\ \underline{61}\downarrow \\ 264 \\ \underline{244} \\ 20 \end{array}$$

12.
$$\begin{array}{r} 7\frac{9}{92} \\ 92\overline{)653} \\ \underline{644} \\ 9 \end{array}$$

SUMMING IT UP

- The system we use for representing numbers is known as the decimal system of numeration. A number is an idea of quantity and a numeral is the symbol that we use to represent that idea.

- Addition is a combining operation. When you add, you are putting together a number of quantities to get a larger quantity.

- If you can add up to 18, you can add any two numbers in the decimal system.

- The multiplication facts, or "times tables," are the basis upon which all multiplication is built.

- Addition and multiplication are combining operations. Both combine smaller quantities to make larger quantities.

- Any form of division (except for the basic division facts) consists of repeating three steps:

  ❶ Divide

  ❷ Multiply

  ❸ Subtract

Fractional Numbers

WHAT IS A FRACTION?

The concepts of fractions and operations with fractions are ones that have caused difficulty for students of mathematics for many years. Indeed, fractions are probably one of the most poorly taught of all the ideas in elementary mathematics. A major cause of the difficulty is the fact that there is no single definition for fraction. Indeed, fractions are used to represent many things, including a part of a whole, a part of a group of things, a comparison (or ratio), a division, and a whole to be further operated upon. Let us examine those definitions one at a time.

A fraction as a part of a whole assumes that there is one whole object, such as a cake or a paycheck, and you are referring to less than the entire object. In order to name the fraction that is being considered, one must look at two components. The first is the number of equal parts that the whole is separated into. The second is the number of those parts being considered. Look at the following:

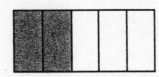

one of four equal parts:
one-fourth $\left(\frac{1}{4}\right)$

two of five equal parts:
two-fifths $\left(\frac{2}{5}\right)$

chapter 4

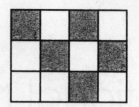

four of seven equal parts:

four-sevenths $\left(\frac{4}{7}\right)$

five of twelve equal parts:

five-twelfths $\left(\frac{5}{12}\right)$

Notice that the part of a fraction that tells the number of parts into which the whole has been separated is at the bottom (known technically as the denominator). The top of the fraction (the numerator) names the number of parts that are being considered (in these cases the number of parts that are shaded).

The following illustrations serve to demonstrate the meaning of fraction as a part of a group:

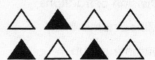

Two of five apples are shaded: $\left(\frac{2}{5}\right)$ Three of eight triangles are shaded: $\left(\frac{3}{8}\right)$

$\left(\frac{4}{7}\right)$ are boys; $\left(\frac{3}{7}\right)$ are girls. $\left(\frac{6}{11}\right)$ are vertical; $\left(\frac{5}{11}\right)$ are horizontal.

There is considerable difference between $\frac{1}{2}$ of an apple and $\frac{1}{2}$ of a group of people. The latter contains only whole people. The former contains no whole apples.

Test Yourself

Directions: Name the fraction indicated by the shaded portion in each diagram.

1.    2.    3.

4.    5.    6.

7. 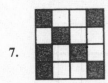   8.   9.

Answers

1. $\dfrac{1}{4}$ 2. $\dfrac{3}{10}$ 3. $\dfrac{2}{3}$ 4. $\dfrac{4}{9}$ 5. $\dfrac{2}{5}$

6. $\dfrac{5}{10}$ or $\dfrac{1}{2}$ 7. $\dfrac{7}{16}$ 8. $\dfrac{8}{15}$ 9. $\dfrac{3}{8}$

There are frequently occasions when fractional notation can be used for making comparisons. If you have 4 dollars and Bill has 7 dollars, then you have $\dfrac{4}{7}$ the amount of money that Bill has. This particular use of fractional notation is called ratio and will be covered separately at a later time.

In algebra, fractions have always been used to represent division. The use of fractional notation for that purpose has become increasingly widespread since the beginning of the popularity of home computers. The fraction line is, in fact, the standard symbol used to indicate division in any computer program. $\dfrac{8}{2}$ means 8 divided by 2. Of course, $\dfrac{8}{2}$ also means (and may be read as) 8 halves. Arithmetically, they are equivalent, since 8 halves or eight divided by 2 equals 4.

Finally, in our examination of the meaning of fractions, consider the fraction as a whole to be further subdivided. Were you to walk into a restaurant and order a piece of pie, it is unlikely that you might notice that the pie (in the case) was cut into eight equal parts. It is even less likely that you would consider ordering "an eighth of pie." Rather, you order a "piece of pie," as if the eighth were indeed a whole entity. If you were not hungry enough to eat the whole piece, you might consider cutting it in half, and sharing it with your dinner companion. Would it ever occur to you that you were eating a sixteenth of the original pie? Most likely, you would be content just to eat "half a piece of pie."

ADDING AND SUBTRACTING FRACTIONS

Adding and Subtracting Fractions with Like Denominators

Fractional numbers, like any other numbers, are capable of being added, subtracted, multiplied, and divided. Addition and subtraction of fractional numbers are approached together here, since the rules that govern those operations are similar. In fact, there is really only one rule that governs the ability of fractions to be added or subtracted:

Fractions may be added or subtracted if they have like denominators.

You may recall, the denominator of a fraction is the numeral on the bottom, and it names the number of parts into which the whole object or group has been separated. If two fractions have the same denominators, they are said to have **like** denominators. It is critical to remember that the denominator names the size of the parts that are being dealt with. So, for example, if an object is divided into eighths, there will be twice as many parts as there would be if the same object were divided into fourths. Each eighth, however, would be half the size of each fourth.

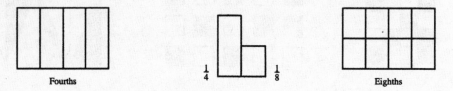

For the same or for equal objects, the more parts there are, the smaller each part is! The diagram below shows the addition of $\frac{1}{5}$ and $\frac{2}{5}$:

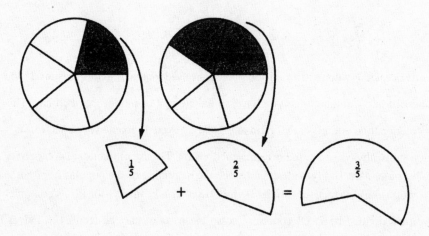

Notice that the numerators (top numerals in the fractions) tell the number of parts being considered. That is, 1 names the first amount of fifths, and 2 names the second amount of fifths. Adding one and two gives three, hence adding $\frac{1}{5}$ and $\frac{2}{5}$ gives $\frac{3}{5}$. Notice that since you started with fifths and ended with fifths, the denominators do not change. Only the numerators are added.

Here is another example:

$$\frac{1}{4} + \frac{2}{4} = \frac{3}{4}$$

Subtracting fractional numbers works in exactly the same way as addition, except that the numerators are subtracted rather than added:

$$\frac{4}{4} - \frac{1}{4} = \frac{3}{4}$$

Test Yourself

Directions: Try your hand at the fractional additions and subtractions below. Pay attention to the signs, so that you know whether to add or subtract.

1. $\frac{2}{5} + \frac{2}{5} =$ _____

2. $\frac{3}{8} + \frac{4}{8} =$ _____

3. $\frac{5}{12} - \frac{2}{12} =$ _____

4. $\frac{9}{15} - \frac{6}{15} =$ _____

5. $\frac{5}{21} + \frac{6}{21} =$ _____

6. $\frac{8}{19} + \frac{5}{19} =$ _____

7. $\frac{4}{9} + \frac{3}{9} =$ _____

8. $\frac{11}{12} - \frac{6}{12} =$ _____

9. $\frac{3}{14} + \frac{8}{14} =$ _____

10. $\frac{15}{16} - \frac{4}{16} =$ _____

11. $\frac{1}{2} + \frac{2}{2} =$ _____

12. $\frac{2}{3} - \frac{1}{3} =$ _____

13. $\frac{9}{18} + \frac{5}{18} =$ _____

14. $\frac{4}{7} - \frac{3}{7} =$ _____

15. $\frac{21}{30} - \frac{11}{30} =$ _____

Answers

1. $\dfrac{4}{5}$ 2. $\dfrac{7}{8}$ 3. $\dfrac{3}{12}$ or $\dfrac{1}{4}$ 4. $\dfrac{3}{15}$ or $\dfrac{1}{5}$

5. $\dfrac{11}{21}$ 6. $\dfrac{13}{19}$ 7. $\dfrac{7}{9}$ 8. $\dfrac{5}{12}$

9. $\dfrac{11}{14}$ 10. $\dfrac{11}{16}$ 11. $\dfrac{3}{2}$ or $1\dfrac{1}{2}$ 12. $\dfrac{1}{3}$

13. $\dfrac{14}{18}$ or $\dfrac{7}{9}$ 14. $\dfrac{1}{7}$ 15. $\dfrac{10}{30}$ or $\dfrac{1}{3}$

Equivalent Fractions

Two fractions that are equal but have different names are called "equivalent fractions." Look at the diagram below:

$\frac{1}{2}$ $\frac{2}{4}$ $\frac{3}{6}$ $\frac{4}{8}$ $\frac{5}{10}$

Each shaded fraction has a different name, as you can see from the numerals beneath each picture. Yet, upon close inspection you should notice that the size of the piece shaded in each of the pictures is the same as the size piece shaded in every other picture. These fractions are equivalent. It can be stated, therefore, that $\frac{1}{2}, \frac{2}{4}, \frac{3}{6}, \frac{4}{8}$, and $\frac{5}{10}$ are all equivalent fractions.

Test Yourself

Directions: Name these equivalent fractions:

1.

2.

3.

Answers

1. $\frac{1}{3}, \frac{2}{6}, \frac{3}{9}, \frac{4}{12}$ 2. $\frac{1}{4}, \frac{2}{8}, \frac{4}{16}, \frac{3}{12}$ 3. $\frac{2}{5}, \frac{4}{10}, \frac{6}{15}, \frac{8}{20}$

Now you are probably saying to yourself, "There must be an easier way to tell whether fractions are equivalent than by drawing boxes." You are right. You can tell equivalent fractions by finding common factors in the numerator *and* denominator of each. Consider the halves in the model problem on the previous page:

$$\frac{2}{4} = \frac{2 \times 1}{2 \times 2} = \frac{2}{2} \times \frac{1}{2} \text{ (but } \frac{2}{2} = 1, \text{ so) } = 1 \times \frac{1}{2} = \frac{1}{2}$$

$$\frac{3}{6} = \frac{3 \times 1}{3 \times 2} = \frac{3}{3} \times \frac{1}{2} \text{ (but } \frac{3}{3} = 1, \text{ so) } = 1 \times \frac{1}{2} = \frac{1}{2}$$

$$\frac{4}{8} = \frac{4 \times 1}{4 \times 2} = \frac{4}{4} \times \frac{1}{2} \text{ (but } \frac{4}{4} = 1, \text{ so) } = 1 \times \frac{1}{2} = \frac{1}{2}$$

$$\frac{5}{10} = \frac{5 \times 1}{5 \times 2} = \frac{5}{5} \times \frac{1}{2} \text{ (but } \frac{5}{5} = 1, \text{ so) } = 1 \times \frac{1}{2} = \frac{1}{2}$$

By finding the common factor that is contained in both the numerator and denominator of each fraction, we are able to write each fraction in lowest possible terms. The lowest possible terms for each of the fractions turns out to be $\frac{1}{2}$.

In case you had not noticed, any fraction that is written as a number in the numerator over the same number in the denominator (e.g. $\frac{2}{2}, \frac{3}{3}, \frac{4}{4}, \frac{5}{5}$) has a value of 1. If you are not sure why, think about it. Two halves make one whole, three thirds make one whole, and so forth. If that is still not clear, remember that any fraction can also be thought of as representing a division. Hence, $\frac{2}{2}$ means 2 divided by 2, which is 1; and $\frac{3}{3}$ means 3 divided by 3, which is 1. There is yet one more key to writing equivalent fractions. Look at the following question:

$$\frac{2}{3} = \frac{?}{9}$$

Ask yourself what the 3 in the first denominator was multiplied by in order to get the 9 of the second denominator. Then multiply the first numerator by the same amount:

$$\frac{2}{3} \times \frac{?}{3} = \frac{?}{9} \quad \therefore^* \quad \frac{2}{3} \times \frac{3}{3} = \frac{6}{9}$$

Notice once again that the first fraction has been multiplied by another name for one $\left(\frac{3}{3}\right)$.

---

* $\therefore$ is mathematical shorthand for "therefore."

Examine the following two examples.

$$\boxed{Q} \quad \frac{3}{5} = \frac{?}{15}$$

$$\boxed{A} \quad \frac{3}{5} \times \frac{?}{3} = \frac{?}{15}$$

$$\frac{3}{5} \times \frac{3}{3} = \frac{9}{15}$$

$$\frac{3}{5} = \frac{9}{15}$$

$$\boxed{Q} \quad \frac{4}{9} = \frac{24}{?}$$

$$\boxed{A} \quad \frac{4}{9} \times \frac{6}{?} = \frac{24}{?}$$

$$\frac{4}{9} \times \frac{6}{6} = \frac{24}{54}$$

$$\frac{4}{9} = \frac{24}{54}$$

Test Yourself

Directions: Complete the following to form equivalent fractions.

1. $\dfrac{1}{4} = \dfrac{}{8} = \dfrac{}{12} = \dfrac{}{16} = \dfrac{}{20} = \dfrac{}{24} = \dfrac{}{28} = \dfrac{}{32} = \dfrac{}{36}$

2. $\dfrac{2}{6} = \dfrac{}{12} = \dfrac{}{18} = \dfrac{}{24} = \dfrac{}{30} = \dfrac{}{36} = \dfrac{}{42} = \dfrac{}{48} = \dfrac{}{54}$

3. $\dfrac{5}{8} = \dfrac{10}{} = \dfrac{15}{} = \dfrac{20}{} = \dfrac{25}{} = \dfrac{30}{} = \dfrac{35}{} = \dfrac{40}{} = \dfrac{45}{}$

4. $\dfrac{7}{12} = \dfrac{}{36}$

5. $\dfrac{9}{11} = \dfrac{36}{}$

6. $\dfrac{12}{15} = \dfrac{}{60}$

7. $\dfrac{5}{18} = \dfrac{}{54}$

8. $\dfrac{8}{9} = \dfrac{72}{}$

9. $\dfrac{5}{7} = \dfrac{30}{}$

10. $\dfrac{8}{13} = \dfrac{}{117}$

Answers

1. 2, 3, 4, 5, 6, 7, 8, 9   2. 4, 6, 8, 10, 12, 14, 16, 18   3. 16, 24, 32, 40, 48, 56, 64, 72

4. 21   5. 44   6. 48

7. 15   8. 81   9. 42

10. 72

Greatest Common Factor and Least Common Multiple

Look at the following group of numbers: {9, 12, 18, 24}. All of the numbers in the group (or set) are divisible by 1 (since all whole numbers are divisible by 1). All of the numbers in the set are also divisible by 3. Can you find another number that is a factor of all the numbers in that set? Since there is none, we say that 3 is the greatest number by which 9, 12, 18, and 24 are divisible. Said differently, 3 is the **Greatest Common Factor** (GCF) of 9, 12, 18, and 24.

Find the greatest common factor of { 10, 20, 30, 45 }. One, of course, is a factor of each number in the set. So is five. Ten divides perfectly into each number in the set except the last. That means the greatest common factor of the four numbers is 5.

Test Yourself

> **Directions:** Find the greatest common factor of each of the following sets.

1. 14, 21, 28 _____
2. 6, 12, 15 _____
3. 8, 12, 20 _____
4. 25, 40 _____
5. 18, 30 _____
6. 24, 36 _____
7. 45, 60 _____
8. 12, 17, 30 _____
9. 12, 16, 36 _____
10. 16, 24, 32 _____

Answers

1. 7	**2.** 3	**3.** 4	**4.** 5	**5.** 6
6. 12	**7.** 15	**8.** 1	**9.** 4	**10.** 8

Another extremely useful concept when dealing with fractional computations is the **Least Common Multiple (LCM).** You will see how it is used in the next section. The notion of LCM is not a complex one. Consider two numbers, say 2 and 4. Now look at some of the multiples of 2 and 4:

> Multiples of 2: 2, 4, 6, 8, 10, 12, 14, 16, 18, 20 ...
> Multiples of 4: 4, 8, 12, 16, 20 ...

Now, circle the common multiples of 2 and 4—that is, the numbers that appear in both sets of multiples. Don't read on until you've done it.* You should have circled 4, 8, 12, 16, and 20. Those are all common multiples of 2 and 4. Now, remember, the LCM is the common multiple with the lowest value (*least* common multiple). Therefore, the LCM for 2 and 4 is 4.

---

* For the technical reader, zero is also considered to be a multiple of all numbers. It has been omitted from the listings of multiples here, however, because its inclusion would have served no purpose. When seeking to find an LCM, we always want the LCM other than zero.

Test Yourself

Directions: List the first 10 multiples for each of the following pairs of numbers.

1. 3 _____   2. 5 _____
    4 _____   3 _____

3. 6 _____   4. 8 _____
    8 _____   4 _____

5. 8 _____   6. 9 _____
    12 _____   8 _____

Directions: Now name the LCM for each of the above.

7. _____   8. _____   9. _____
10. _____   11. _____   12. _____

Directions: Find the LCMs for the following pairs of numbers.

13.	7, 11	14.	8, 3	15.	5, 4	16.	4, 6	17.	9, 12
18.	8, 2	19.	5, 17	20.	2, 18	21.	3, 20	22.	18, 4
23.	6, 11	24.	9, 17	25.	4, 10	26.	10, 14	27.	6, 15
28.	6, 14	29.	2, 9	30.	15, 3	31.	4, 14	32.	8, 20

Answers

1. 3, 6, 9, 12, 15, 18, 21, 24, 27, 30
    4, 8, 12, 16, 20, 24, 28, 32, 36, 40

2. 5, 10, 15, 20, 25, 30, 35, 40, 45, 50
    3, 6, 9, 12, 15, 18, 21, 24, 27, 30

3. 6, 12, 18, 24, 30, 36, 42, 48, 54, 60
    8, 16, 24, 32, 40, 48, 56, 64, 72, 80

4. 8, 16, 24, 32, 40, 48, 56, 64, 72, 80
    4, 8, 12, 16, 20, 24, 28, 32, 36, 40

5. 8, 16, 24, 32, 40, 48, 56, 64, 72, 80
    12, 24, 36, 48, 60, 72, 84, 96, 108, 120

6. 9, 18, 27, 36, 45, 54, 63, 72, 81, 90
    8, 16, 24, 32, 40, 48, 56, 64, 72, 80

7.	12	8.	15	9.	24	10.	8	11.	24
12.	72	13.	77	14.	24	15.	20	16.	12
17.	36	18.	8	19.	85	20.	18	21.	60
22.	36	23.	66	24.	153	25.	20	26.	70
27.	30	28.	42	29.	18	30.	15	31.	28
32.	40								

Using Equivalent Fractions to Add and Subtract

When we first looked at addition and subtraction of fractional numbers, it was emphatically stated that addition and subtraction cannot be performed unless the denominators are the same. Unfortunately, the likelihood of all fractions that one wishes to add having identical denominators is slim. Fractions just are not generally that obliging. That fact, in case you were wondering, is responsible for the three sections that came between our first look at addition and subtraction of fractional numbers and this one. Consider the following example.

Q Mr. Anderson painted $\frac{3}{8}$ of his garage. Mrs. Anderson then painted another $\frac{1}{6}$. How much of the garage was painted?

A Since one painted a part of the garage, and then the other did another part of the same job, to find out how much of the job was done altogether, addition is required:

$$\frac{3}{8} + \frac{1}{6} = ?$$

You will recall that this addition, as it stands, is impossible to do. The denominators are not the same.

Now, however, you know that for any fraction, equivalent fractions can be written. Look, then, at the denominators of the two fractions in question. They are 8 and 6. We find that the least common multiple of 8 and 6 is 24.

$$\frac{3}{8} = \frac{?}{24}$$
$$\frac{1}{6} = \frac{?}{24}$$

$$\frac{}{8} + \frac{}{6}$$
$$\text{LCM} = 24$$

Having found that, we proceed to write equivalent fractions for $\frac{3}{8}$ and $\frac{1}{6}$, using 24 as the **Least Common Denominator** (abbreviated LCD).

Once the equivalent fractions are written, they may be added to find the sum.

$$\frac{3}{8} = \frac{9}{24}$$
$$+\frac{1}{6} = \frac{4}{24}$$
$$\overline{\phantom{+\frac{1}{6} =} \frac{13}{24}}$$

Now let's review the steps for adding fractions with unlike denominators:

1. Find the LCM of the denominators.
2. Make the LCM the least common denominator to use for writing equivalent fractions.
3. Write equivalent fractions with common denominators.
4. Add the equivalent fractions to find the sum.

The two examples below are done for you to show two different styles of notation. Decide on the style that will work best for you, and then stick with it.

Q $\dfrac{1}{3} + \dfrac{2}{5} = ?$

A LCM for 3 and 5 = 15

LCD = 15ths

$$\dfrac{1}{3} = \dfrac{5}{15}$$
$$+\dfrac{2}{5} = \dfrac{6}{15}$$
$$\dfrac{11}{15}$$

Q $\dfrac{4}{7} + \dfrac{1}{5} = ?$

A LCM for 7 and 5 = 35

LCD = 35ths

Write equivalent fractions:

$$\dfrac{}{35} + \dfrac{}{35} = \dfrac{}{35}$$
$$\dfrac{20}{35} + \dfrac{7}{35} = \dfrac{27}{35}$$

Subtraction of fractional numbers is, once again, performed in a way that is similar to addition. Find common denominators for the two fractions and, once the denominators are the same, simply subtract the numerators.

Test Yourself

Directions: Add or subtract.

1. $\dfrac{2}{3} + \dfrac{1}{4} = ?$

2. $\dfrac{1}{3} + \dfrac{1}{5} = ?$

3. $\dfrac{7}{8} - \dfrac{3}{5} = ?$

4. $\dfrac{9}{10} - \dfrac{5}{6} = ?$

5. $\dfrac{3}{8} + \dfrac{5}{12} = ?$

6. $\dfrac{3}{7} + \dfrac{5}{14} = ?$

7. $\dfrac{1}{2} + \dfrac{1}{9} = ?$

8. $\dfrac{3}{4} - \dfrac{7}{12} = ?$

9. $\dfrac{9}{13} - \dfrac{1}{2} = ?$

10. $\dfrac{3}{11} + \dfrac{5}{44} = ?$

11. $\dfrac{2}{5} + \dfrac{4}{9} = ?$

12. $\dfrac{17}{18} - \dfrac{11}{12} = ?$

Answers

1. $\dfrac{11}{12}$

2. $\dfrac{8}{15}$

3. $\dfrac{11}{40}$

4. $\dfrac{2}{30}$ or $\dfrac{1}{15}$

5. $\dfrac{19}{24}$

6. $\dfrac{11}{14}$

7. $\dfrac{11}{18}$

8. $\dfrac{2}{12}$ or $\dfrac{1}{6}$

9. $\dfrac{5}{26}$

10. $\dfrac{17}{44}$

11. $\dfrac{38}{45}$

12. $\dfrac{1}{36}$

Expressing Fractions in Lowest Terms

We have already discussed that the same fractional number can be written in many different ways (equivalent fractions). In many applications—and especially on the GED test—it is often necessary to express a fraction in its lowest terms. Writing a fraction in lowest terms means writing it with the smallest numerator and denominator possible. Here is an example:

$\dfrac{24}{32}$ can be written as an equivalent fraction by dividing the numerator and the denominator by 2. Doing so, we get

$\dfrac{12}{16}$ But this fraction can have its numerator and denominator divided by 2, in which case we get

$\dfrac{6}{8}$ You will note that 2 is a factor of both 6 and 8. Dividing both by 2, we get a result of

$\dfrac{3}{4}$ There is no number, except for 1, that can be divided exactly into 3 and into 4. That means that $\dfrac{3}{4}$ is the lowest terms in which $\dfrac{24}{32}$ can be expressed.

The method described above required finding a common factor in the numerator and denominator and then dividing both by that amount. The process was repeated and repeated until no common factor of both the numerator and denominator remained. This method works fine but may be unnecessarily time consuming. Do you remember what is meant by greatest common factor? Let's go back to the original fraction, $\dfrac{24}{32}$. What is the GCF for 24 and 32? If you calculated it to be 8, you are correct. 8 is the largest number by which both 24 and 32 can be divided exactly.

Having found the greatest common factor, divide both numerator and denominator by it:

$$\frac{24 \div 8}{32 \div 8} = \frac{3}{4}$$

Remember, when a number is divided by 1 (expressed as $\dfrac{8}{8}$ in this case), its value does not change. That means that $\dfrac{3}{4}$ is equivalent to $\dfrac{24}{32}$ and is, in fact, that fraction expressed in lowest (or simplest) terms.

Test Yourself

Directions: Express the following fractions in lowest terms.

1. $\dfrac{8}{10} =$ _____
2. $\dfrac{16}{32} =$ _____
3. $\dfrac{15}{20} =$ _____
4. $\dfrac{18}{24} =$ _____

5. $\dfrac{6}{15} =$ _____
6. $\dfrac{12}{18} =$ _____
7. $\dfrac{24}{30} =$ _____
8. $\dfrac{25}{45} =$ _____

9. $\dfrac{40}{50} =$ _____
10. $\dfrac{32}{36} =$ _____
11. $\dfrac{54}{63} =$ _____
12. $\dfrac{21}{28} =$ _____

13. $\dfrac{56}{80} =$ _____
14. $\dfrac{45}{75} =$ _____
15. $\dfrac{25}{50} =$ _____
16. $\dfrac{75}{100} =$ _____

Answers

1. $\dfrac{4}{5}$
2. $\dfrac{1}{2}$
3. $\dfrac{3}{4}$
4. $\dfrac{3}{4}$
5. $\dfrac{2}{5}$

6. $\dfrac{2}{3}$
7. $\dfrac{4}{5}$
8. $\dfrac{5}{9}$
9. $\dfrac{4}{5}$
10. $\dfrac{8}{9}$

11. $\dfrac{6}{7}$
12. $\dfrac{3}{4}$
13. $\dfrac{7}{10}$
14. $\dfrac{3}{5}$
15. $\dfrac{1}{2}$

16. $\dfrac{3}{4}$

It is customary for solutions to fractional arithmetic problems to be expressed in lowest terms. Some books refer to expressing fractions in lowest terms as "reducing." Even though that term was widely used not so long ago, be wary. When fractions are expressed in lowest terms, they are not reduced (made smaller). Rather, the sizes of the numerator and denominator are reduced. The fractional part represented, however, stays the same size.

Fractional arithmetic answers on the GED examination are usually expressed in lowest terms. It is therefore a good idea to get into the habit of always expressing the answers to any computations that you do in that form.

Test Yourself

Directions: Solve and express your answers in lowest terms.

1. $\dfrac{2}{4}+\dfrac{1}{3}=$ _____

2. $\dfrac{6}{8}+\dfrac{1}{4}=$ _____

3. $\dfrac{1}{3}+\dfrac{2}{6}=$ _____

4. $\dfrac{2}{6}+\dfrac{1}{9}=$ _____

5. $\dfrac{5}{12}+\dfrac{3}{12}=$ _____

6. $\dfrac{3}{15}+\dfrac{6}{15}=$ _____

7. $\dfrac{11}{12}-\dfrac{1}{4}=$ _____

8. $\dfrac{17}{20}-\dfrac{1}{4}=$ _____

9. $\dfrac{7}{18}-\dfrac{1}{6}=$ _____

10. $\dfrac{1}{3}+\dfrac{2}{7}=$ _____

11. $\dfrac{19}{20}-\dfrac{3}{4}=$ _____

12. $\dfrac{1}{6}+\dfrac{1}{2}=$ _____

13. $\dfrac{3}{5}-\dfrac{4}{15}=$ _____

14. $\dfrac{23}{24}-\dfrac{1}{8}=$ _____

15. $\dfrac{17}{18}-\dfrac{1}{2}=$ _____

Answers

1. $\dfrac{5}{6}$

2. 1

3. $\dfrac{2}{3}$

4. $\dfrac{4}{9}$

5. $\dfrac{2}{3}$

6. $\dfrac{3}{5}$

7. $\dfrac{2}{3}$

8. $\dfrac{3}{5}$

9. $\dfrac{2}{9}$

10. $\dfrac{13}{21}$

11. $\dfrac{1}{5}$

12. $\dfrac{2}{3}$

13. $\dfrac{1}{3}$

14. $\dfrac{5}{6}$

15. $\dfrac{4}{9}$

MULTIPLYING FRACTIONS

Multiplication of fractional numbers does not at first appear to be analogous to multiplication of whole numbers, even though it is performed in almost the same manner. That is because when whole numbers are multiplied together, the product (answer) is usually greater than those numbers. When fractional numbers are multiplied together, however, the product often becomes smaller than either of the fractions. To understand this fact, substitute the word "of" for the times sign in a whole number multiplication and then in a fractional one:

3×4 • • • •   Look at the four. Change the $\times$ to *of*.

↓   ↓

3 of 4   • • • •   Here are three of them.

           • • • •

           • • • •

           ↓

3 4's $=$ 12   Three fours make twelve.

Now see what happens with fractional numbers:

$\frac{1}{2} \times \frac{1}{2}$ Look at the second one half. Change the $\times$ to *of*.

$\frac{1}{2}$ of $\frac{1}{2}$ Here is one half of one half.

$\frac{1}{2}$ of $\frac{1}{2} = \frac{1}{4}$ Half of one half is one fourth.

As you can see from the illustration, fractional multiplication may be thought of as taking a part of a part of something. When taking a part of a part of something, it is only to be expected that the resulting part will be smaller than the original part.

Multiplication of fractional numbers is accomplished by multiplying numerator by numerator and denominator by denominator. We do not have to worry about finding common denominators before we do this. Examine the example below.

Q $\frac{2}{3} \times \frac{5}{7} =$

A $\frac{2 \times 5}{3 \times 7} = \frac{10}{21}$

Test Yourself

Directions: Multiply the following, and express the products in lowest terms.

1. $\frac{2}{3} \times \frac{3}{4} =$ _____   2. $\frac{5}{8} \times \frac{4}{5} =$ _____   3. $\frac{3}{7} \times \frac{4}{9} =$ _____

4. $\frac{5}{8} \times \frac{4}{7} =$ _____   5. $\frac{3}{8} \times \frac{1}{3} =$ _____   6. $\frac{3}{4} \times \frac{5}{8} =$ _____

7. $\frac{2}{11} \times \frac{6}{7} =$ _____   8. $\frac{1}{7} \times \frac{5}{6} =$ _____   9. $\frac{3}{10} \times \frac{5}{6} =$ _____

10. $\frac{2}{13} \times \frac{5}{11} =$ _____   11. $\frac{4}{5} \times \frac{3}{8} =$ _____   12. $\frac{1}{2} \times \frac{3}{4} =$ _____

13. $\frac{5}{6} \times \frac{1}{3} =$ _____   14. $\frac{2}{7} \times \frac{4}{11} =$ _____   15. $\frac{5}{8} \times \frac{1}{6} =$ _____

Answers

1. $\dfrac{1}{2}$ 2. $\dfrac{1}{2}$ 3. $\dfrac{4}{21}$ 4. $\dfrac{5}{14}$ 5. $\dfrac{1}{8}$

6. $\dfrac{15}{32}$ 7. $\dfrac{12}{77}$ 8. $\dfrac{5}{42}$ 9. $\dfrac{1}{4}$ 10. $\dfrac{10}{143}$

11. $\dfrac{3}{10}$ 12. $\dfrac{3}{8}$ 13. $\dfrac{5}{18}$ 14. $\dfrac{8}{77}$ 15. $\dfrac{5}{48}$

Canceling

A technique that makes it somewhat easier to multiply two fractions and express the result in lowest terms is known as **canceling**. Canceling is really a device that permits you to assure that the product of two fractions will be in lowest terms before even multiplying them. Here is how it works:

Example 1

Q $\dfrac{3}{8} \times \dfrac{4}{5}$

A $\dfrac{3}{\underset{2}{8}} \times \dfrac{\overset{1}{4}}{5}$

$\downarrow$

$\dfrac{3}{2} \times \dfrac{1}{5} = \dfrac{3}{10}$

Pretend that you are attempting to express a fraction in lowest terms. You would then seek to find the greatest common factor in the numerator and the denominator. Having found it, you would divide numerator and denominator by that number. When fractions are being multiplied together, you may seek to divide common factors out of either or all numerators and denominators, whether or not they are parts of the same fraction.

Example 2

Q $\dfrac{3}{7} \times \dfrac{21}{6}$

A $\dfrac{\overset{1}{3}}{7} \times \dfrac{21}{\underset{2}{6}}$ 3 is the GCF for 3 and 6…

$\downarrow$

$\dfrac{1}{\underset{1}{7}} \times \dfrac{\overset{3}{21}}{2}$

7 is the GCF for 7 and 21.

$\downarrow$

$\dfrac{1}{1} \times \dfrac{3}{2} = \dfrac{3}{2} = 1\dfrac{1}{2}$

Test Yourself

Directions: Cancel before multiplying, if possible. If your product is not in lowest terms, check back to see what else you could have canceled.

1. $\dfrac{3}{4} \times \dfrac{5}{9} =$ _____   2. $\dfrac{4}{7} \times \dfrac{3}{8} =$ _____   3. $\dfrac{5}{11} \times \dfrac{3}{10} =$ _____

4. $\dfrac{7}{12} \times \dfrac{6}{11} =$ _____   5. $\dfrac{5}{9} \times \dfrac{3}{8} =$ _____   6. $\dfrac{8}{15} \times \dfrac{5}{12} =$ _____

7. $\dfrac{8}{14} \times \dfrac{7}{10} =$ _____   8. $\dfrac{9}{16} \times \dfrac{4}{6} =$ _____   9. $\dfrac{15}{24} \times \dfrac{3}{10} =$ _____

10. $\dfrac{3}{7} \times \dfrac{14}{21} =$ _____   11. $\dfrac{9}{32} \times \dfrac{16}{12} =$ _____   12. $\dfrac{7}{18} \times \dfrac{9}{21} =$ _____

13. $\dfrac{18}{25} \times \dfrac{5}{9} \times \dfrac{5}{2} =$ _____   14. $\dfrac{4}{9} \times \dfrac{3}{6} \times \dfrac{3}{2} =$ _____   15. $\dfrac{13}{24} \times \dfrac{8}{26} \times \dfrac{3}{5} =$ _____

Answers

1. $\dfrac{5}{12}$ 2. $\dfrac{3}{14}$ 3. $\dfrac{3}{22}$ 4. $\dfrac{7}{22}$ 5. $\dfrac{5}{24}$

6. $\dfrac{2}{9}$ 7. $\dfrac{2}{5}$ 8. $\dfrac{3}{8}$ 9. $\dfrac{3}{16}$ 10. $\dfrac{2}{7}$

11. $\dfrac{3}{8}$ 12. $\dfrac{1}{6}$ 13. 1 14. $\dfrac{1}{3}$ 15. $\dfrac{1}{10}$

DIVISION OF FRACTIONS

Division of fractional numbers takes advantage of the fact that division and multiplication are reciprocal operations. To understand what that means, it is necessary to consider the meaning of reciprocals. The reciprocal of a number is the number by which you must multiply to get a product of one. The reciprocal of 2 is $\dfrac{1}{2}$ since $2 \times \dfrac{1}{2} = 1$. The reciprocal of $\dfrac{1}{4}$ is 4 since $\dfrac{1}{4} \times 4 = 1$.

Test Yourself

Directions: Name the reciprocal.

1. 7	2. 9	3. 23	4. 67	5. $\frac{1}{8}$
6. $\frac{1}{7}$	7. $\frac{1}{25}$	8. $\frac{1}{235}$	9. $\frac{2}{3}$	10. $\frac{3}{4}$
11. $\frac{5}{17}$	12. $\frac{7}{36}$	13. $\frac{6}{93}$		

Answers

1. $\frac{1}{7}$	2. $\frac{1}{9}$	3. $\frac{1}{23}$	4. $\frac{1}{67}$	5. 8
6. 7	7. 25	8. 235	9. $\frac{3}{2}$	10. $\frac{4}{3}$
11. $\frac{17}{5}$	12. $\frac{36}{7}$	13. $\frac{93}{6}$		

Any division example is capable of being solved as a multiplication, where the reciprocal of the divisor is used to multiply the dividend. Consider the two division examples below. Each has been rewritten as a multiplication by the reciprocal of the divisor.

NOTE

It is always the number being *divided by* whose reciprocal is used.

Q $8 \div 2$

A $8 \times \frac{1}{2}$

$\downarrow$

$\frac{8}{1} \times \frac{1}{2} = \frac{8}{2} = 4$

Q $5\overline{)20}$

A $20 \div 5$

$\downarrow$

$20 \times \frac{1}{5}$

$\downarrow$

$\frac{20}{1} \times \frac{1}{5} = \frac{20}{5} = 4$

Look at the divisions below. Change each to a reciprocal multiplication and solve. The numbers are familiar ones, so that you should have no difficulty in determining whether your answers are correct. It is the method that is of concern here.

Test Yourself

Directions: Solve the following.

1. $12 \div 3 = \dfrac{12}{1} \times \underline{\hspace{1cm}} = \underline{\hspace{1cm}}$

2. $6\overline{)30} = 30 \div \underline{\hspace{1cm}} = \dfrac{30}{1} \times \underline{\hspace{1cm}} = \underline{\hspace{1cm}}$

3. $15 \div 5 = \dfrac{15}{1} \times \underline{\hspace{1cm}} = \underline{\hspace{1cm}}$

4. $18 \div 9 = \dfrac{18}{1} \times \underline{\hspace{1cm}} = \underline{\hspace{1cm}}$

5. $32 \div 4 = \underline{\hspace{1cm}} \times \underline{\hspace{1cm}} = \underline{\hspace{1cm}}$

6. $21 \div 3 = \underline{\hspace{1cm}} \times \underline{\hspace{1cm}} = \underline{\hspace{1cm}}$

7. $63 \div 7 = \underline{\hspace{1cm}} \times \underline{\hspace{1cm}} = \underline{\hspace{1cm}}$

Answers

1. $\dfrac{1}{3}, 4$

2. $6, \dfrac{1}{6}, 5$

3. $\dfrac{1}{5}, 3$

4. $\dfrac{1}{9}, 2$

5. $\dfrac{32}{1}, \dfrac{1}{4}, 8$

6. $\dfrac{21}{1}, \dfrac{1}{3}, 7$

7. $\dfrac{63}{1}, \dfrac{1}{7}, 9$

Fractional division is identical to the reciprocal multiplication form shown earlier. In order to divide one fraction by another, we multiply by the reciprocal of the divisor. Note that in the standard form in which fractional multiplications are written, the divisor is always the second numeral. That means that it is always the second numeral whose reciprocal you must use to multiply by.

Q $\dfrac{3}{4} \div \dfrac{2}{3}$

A $\dfrac{3}{4} \times \dfrac{3}{2}$

$\downarrow$

$\dfrac{3}{4} \times \dfrac{3}{2} = \dfrac{9}{8} = 1\dfrac{1}{8}$

Q $\dfrac{5}{14} \div \dfrac{1}{2}$

A $\dfrac{5}{14} \times \dfrac{2}{1}$

$\downarrow$

$\dfrac{5}{14_7} \times \dfrac{\overset{1}{2}}{1} = \dfrac{5}{7}$

Note that in the second example, we can cancel as usual after we rewrite the division as a multiplication. However, we cannot cancel numbers in a division problem. Note that in the first line of the first example, the 3s cannot be cancelled.

Test Yourself

Directions: Divide. Express the quotients in lowest terms.

1. $\dfrac{1}{2} \div \dfrac{1}{4} = $ _____

2. $\dfrac{3}{5} \div \dfrac{2}{7} = $ _____

3. $\dfrac{6}{11} \div \dfrac{4}{5} = $ _____

4. $\dfrac{3}{8} \div \dfrac{3}{4} = $ _____

5. $\dfrac{6}{7} \div \dfrac{12}{14} = $ _____

6. $\dfrac{5}{9} \div \dfrac{4}{5} = $ _____

7. $\dfrac{7}{12} \div \dfrac{3}{4} = $ _____

8. $\dfrac{11}{16} \div \dfrac{22}{32} = $ _____

9. $\dfrac{5}{6} \div \dfrac{10}{11} = $ _____

10. $\dfrac{9}{16} \div \dfrac{2}{3} = $ _____

11. $\dfrac{7}{13} \div \dfrac{5}{6} = $ _____

12. $\dfrac{2}{3} \div \dfrac{1}{4} = $ _____

13. $\dfrac{5}{4} \div \dfrac{1}{3} = $ _____

14. $\dfrac{2}{3} \div \dfrac{3}{4} = $ _____

15. $\dfrac{1}{5} \div \dfrac{3}{8} = $ _____

16. $\dfrac{9}{10} \div \dfrac{11}{12} =$ _____ 17. $\dfrac{5}{9} \div \dfrac{6}{7} =$ _____ 18. $\dfrac{1}{2} \div \dfrac{5}{7} =$ _____

19. $\dfrac{2}{7} \div \dfrac{5}{6} =$ _____ 20. $\dfrac{1}{4} \div \dfrac{3}{8} =$ _____ 21. $\dfrac{4}{9} \div \dfrac{3}{5} =$ _____

Answers

1. 2 **2.** $2\dfrac{1}{10}$ **3.** $\dfrac{15}{22}$ **4.** $\dfrac{1}{2}$ **5.** 1

6. $\dfrac{25}{36}$ **7.** $\dfrac{7}{9}$ **8.** 1 **9.** $\dfrac{11}{12}$ **10.** $\dfrac{27}{32}$

11. $\dfrac{42}{65}$ **12.** $2\dfrac{2}{3}$ **13.** $3\dfrac{3}{4}$ **14.** $\dfrac{8}{9}$ **15.** $\dfrac{8}{15}$

16. $\dfrac{54}{55}$ **17.** $\dfrac{35}{54}$ **18.** $\dfrac{7}{10}$ **19.** $\dfrac{12}{35}$ **20.** $\dfrac{2}{3}$

21. $\dfrac{20}{27}$

MIXED NUMERALS

A numeral consisting of a digit (or several digits) and a fraction is known as a mixed numeral. 2 is a digit. $\frac{1}{2}$ is a fraction. $2\frac{1}{2}$ (read two and one half) is a mixed numeral. Any fraction whose denominator is smaller than its numerator may be written as a whole or as a mixed numeral. The transformation is accomplished by dividing the numerator by the denominator. Look at the model examples.

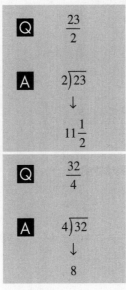

Adding with Mixed Numerals

Since mixed numerals contain both a whole number portion and a fractional number portion, the technique used in adding them consists of treating each component separately. First the fractions are combined in the usual manner. Then the whole numbers are added. It is important that the addition take place in that order: **fractions first, then whole numbers**. That is so if the fractional addition yields a mixed numeral, the whole number portion of it may be incorporated in the addition of the whole numbers.

Q $3\frac{1}{8} + 4\frac{3}{8}$

A

$$\begin{array}{r} 3\frac{1}{8} \\ +4\frac{3}{8} \\ \hline \frac{4}{8} \end{array}$$

$$\downarrow$$

$$\begin{array}{r} 3\frac{1}{8} \\ +4\frac{3}{8} \\ \hline 7\frac{4}{8} \end{array} \rightarrow 7\frac{1}{2}$$

Q $4\frac{7}{16} + 3\frac{11}{16}$

A

$$\begin{array}{r} 4\frac{7}{16} \\ +3\frac{11}{16} \\ \hline \frac{18}{16} \end{array} \longrightarrow 1\frac{1}{8}$$

$$\begin{array}{r} {}^{1} \\ 4\frac{7}{16} \\ +3\frac{11}{16} \\ \hline 8\frac{1}{8} \end{array}$$

Q $5\dfrac{1}{8}$ → $5\dfrac{2}{16}$ → $5\dfrac{2}{16}$

$+3\dfrac{7}{16}$ → $3\dfrac{7}{16}$ → $3\dfrac{7}{16}$

$\dfrac{9}{16}$ $8\dfrac{9}{16}$

A $8\dfrac{9}{16}$

Q $3\dfrac{9}{10}$ → $3\dfrac{18}{20}$ → $3\dfrac{18}{20}$ → $^{1}3\dfrac{18}{20}$ → $^{1}3\dfrac{18}{20}$

$+4\dfrac{17}{20}$ → $4\dfrac{17}{20}$ → $4\dfrac{17}{20}$ → $4\dfrac{17}{20}$ → $4\dfrac{17}{20}$

$\dfrac{35}{20}$ $=$ $1\dfrac{3}{4}$ $\dfrac{3}{4}$ $8\dfrac{3}{4}$

A $8\dfrac{3}{4}$

Test Yourself

Directions: Add. Express the sum as a mixed numeral in lowest terms.

1. $4\dfrac{1}{3}$
    $+2\dfrac{1}{3}$

2. $5\dfrac{2}{5}$
    $+2\dfrac{2}{5}$

3. $3\dfrac{2}{7}$
    $+4\dfrac{3}{7}$

4. $1\dfrac{5}{8}$
    $+6\dfrac{1}{8}$

5. $3\dfrac{1}{6}$
    $+2\dfrac{2}{3}$

6. $4\dfrac{1}{5}$
    $+3\dfrac{3}{10}$

7. $6\dfrac{1}{7}$
    $+4\dfrac{5}{21}$

8. $3\dfrac{1}{8}$
    $+4\dfrac{5}{16}$

9. $2\dfrac{1}{4}$
    $+3\dfrac{1}{6}$

10. $4\dfrac{5}{9}$
    $+3\dfrac{1}{4}$

11. $6\dfrac{1}{3}$
    $+5\dfrac{1}{4}$

12. $4\dfrac{2}{7}$
    $+3\dfrac{1}{5}$

13. $3\dfrac{5}{8}$ 14. $2\dfrac{7}{9}$ 15. $7\dfrac{3}{4}$ 16. $4\dfrac{7}{8}$

$+2\dfrac{1}{2}$ $+6\dfrac{5}{6}$ $+5\dfrac{2}{3}$ $+3\dfrac{5}{6}$

17. $3\dfrac{8}{11}$ 18. $5\dfrac{8}{9}$ 19. $2\dfrac{5}{6}$ 20. $4\dfrac{9}{10}$

$+2\dfrac{3}{4}$ $+6\dfrac{7}{8}$ $+3\dfrac{3}{4}$ $+2\dfrac{7}{8}$

Answers

1. $6\dfrac{2}{3}$ 2. $7\dfrac{4}{5}$ 3. $7\dfrac{5}{7}$ 4. $7\dfrac{3}{4}$ 5. $5\dfrac{5}{6}$

6. $7\dfrac{1}{2}$ 7. $10\dfrac{8}{21}$ 8. $7\dfrac{7}{16}$ 9. $5\dfrac{5}{12}$ 10. $7\dfrac{29}{36}$

11. $11\dfrac{7}{12}$ 12. $7\dfrac{17}{35}$ 13. $6\dfrac{1}{8}$ 14. $9\dfrac{11}{18}$ 15. $13\dfrac{5}{12}$

16. $8\dfrac{17}{24}$ 17. $6\dfrac{21}{44}$ 18. $12\dfrac{55}{72}$ 19. $6\dfrac{7}{12}$ 20. $7\dfrac{31}{40}$

Subtracting with Mixed Numerals

Up to a point, subtraction of mixed numerals is the same as addition of mixed numerals, except, of course, that you subtract instead of add. Try the following subtractions, subtracting the fractions first, and then the whole numbers.

Test Yourself

Directions: Solve the following.

1. $5\dfrac{7}{8}$ 2. $9\dfrac{4}{5}$ 3. $8\dfrac{3}{4}$ 4. $6\dfrac{11}{16}$

$-3\dfrac{1}{8}$ $-3\dfrac{1}{6}$ $-5\dfrac{2}{3}$ $-2\dfrac{3}{8}$

5. $8\dfrac{3}{5}$ 6. $10\dfrac{9}{10}$ 7. $6\dfrac{11}{12}$ 8. $9\dfrac{5}{6}$

$-4\dfrac{1}{7}$ $-4\dfrac{3}{8}$ $-2\dfrac{1}{2}$ $-4\dfrac{1}{3}$

Answers

1. $2\dfrac{3}{4}$ 2. $6\dfrac{19}{30}$ 3. $3\dfrac{1}{12}$ 4. $4\dfrac{5}{16}$ 5. $4\dfrac{16}{35}$

6. $6\dfrac{21}{40}$ 7. $4\dfrac{5}{12}$ 8. $5\dfrac{1}{2}$

Not all fractional subtractions, however, are quite this straightforward. Consider, for example, the following example.

$$5\dfrac{1}{4}$$
$$-2\dfrac{3}{4}$$

You should notice that since $\dfrac{3}{4}$ is larger than $\dfrac{1}{4}$, it cannot be subtracted from it.*

In order to be able to subtract one fraction from the other, it is necessary to take the first fraction and rename one from the whole number five as a fraction, $\dfrac{4}{4}$.

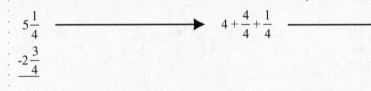

$$5\dfrac{1}{4} \longrightarrow 4 + \dfrac{4}{4} + \dfrac{1}{4} \longrightarrow \begin{array}{c} 4\dfrac{5}{4} \\ -2\dfrac{3}{4} \\ \hline \downarrow \end{array}$$

$$\begin{array}{c} 4\dfrac{5}{4} \\ -2\dfrac{3}{4} = 2\dfrac{1}{2} \end{array}$$

Now, of course, since there is a larger fraction on top than on the bottom, it is possible to subtract:

---

*At least, it cannot be subtracted from it and yield a positive value, which at this point is what we are interested in maintaining. We will consider negative numbers later in this book.

Test Yourself

Directions: Using the previous example as your guide, subtract the following.

1. $6\dfrac{3}{8}$ $5+\dfrac{8}{8}+\dfrac{3}{8}$ $5\dfrac{11}{8}$

    $-2\dfrac{5}{8}$ $-2\dfrac{5}{8}$ $-2\dfrac{5}{8}$

    $\phantom{-2\dfrac{5}{8}\quad-2\dfrac{5}{8}}\;3\dfrac{6}{8}=$ _____

2. $7\dfrac{2}{5}$

    $-1\dfrac{4}{5}$

3. $6\dfrac{7}{16}$

    $-3\dfrac{11}{16}$

4. $9\dfrac{1}{3}$

    $-5\dfrac{2}{3}$

5. $6\dfrac{3}{7}$

    $-3\dfrac{6}{7}$

6. $8\dfrac{1}{4}$

    $-5\dfrac{3}{4}$

7. $7\dfrac{1}{6}$

    $-2\dfrac{5}{6}$

8. $6\dfrac{1}{9}$

    $-4\dfrac{8}{9}$

9. $9\dfrac{2}{5}$

    $-8\dfrac{3}{5}$

Answers

1. $3\dfrac{3}{4}$
2. $5\dfrac{3}{5}$
3. $2\dfrac{3}{4}$
4. $3\dfrac{2}{3}$
5. $2\dfrac{4}{7}$

6. $2\dfrac{1}{2}$
7. $4\dfrac{1}{3}$
8. $1\dfrac{2}{9}$
9. $\dfrac{4}{5}$

Of course, unlike the subtractions shown previously, there is also no guarantee that the denominators of the fractional parts of the mixed numerals are going to be identical. The following examples put together all of the steps that may be encountered in a mixed numeral subtraction.

Example

$$7\dfrac{2}{5} \longrightarrow 7\dfrac{16}{40} \longrightarrow 6\dfrac{40}{40} + \dfrac{16}{40} \longrightarrow 6\dfrac{56}{40}$$
$$-3\dfrac{5}{8} \longrightarrow -3\dfrac{25}{40} \xrightarrow{\hspace{4cm}} -3\dfrac{25}{40}$$
$$3\dfrac{31}{40}$$

Example

$$8\dfrac{2}{3} \longrightarrow 8\dfrac{8}{12} \longrightarrow 7\dfrac{12}{12} + \dfrac{8}{12} \longrightarrow 7\dfrac{20}{12}$$
$$-5\dfrac{3}{4} \longrightarrow -5\dfrac{9}{12} \xrightarrow{\hspace{4cm}} -5\dfrac{9}{12}$$
$$2\dfrac{11}{12}$$

Test Yourself

Directions: Solve the following mixed numeral subtractions.

1. $4\frac{7}{8}$
   $-2\frac{3}{8}$

2. $6\frac{5}{6}$
   $-2\frac{3}{4}$

3. $8\frac{2}{3}$
   $-5\frac{1}{4}$

4. $5\frac{3}{8}$
   $-2\frac{2}{3}$

5. $6\frac{1}{4}$
   $-4\frac{1}{2}$

6. $7\frac{1}{5}$
   $-3\frac{1}{2}$

7. $6\frac{3}{8}$
   $-2\frac{3}{4}$

8. $11\frac{2}{7}$
   $-8\frac{4}{5}$

9. $12\frac{1}{3}$
   $-5\frac{7}{12}$

10. $5\frac{4}{9}$
    $-3\frac{7}{8}$

11. $6\frac{1}{6}$
    $-4\frac{9}{16}$

12. $12\frac{1}{2}$
    $-11\frac{7}{12}$

13. $8\frac{3}{10}$
    $-4\frac{5}{8}$

14. $9\frac{5}{13}$
    $-8\frac{3}{4}$

15. $7\frac{1}{3}$
    $-6\frac{5}{7}$

Answers

1. $2\frac{1}{2}$

2. $4\frac{1}{12}$

3. $3\frac{5}{12}$

4. $2\frac{17}{24}$

5. $1\frac{3}{4}$

6. $3\frac{7}{10}$

7. $3\frac{5}{8}$

8. $2\frac{17}{35}$

9. $6\frac{3}{4}$

10. $1\frac{41}{72}$

11. $1\frac{29}{48}$

12. $\frac{11}{12}$

13. $3\frac{27}{40}$

14. $\frac{33}{52}$

15. $\frac{13}{21}$

Multiplying and Dividing with Mixed Numerals

Unlike addition and subtraction, multiplication and division of mixed numerals is not readily carried out by working first with the fractions and then with the whole numbers. Rather, it is necessary to first convert the entire mixed numeral to a fraction. Then, the fractions are multiplied or divided in the normal fashion.

To change a mixed numeral to a fraction, follow the formula below to obtain the numerator of the fraction:

Denominator × Whole Number + Numerator

Change $3\frac{5}{8}$ to a fraction:

$$8 \times 3 = 24$$
$$24 + 5 = 29$$

$$\therefore \ 3\frac{5}{8} = \frac{29}{8}$$

Note: The denominator is unchanged.

First the denominator (8) is multiplied by the whole number (3). Then the numerator (5) is added to the product (24). That determines the new numerator (29). The denominator for the new fraction is the same as the old denominator.

Here is another example:

$$2\frac{1}{3} \longrightarrow 3 \times 2 + 1 \longrightarrow 6 + 1 \longrightarrow \frac{7}{3}$$

Numerator

Integer

Denominator

$$\therefore \ 2\frac{1}{3} = \frac{7}{3}$$

Test Yourself

Directions: Express the following mixed numerals as fractions. Refer to the examples above for guidance, if needed.

1. $2\frac{1}{2} =$ _____

2. $3\frac{1}{4} =$ _____

3. $4\frac{1}{5} =$ _____

4. $5\frac{2}{3} =$ _____

5. $6\frac{3}{8} =$ _____

6. $7\frac{4}{5} =$ _____

7. $8\frac{3}{4} =$ _____

8. $9\frac{2}{7} =$ _____

9. $6\frac{7}{8} =$ _____

10. $3\frac{4}{5} =$ _____

11. $5\frac{6}{7} =$ _____

12. $4\frac{7}{12} =$ _____

13. $5\frac{8}{9} =$ _____

14. $2\frac{3}{16} =$ _____

15. $3\frac{5}{9} =$ _____

16. $4\frac{5}{24} =$ _____

17. $4\frac{7}{11} =$ _____

18. $3\frac{7}{18} =$ _____

19. $2\frac{5}{23} =$ _____

20. $6\frac{7}{20} =$ _____

Answers

1. $\dfrac{5}{2}$	2. $\dfrac{13}{4}$	3. $\dfrac{21}{5}$	4. $\dfrac{17}{3}$	5. $\dfrac{51}{8}$
6. $\dfrac{39}{5}$	7. $\dfrac{35}{4}$	8. $\dfrac{65}{7}$	9. $\dfrac{55}{8}$	10. $\dfrac{19}{5}$
11. $\dfrac{41}{7}$	12. $\dfrac{55}{12}$	13. $\dfrac{53}{9}$	14. $\dfrac{35}{16}$	15. $\dfrac{32}{9}$
16. $\dfrac{101}{24}$	17. $\dfrac{51}{11}$	18. $\dfrac{61}{18}$	19. $\dfrac{51}{23}$	20. $\dfrac{127}{20}$

When multiplying or dividing mixed numerals, as previously noted, you must first change those mixed numerals to fractions. Since the fractions' terms will often be large, always try to cancel (if possible) before multiplying.

Test Yourself

Directions: Change to fractions, then multiply or divide (as indicated)*.

1. $2\dfrac{1}{2} \times 3\dfrac{1}{2} = $ _____ 2. $3\dfrac{1}{4} \times 2\dfrac{2}{3} = $ _____ 3. $4\dfrac{1}{3} \div 5\dfrac{1}{2} = $ _____

4. $3\dfrac{2}{3} \div 2\dfrac{1}{4} = $ _____ 5. $5\dfrac{5}{6} \times 4\dfrac{3}{8} = $ _____ 6. $2\dfrac{3}{16} \div 1\dfrac{5}{8} = $ _____

7. $3\dfrac{5}{7} \div 2\dfrac{8}{21} = $ _____ 8. $4\dfrac{5}{8} \times 3\dfrac{4}{7} = $ _____ 9. $4\dfrac{3}{8} \div 3\dfrac{7}{12} = $ _____

10. $2\dfrac{5}{9} \times 3\dfrac{3}{7} = $ _____ 11. $5\dfrac{3}{4} \times 4\dfrac{5}{9} = $ _____ 12. $3\dfrac{5}{16} \div 4\dfrac{15}{24} = $ _____

Answers

1. $8\dfrac{3}{4}$	2. $8\dfrac{2}{3}$	3. $\dfrac{26}{33}$	4. $1\dfrac{17}{27}$	5. $25\dfrac{25}{48}$
6. $1\dfrac{9}{26}$	7. $1\dfrac{14}{25}$	8. $16\dfrac{29}{56}$	9. $1\dfrac{19}{86}$	10. $8\dfrac{16}{21}$
11. $26\dfrac{7}{36}$	12. $\dfrac{53}{74}$			

*If you are unsure of how to divide here, review the section on dividing fractions earlier in this chapter.

EXERCISES: EQUIVALENT FRACTIONS

Directions: Complete each to form an equivalent fraction.

1. $\dfrac{3}{5} = \dfrac{-}{30}$

2. $\dfrac{4}{7} = \dfrac{12}{-}$

3. $\dfrac{3}{4} = \dfrac{75}{-}$

4. $\dfrac{5}{9} = \dfrac{-}{81}$

5. $\dfrac{1}{5} = \dfrac{-}{20}$

6. $\dfrac{3}{11} = \dfrac{15}{-}$

7. $\dfrac{5}{8} = \dfrac{-}{32}$

8. $\dfrac{11}{12} = \dfrac{-}{48}$

9. $\dfrac{6}{13} = \dfrac{-}{39}$

10. $\dfrac{2}{3} = \dfrac{12}{-}$

11. $\dfrac{4}{15} = \dfrac{-}{60}$

12. $\dfrac{7}{18} = \dfrac{-}{54}$

ANSWERS

1. 18	**2.** 21	**3.** 100	**4.** 45	**5.** 4
6. 55	**7.** 20	**8.** 44	**9.** 18	**10.** 18
11. 16	**12.** 21			

EXERCISES: ADDING AND SUBTRACTING FRACTIONS

Directions: Solve. Express your answer in lowest terms.

1. $\dfrac{4}{8} + \dfrac{2}{6} =$

2. $\dfrac{11}{15} - \dfrac{1}{5} =$

3. $\dfrac{1}{2} + \dfrac{3}{4} =$

4. $\dfrac{29}{30} - \dfrac{1}{6} =$

5. $\dfrac{7}{8} - \dfrac{1}{12} =$

6. $\dfrac{15}{40} - \dfrac{4}{16} =$

7. $\dfrac{2}{5} + \dfrac{3}{8} =$

8. $\dfrac{2}{6} + \dfrac{2}{8} =$

9. $\dfrac{4}{14} + \dfrac{2}{6} =$

ANSWERS

1. $\dfrac{5}{6}$ 2. $\dfrac{8}{15}$ 3. $1\dfrac{1}{4}$ 4. $\dfrac{4}{5}$ 5. $\dfrac{19}{24}$

6. $\dfrac{1}{8}$ 7. $\dfrac{31}{40}$ 8. $\dfrac{7}{12}$ 9. $\dfrac{13}{21}$

EXERCISES : MULTIPLYING AND DIVIDING FRACTIONS

Directions: Cancel, then multiply.

1. $\dfrac{4}{9} \times \dfrac{3}{4} =$ 2. $\dfrac{7}{18} \times \dfrac{9}{14} =$ 3. $\dfrac{10}{24} \times \dfrac{8}{15} =$ 4. $\dfrac{12}{17} \times \dfrac{1}{8} =$

5. $\dfrac{15}{36} \times \dfrac{18}{20} =$ 6. $\dfrac{27}{40} \times \dfrac{12}{27} =$ 7. $\dfrac{35}{42} \times \dfrac{14}{70} =$ 8. $\dfrac{24}{34} \times \dfrac{17}{48} =$

9. $\dfrac{21}{36} \times \dfrac{9}{14} =$ 10. $\dfrac{15}{24} \times \dfrac{8}{25} =$ 11. $\dfrac{21}{56} \times \dfrac{28}{35} =$ 12. $\dfrac{54}{72} \times \dfrac{45}{63} =$

13. $\dfrac{2}{3} \div \dfrac{1}{4} =$ 14. $\dfrac{5}{8} \div \dfrac{3}{12} =$ 15. $\dfrac{2}{7} \div \dfrac{3}{14} =$ 16. $\dfrac{5}{9} \div \dfrac{1}{3} =$

17. $\dfrac{4}{7} \div \dfrac{6}{14} =$ 18. $\dfrac{3}{8} \div \dfrac{3}{4} =$ 19. $\dfrac{5}{12} \div \dfrac{10}{18} =$ 20. $\dfrac{1}{2} \div \dfrac{1}{3} =$

21. $\dfrac{3}{4} \div \dfrac{2}{3} =$ 22. $\dfrac{8}{19} \div \dfrac{4}{38} =$ 23. $\dfrac{4}{38} \div \dfrac{8}{19} =$ 24. $\dfrac{10}{18} \div \dfrac{5}{12} =$

ANSWERS

1. $\dfrac{1}{3}$ 2. $\dfrac{1}{4}$ 3. $\dfrac{2}{9}$ 4. $\dfrac{3}{34}$ 5. $\dfrac{3}{8}$ 6. $\dfrac{3}{10}$

7. $\dfrac{1}{6}$ 8. $\dfrac{1}{4}$ 9. $\dfrac{3}{8}$ 10. $\dfrac{1}{5}$ 11. $\dfrac{3}{10}$ 12. $\dfrac{15}{28}$

13. $\dfrac{8}{3}$ or $2\dfrac{2}{3}$ 14. $\dfrac{5}{2}$ or $2\dfrac{1}{2}$ 15. $\dfrac{4}{3}$ or $1\dfrac{1}{3}$ 16. $\dfrac{5}{3}$ or $1\dfrac{2}{3}$ 17. $\dfrac{4}{3}$ or $1\dfrac{1}{3}$ 18. $\dfrac{1}{2}$

19. $\dfrac{3}{4}$ 20. $\dfrac{3}{2}$ or $1\dfrac{1}{2}$ 21. $\dfrac{9}{8}$ or $1\dfrac{1}{8}$ 22. 4 23. $\dfrac{1}{4}$ 24. $\dfrac{4}{3}$ or $1\dfrac{1}{3}$

EXERCISES: SUBTRACTING MIXED NUMERALS

Directions: Subtract and express the differences in lowest terms.

1. $4\dfrac{3}{5}$ 2. $5\dfrac{1}{3}$ 3. $4\dfrac{2}{7}$ 4. $6\dfrac{1}{16}$ 5. $9\dfrac{1}{4}$ 6. $10\dfrac{1}{3}$

$-2\dfrac{4}{5}$ $-2\dfrac{3}{8}$ $-3\dfrac{7}{8}$ $-2\dfrac{3}{8}$ $-7\dfrac{5}{6}$ $-7\dfrac{2}{3}$

7. $8\dfrac{3}{8} - 4\dfrac{2}{3} =$ ___

8. $6\dfrac{1}{5} - 3\dfrac{5}{6} =$ ___

9. $12\dfrac{1}{3} - 5\dfrac{7}{8} =$ ___

10. $7\dfrac{2}{5} - 3\dfrac{11}{15} =$ ___

ANSWERS

1. $1\frac{4}{5}$
2. $2\frac{23}{24}$
3. $\frac{23}{56}$
4. $3\frac{11}{16}$
5. $1\frac{5}{12}$
6. $2\frac{2}{3}$
7. $3\frac{17}{24}$
8. $2\frac{11}{30}$
9. $6\frac{11}{24}$
10. $3\frac{2}{3}$

SUMMING IT UP

- A fraction is a part of a unit.

- A fraction has a numerator and a denominator.

- Fractions cannot be added or subtracted unless the denominators are all the same.

- To be multiplied, fractions need not have the same denominators.

- Most fraction problems can be arranged in the form: "What fraction of a number is another number?"

Decimal Fractions

OVERVIEW

- What is a decimal fraction?
- Expressing common fractions as decimal fractions
- Rounding decimal fractions
- Adding decimal fractions
- Subtracting decimal fractions
- Multiplying decimal fractions
- Multiplying and dividing by powers of 10
- Dividing decimal fractions
- Summing it up

WHAT IS A DECIMAL FRACTION?

Having completed our study of common fractions, perhaps you thought that you were through seeing the word fraction for awhile. In fact, even more important to the mathematics of today is a different kind of fraction—the decimal fraction. What makes decimal fractions so important is that they are a logical rightward extension of the decimal system in which we express all whole numbers. Decimal fractions follow all the rules of place value numeration and can be easily fit together with whole number operations. Two rather recent innovations make decimal fractions particularly significant in our day-to-day lives. One is the fact that pocket calculators, as well as computers, are capable of readily handling fractions when they are in decimal form. Second is that the universal adoption of the metric system for calculating weights and measures may soon make common fractions all but obsolete, with computations taking place with decimals exclusively. Finally, as an additional sidelight, it should be noted that our monetary system is based upon decimal notation, with $0.25 being a quarter of a dollar, $.50 being a half dollar, and so forth.

The word **decimal** means tenth. As you move from left to right across a place-value chart, each place is worth one tenth as much as the one to its immediate left:

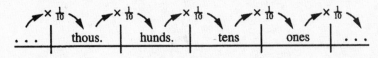

chapter 5

You will notice in the preceding chart that there is no beginning and there is no end. As you move along the chart, each place is multiplied by one tenth to find the value of the place to the right. After the ones' place, a decimal point is placed. That decimal point is used to separate the whole numbers from the fractions:

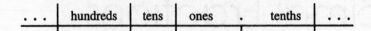

All digits to the left of the decimal point are whole numbers. All digits to the right of the decimal point are fractional numbers. Consider the value of each 3 in the place value chart below.

Write	. . .100's	10's	1's	10ths	100ths	1000ths . . .	Value
300	3						300
30		3					30
3			3				3
.3				3			$\frac{3}{10}$
.03					3		$\frac{3}{100}$
.003						3	$\frac{3}{1000}$

NOTE

Each denominator is a multiple of ten. The number of zeroes in the denominator corresponds to the number of places to the right of the decimal point.

Examples

1. $0.5 = \dfrac{5}{10}$

2. $0.05 = \dfrac{5}{100}$

3. $0.005 = \dfrac{5}{1000}$

4. $0.0005 = \dfrac{5}{10,000}$

Test Yourself

Directions: Write the fraction or mixed numeral that means the same thing.

1. .4 = _____   2. .8 = _____   3. .02 = _____

4. .05 = _____   5. .006 = _____ 6. .009 = _____

7. .00001 = _____   8. .00004 = _____

Directions: Write the decimal fraction that means the same thing.

9. $\dfrac{5}{10}$ = _____

10. $\dfrac{7}{10}$ = _____

11. $\dfrac{3}{100}$ = _____

12. $\dfrac{1}{1000}$ = _____

13. $\dfrac{9}{100}$ = _____

14. $\dfrac{5}{1000}$ = _____

15. $\dfrac{9}{10}$ = _____

16. $\dfrac{6}{10,000}$ = _____

17. $\dfrac{9}{1000}$ = _____

18. $\dfrac{7}{10,000}$ = _____

19. $\dfrac{4}{100,000}$ = _____

21. $\dfrac{8}{1,000,000}$ = _____

Answers

1. $\dfrac{4}{10}$

2. $\dfrac{8}{10}$

3. $\dfrac{2}{100}$

4. $\dfrac{5}{100}$

5. $\dfrac{6}{1000}$

6. $\dfrac{9}{1000}$

7. $\dfrac{1}{100,000}$

8. $\dfrac{4}{100,000}$

9. .5

10. .7

11. .03

12. .001

13. .09

14. .005

15. .9

16. .0006

17. .009

18. .0007

19. .00004

20. .000008

When a decimal fraction contains more than a single digit, it is named by the place occupied by the digit farthest to the right. So, for example, .17 would be read as 17 hundredths, .23 is 23 hundredths, and .024 is 24 thousandths. Similarly, .345 is read as three hundred forty-five thousandths, and .00216 is two hundred sixteen hundred-thousandths.

Test Yourself

Directions: Name these decimal fractions.

1. .24 _____

2. .68 _____

3. .069 _____

4. .053 _____

5. .437 _____

6. .619 _____

7. .0085 _____

8. .0136 _____

9. .00148 _____

10. 7.00045 _____

Answers

1. 24 hundredths

2. 68 hundredths

3. 69 thousandths

4. 53 thousandths

5. 437 thousandths

6. 619 thousandths

7. 85 ten-thousandths

8. 136 ten-thousandths

9. 148 hundred-thousandths

10. 7 and 45 hundred-thousandths

EXPRESSING COMMON FRACTIONS AS DECIMAL FRACTIONS

Certain common fractions are easily expressed as decimal fractions. That is particularly true of common fractions whose denominators are multiples of ten, such as tenths, hundredths, and thousandths. Other common fractions can be changed to decimal fractions by first writing equivalent fractions whose denominators are multiples of ten. See the examples below.

Examples

1. $\dfrac{1}{4} = \dfrac{}{100} \rightarrow \dfrac{1}{4} = \dfrac{25}{100} = .25$

   4 does not divide perfectly into 10, but it goes into 100 evenly, 25 times.

2. $\dfrac{3}{5} = \dfrac{}{10} \rightarrow \dfrac{3}{5} = \dfrac{6}{10} = .6$

   5 divides perfectly into 10.

3. $\dfrac{3}{8} = \dfrac{}{1000} \rightarrow \dfrac{3}{8} = \dfrac{375}{1000} = .375$

   8 divides perfectly into neither 10 nor 100. It does fit into 1000 evenly, 125 times.

Test Yourself

Directions: Express the following common fractions as decimal fractions.

1. $\dfrac{3}{4} = \dfrac{}{100} = \underline{}$　　2. $\dfrac{1}{2} = \dfrac{}{10} = \underline{}$　　3. $\dfrac{2}{5} = \dfrac{}{10} = \underline{}$

4. $\dfrac{5}{8} = \dfrac{}{1000} = \underline{}$　　5. $\dfrac{7}{20} = \dfrac{}{100} = \underline{}$　　6. $\dfrac{7}{8} = \dfrac{}{1000} = \underline{}$

Answers

1. $\dfrac{75}{100}, .75$　　2. $\dfrac{5}{10}, .5$　　3. $\dfrac{4}{10}, .4$

4. $\dfrac{625}{1000}, .625$　　5. $\dfrac{35}{100}, .35$　　6. $\dfrac{875}{1000}, .875$

As you may have guessed, not all fractions are as obliging as the ones listed above. That is, not all of them will convert quite so readily to equivalent fractions with denominators that are multiples of ten. Fortunately, however, there is a technique that can be employed that will allow any common fraction to be expressed as a decimal. That technique requires dividing the numerator of any fraction by its denominator.

Examples

1. $\frac{1}{4} \rightarrow 4\overline{)1} \rightarrow 4\overline{)1.00} \rightarrow 4\overline{)1.00}^{.25}$

   To find the decimal equivalent of $\frac{1}{4}$, 1 must be divided by 4. To do so, a decimal point is placed after the 1, and two zeroes are added. (1.00 = 1) Place a decimal point in the quotient over the decimal point in the dividend and get the decimal .25.

2. $\frac{3}{5} \rightarrow 5\overline{)3} \rightarrow 5\overline{)3.0} \rightarrow 5\overline{)3.0}^{.6}$

   To divide 3 by 5, one zero after the decimal point is enough, although adding a second zero after the decimal point will yield the same answer (since .60 = .6). Proceed as in model example 1.

3. $\frac{1}{3} \rightarrow 3\overline{)1} \rightarrow 3\overline{)1.00} \rightarrow 3\overline{)1.00}^{.3\overline{3}}$

   No matter how many zeroes are added after the decimal point, 3s will keep appearing in the quotient. The fraction $\frac{1}{3}$ translates to what is known as a "repeating decimal." The bar over the second 3 in the quotient indicates that the 3 keeps repeating.

ROUNDING DECIMAL FRACTIONS

It is often either inconvenient or unnecessary to express a scrupulously accurate quantity. When, for example, estimating the crowd at a baseball game, the fact that 48,143 persons attended is generally an excessively accurate piece of information. It is usually sufficient to say that there were about 48,000 in attendance. 48,000 is the attendance figure rounded to the nearest thousand.

In order to round a number, there are two things that must be considered. First is the place to which the number is to be rounded. Second is the figure in the place immediately to the right of the place to which you wish to round the number.

Examples

1. Round 34,567 to the nearest hundred.

   34,567 $\rightarrow$ 34,567 $\rightarrow$ 34,600

   $\uparrow$ $\qquad\qquad\qquad$ $\uparrow$ $\qquad\qquad\qquad$ $\uparrow$

   Place to be rounded to. $\quad$ Five or larger. $\quad$ Rounded up to 6.

2. Round 34,567 to the nearest 10,000.

   34,567 $\rightarrow$ 34,567 $\rightarrow$ 30,000

   $\uparrow$ $\qquad\qquad\qquad$ $\uparrow$ $\qquad\qquad\qquad$ $\uparrow$

   Place to be rounded to. $\quad$ Less than five. $\quad$ Stays a 3.

3. Round 27.635 to the nearest tenth.

27.635   →   27.635   →   27.6

↑   ↑   ↑

Place to be rounded to.   Less than five.   Stays a 6.

4. Round 27.635 to the nearest hundredth.

27.635   →   27.635   →   27.64

↑   ↑   ↑

Place to be rounded to.   Five or larger.   Rounded up to 4.

You may have noticed that when rounding to a whole number, the place rounded to is the last one in which a non-zero digit appears. Each place to the right contains a zero (note that 34,567 to the nearest 10,000 became 30,000). When rounding a decimal fraction, however, the place to which you round is the last digit that will appear in the rounded numeral. No zeroes are inserted to follow it (see examples 3 and 4).

Test Yourself

Directions: Try the following exercises in rounding off. If needed, use the model examples (above) as your guide.

Round to the nearest	hundred	ten	tenth	hundredth
1. 4562.738	_____	_____	_____	_____
2. 328.4929	_____	_____	_____	_____
3. 255.555	_____	_____	_____	_____
4. 8134.8134	_____	_____	_____	_____
5. 2121.9192	_____	_____	_____	_____

Answers

1. 4600, 4560, 4562.7, 4562.74   **2.** 300, 330, 328.5, 328.49

3. 300, 260, 255.6, 255.56   **4.** 8100, 8130, 8134.8, 8134.81

5. 2100, 2120, 2121.9, 2121.92

Now, in case you have been wondering what a discussion of rounding was doing in the middle of a discussion of changing common fractions to decimals, here is the answer. Since many fractions do not translate evenly into decimal fractions, it is often necessary to round them off. You already saw that $\frac{1}{3}$ translates to $.33333\overline{3}$ when expressed as a decimal. It is far from being a unique case.

It is a good rule of thumb, therefore, to round off decimal fractions that do not work out evenly to the nearest hundredth. In order to round off to the nearest hundredth, however, you must work a decimal out to the thousandths' place. It is, after all, the digit in the thousandths' place that will tell you whether to leave the digit in the hundredths' place as it is or to round it up to the next higher digit. Bear this in mind when working the following exercises.

Test Yourself

> **Directions:** Express each common fraction as a decimal fraction. Round to the nearest hundredth when there are more than 2 decimal places.

1. $\dfrac{5}{16} =$ _____ 2. $\dfrac{5}{9} =$ _____ 3. $\dfrac{2}{3} =$ _____ 4. $\dfrac{4}{5} =$ _____

5. $\dfrac{8}{11} =$ _____ 6. $\dfrac{7}{12} =$ _____ 7. $\dfrac{4}{7} =$ _____ 8. $\dfrac{5}{8} =$ _____

9. $\dfrac{11}{20} =$ _____ 10. $\dfrac{17}{35} =$ _____ 11. $\dfrac{15}{40} =$ _____ 12. $\dfrac{9}{17} =$ _____

13. $\dfrac{9}{14} =$ _____ 14. $\dfrac{3}{4} =$ _____ 15. $\dfrac{5}{12} =$ _____ 16. $\dfrac{1}{2} =$ _____

17. $\dfrac{9}{15} =$ _____ 18. $\dfrac{1}{11} =$ _____ 19. $\dfrac{2}{13} =$ _____ 20. $\dfrac{1}{6} =$ _____

Answers

1. .31	**2.** .56	**3.** .67	**4.** .80	**5.** .73
6. .58	**7.** .57	**8.** .63	**9.** .55	**10.** .49
11. .38	**12.** .53	**13.** .64	**14.** .75	**15.** .42
16. .50	**17.** .60	**18.** .09	**19.** .15	**20.** .17

ADDING DECIMAL FRACTIONS

There is one and only one key to the addition of decimal fractions and mixed numerals—the decimal points must be lined up one beneath another. The decimal point in the sum is placed beneath the decimal points in the addition, and then the numbers are added in the usual manner. If no decimal point is shown in a numeral, then it is understood that the decimal point appears to the right of the last digit in the numeral.

Examples

1. $2.4 + 1.8 + 3.72 \quad \rightarrow$

$$
\begin{array}{r}
2.4 \\
1.8 \\
\underline{3.72}
\end{array}
\qquad \rightarrow \qquad
\begin{array}{r}
2.4 \\
1.8 \\
\underline{3.72} \\
7.92
\end{array}
$$

(First align the decimal points)   (Then add)

2. $3.6 + .074 + 59 + 16.08 \rightarrow$

3.6	3.6
.074 $\rightarrow$	.074
59.	59.
16.08	16.08
	78.754

Test Yourself

Directions: Now try solving the additions below. Bear in mind that the decimal points must be aligned one beneath the other.

1.
3.4
2.86
12.5

2.
5.61
.007
21.0

3.
6.81
.009
5.73
24.0

4. $6.2 + 7.5 + 3.82$

5. $9.4 + 3.56 + 21.5$

6. $18.7 + 17.39 + 6.25$

7. $25 + .08 + 3.5$

8. $342 + .007$

9. $46 + .38 + 2.592$

10. $17.2 + 8.09 + 4.6$

11. $61.9 + 3.85 + 596 + .35$

12. $63 + .34 + 5.21 + .098 + 351.7 + 84.2$

13. $19.6 + .342 + 8.29 + .074 + 39$

14. $.159 + 28.6 + 3.14 + 9 + .8 + .216$

15. $4.72 + 15.3 + 69.25 + 47 + .356 + 9 + .05$

Answers

1. 18.76	**2.** 26.617	**3.** 36.549	**4.** 17.52	**5.** 34.46
6. 42.34	**7.** 28.58	**8.** 342.007	**9.** 48.972	**10.** 29.89
11. 662.1	**12.** 504.548	**13.** 67.306	**14.** 41.915	**15.** 145.676

SUBTRACTING DECIMAL FRACTIONS

As with addition of decimals, subtraction is accomplished by lining the decimal points up, one beneath the other. With one exception, subtraction would then take place in the usual manner. The exception is illustrated in the example below.

Example

$$8.6 - 3.425$$
$$\downarrow$$

$$8.6$$
$$\underline{-3.425}$$
$$\downarrow$$

$$8.600$$
$$\underline{-3.425}$$
$$\downarrow$$

$$8.600$$
$$\underline{-3.425}$$

$$8.600$$
$$\underline{-3.425}$$
$$5.175$$

The subtraction is first rewritten in column form, lining up the decimal points one below the other.

... but from what are the 2 and the 5 to be subtracted?

Zeroes are added on the top numeral in order to fill the empty places. **The values of 8.6 and 8.600 are identical.**

Now subtraction can take place. Note the decimal point's placement in the difference.

TIP

Adding zeroes after the last significant figure to the right of the decimal point does not change the number's value. Consider that .6, .60, and .600 are equivalent to $\frac{6}{10}$, $\frac{60}{100}$, and $\frac{600}{1000}$ respectively. All three of those fractions expressed in lowest terms are equal to $\frac{3}{5}$.

Test Yourself

Directions: Complete the following.

1. $\begin{array}{r} 8.69 \\ \underline{-3.54} \end{array}$	**2.** $\begin{array}{r} 15.7 \\ \underline{-5.9} \end{array}$	**3.** $\begin{array}{r} 36.4 \\ \underline{-7.8} \end{array}$	**4.** $\begin{array}{r} 57.9 \\ \underline{-24.6} \end{array}$
5. $\begin{array}{r} 11.35 \\ \underline{-8.59} \end{array}$	**6.** $\begin{array}{r} 24.21 \\ \underline{-19.8} \end{array}$	**7.** $\begin{array}{r} 81.6 \\ \underline{-49.83} \end{array}$	**8.** $\begin{array}{r} 64.2 \\ \underline{-38.97} \end{array}$

9. 29.8 - 15.6   **10.** 34.7 - 8.9   **11.** 19.5 - 12.71   **12.** 59.2 - 23.84

13. 125 - .02   **14.** 67.1 - .875   **15.** 143.5 - 68.9   **16.** 234.2 - 7.895

17. 4.60 - .349   **18.** 7 - .23   **19.** 14 - 1.761   **20.** 11 - .2487

Answers

1. 5.15	**2.** 9.8	**3.** 28.6	**4.** 33.3	**5.** 2.76
6. 4.41	**7.** 31.77	**8.** 25.23	**9.** 14.2	**10.** 25.8
11. 6.79	**12.** 35.36	**13.** 124.98	**14.** 66.225	**15.** 74.6
16. 226.305	**17.** 4.251	**18.** 6.77	**19.** 12.239	**20.** 10.7513

MULTIPLYING DECIMAL FRACTIONS

The rationale for multiplication of decimal fractions can be found in an examination of the multiplication of common fractions whose denominators are multiples of ten. A pattern can be seen by examining the table below. Note that each line contains a single multiplication, first expressed in common fractions and then in decimal fractions.

Multiplication	Product	Multiplication	Product
$\frac{1}{10} \times \frac{1}{10}$	$\frac{1}{100}$	$.1 \times .1$	$.01$
$\frac{1}{10} \times \frac{1}{100}$	$\frac{1}{1000}$	$.1 \times .01$	$.001$
$\frac{1}{100} \times \frac{1}{100}$	$\frac{1}{10,000}$	$.01 \times .01$	$.0001$
$\frac{1}{10} \times \frac{1}{1000}$	$\frac{1}{10,000}$	$.1 \times .001$	$.0001$
$\frac{1}{100} \times \frac{1}{1000}$	$\frac{1}{100,000}$	$.01 \times .001$	$.00001$
$\frac{1}{1000} \times \frac{1}{1000}$	$\frac{1}{1,000,000}$	$.001 \times .001$	$.000001$
$\frac{1}{10} \times \frac{1}{10,000}$	$\frac{1}{100,000}$	$.1 \times .0001$	$.00001$

Count up the number of digits to the right of the decimal point in each of the multiplications above. Then count up the number of digits to the right of the decimal point in each product. Do you see a pattern? When multiplying decimals, the number of digits to the right of the decimal point in the product will equal the total number of digits to the right of the points in the two numbers involved in the multiplication.

In order to see how the procedure of counting digits to the right of the decimal point works in actual usage, examine the model problems below. Notice that no attention whatsoever is paid to the decimal points while the multiplication is taking place.

Examples

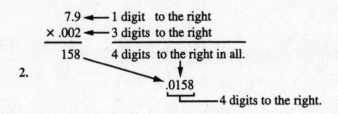

$$
\begin{array}{r}
2.5 \quad \longleftarrow \text{1 digit to the right} \\
\times\ 1.2 \quad \longleftarrow \text{1 digit to the right} \\
\hline
50 \qquad \text{2 digits to the right in all.} \\
250 \\
\hline
300 \quad \longrightarrow 3.00
\end{array}
$$

1. ⌐— Two digits to the right.

$$
\begin{array}{r}
7.9 \quad \longleftarrow \text{1 digit to the right} \\
\times\ .002 \quad \longleftarrow \text{3 digits to the right} \\
\hline
158 \qquad \text{4 digits to the right in all.}
\end{array}
$$

2. .0158 — 4 digits to the right.

$$
\begin{array}{r}
.006 \quad \\
\times\ .002 \quad \text{6 digits to the right in all.}
\end{array}
$$

3. 12 ⟶ .000012 — 6 digits to the right.

Test Yourself

Directions: The following multiplications have been computed for you, but the decimal point has not been placed in the product. Place the decimal point in the correct place in each product.

1. 12.4
    ×.01
    124

2. 3.56
    ×.12
    4272

3. 4.7
    ×1.6
    752

4. 28
    ×.3
    84

5. 1.56
    ×.04
    624

6. 124.3
    ×1.5
    18645

7. 697
    × 0.3
    2091

8. 7.55
    ×.13
    9815

9. .214
    ×.001
    214

10. .715
    ×.008
    5720

11. .3641
    ×.0003
    10923

12. .00005
    ×.0002
    10

Answers

1. .124   2. .4272   3. 7.52   4. 8.4   5. .0624

6. 186.45   7. 209.1   8. .9815   9. .000214   10. .005720

11. .00010923   12. .00000001

If the answer to number 12 (above) confused you, remember that a zero after the last significant figure to the right of the decimal point has no meaning, just as a zero to the left of the first significant figure to the left of a decimal point has no meaning. (.6 = .60; 025.3 = 25.3) The answer, thus, also could have been written .000000010.

Now that you have had practice placing the decimal point, it is time to try doing the entire multiplication from scratch. Once more, remember that the multiplication proceeds as any whole number multiplication would. It is only after the computation has been completed that the decimal point is placed.

Test Yourself

Directions: Multiply. Then place the decimal point correctly.

1. $\begin{array}{r} 2.6 \\ \times 4 \\ \hline \end{array}$	**2.** $\begin{array}{r} 3.7 \\ \times .5 \\ \hline \end{array}$	**3.** $\begin{array}{r} 4.9 \\ \times .6 \\ \hline \end{array}$	**4.** $\begin{array}{r} 5.8 \\ \times .7 \\ \hline \end{array}$	**5.** $\begin{array}{r} .35 \\ \times 2 \\ \hline \end{array}$
6. $\begin{array}{r} .48 \\ \times 5 \\ \hline \end{array}$	**7.** $\begin{array}{r} .45 \\ \times .7 \\ \hline \end{array}$	**8.** $\begin{array}{r} .63 \\ \times .6 \\ \hline \end{array}$	**9.** $\begin{array}{r} .73 \\ \times .04 \\ \hline \end{array}$	**10.** $\begin{array}{r} .89 \\ \times .62 \\ \hline \end{array}$
11. $\begin{array}{r} .96 \\ \times 2.7 \\ \hline \end{array}$	**12.** $\begin{array}{r} .085 \\ \times .31 \\ \hline \end{array}$	**13.** $\begin{array}{r} .058 \\ \times .012 \\ \hline \end{array}$	**14.** $\begin{array}{r} 1.79 \\ \times 1.3 \\ \hline \end{array}$	**15.** $\begin{array}{r} 38.4 \\ \times .05 \\ \hline \end{array}$
16. $\begin{array}{r} 67.3 \\ \times 1.7 \\ \hline \end{array}$	**17.** $\begin{array}{r} 4.9 \\ \times 5.6 \\ \hline \end{array}$	**18.** $\begin{array}{r} 7.4 \\ \times .58 \\ \hline \end{array}$	**19.** $\begin{array}{r} 9.20 \\ \times .35 \\ \hline \end{array}$	**20.** $\begin{array}{r} .0064 \\ \times .037 \\ \hline \end{array}$

Answers

1. 10.4	**2.** 1.85	**3.** 2.94	**4.** 4.06	**5.** .7
6. 2.4	**7.** .315	**8.** .378	**9.** .0292	**10.** .5518
11. 2.592	**12.** .02635	**13.** .000696	**14.** 2.327	**15.** 1.92
16. 114.41	**17.** 27.44	**18.** 4.292	**19.** 3.22	**20.** .0002368

MULTIPLYING AND DIVIDING BY POWERS OF 10

Any whole number may be multiplied by 10 simply by placing a zero at the end of it. For example, 10×2 is 20, $10 \times 8 = 80$, $10 \times 35 = 350$, and $10 \times 237 = 2370$. How do you think you might multiply a number by 100? By 1000? By 10,000?

Just in case you hadn't guessed, to multiply a whole number by 100, attach two zeroes ($3 \times 100 = 300$; $46 \times 100 = 4600$; $342 \times 100 = 34,200$). To multiply a whole number by 1000, attach three zeroes ($6 \times 1000 = 6000$; $57 \times 1000 = 57,000$; $416 \times 1000 = 416,000$). To multiply by 10,000, attach four zeroes; to multiply by 100,000 attach five zeroes, etc.

This formula (attaching the same number of zeroes as there are in the multiple of ten by which you are multiplying) is a convenient mechanism for dealing with multiplying whole numbers by powers of ten. Now that you are familiar with the meaning of a decimal point in a numeral, consider the following alternative explanation for multiplying by powers of ten:

> To multiply by 10, move the decimal point 1 place to the right.
>
> To multiply by 100, move the decimal point 2 places to the right.
>
> To multiply by 1000, move the decimal point 3 places to the right.
>
> To multiply a number by any power of ten, move the point a number of spaces to the right equal to the number of zeroes in the power of ten by which you are multiplying.

This explanation demonstrates the actual process that occurs when zeroes are attached and also is applicable to numbers containing decimal fractions. Study the table below.

Number	× 10	× 100	× 1000	× 10,000	× 100,000
23	230	2300	23,000	230,000	2,300,000
3.4	34	340	3400	34,000	340,000
.59	5.9	59	590	5900	59,000
.028	.28	2.8	28	280	2800
.0046	.046	.46	4.6	46	460
.00081	.0081	.081	.81	8.1	81

Now, the beauty of this approach to multiplying by powers of 10 becomes even more evident when we consider division by powers of 10. Since division is the undoing of multiplication, then it is to be expected that when dividing by powers of ten, we do the opposite from what we do when multiplying. That is, the decimal point must be moved to the left the same number of places as there are zeroes in the multiple of ten by which you are dividing. Look at the following table.

Number	÷ 10	÷ 100	÷ 1000	÷ 10,000	÷ 100,000
23	2.3	.23	.023	.0023	.00023
3.4	.34	.034	.0034	.00034	.000034
567	56.7	5.67	.567	.0567	.00567
8936	893.6	89.36	8.936	.8936	.08936
45,382	4538.2	453.82	45.382	4.5382	.45382
732,971	73,297.1	7329.71	732.971	73.2971	7.32971

Test Yourself

> **Directions:** Multiply each number in your head by the multiplier specified.

10 ×

1. 3.4
2. .62
3. 41.7
4. .018
5. 6.91
6. 34

100 ×

7. 5.9
8. .82
9. 63.4
10. .052
11. 3.97
12. 78

1000 ×

13. 9.3
14. .93
15. 82.7
16. .095
17. 9.36
18. 5.1

> **Directions:** Divide each number in your head by the divisor specified.

10)

19. 3.4
21. .62
21. 41.7
22. .018
23. 6.91
24. 34

100)

25. 5.9
26. .82
27. 63.4
28. .052
29. 3.97
30. 78

1000)

31. 9.3
32. .93
33. 82.7
34. .095
35. 9.36
36. 5.1

Answers

1. 34	2. 6.2	3. 417	4. .18	5. 69.1
6. 340	7. 590	8. 82	9. 6340	10. 5.2
11. 397	12. 7800	13. 9300	14. 930	15. 82,700
16. 95	17. 9360	18. 5100	19. .34	20. .062
20. 4.17	22. .0018	23. .691	24. 3.4	25. .059
26. .0082	27. .634	28. .00052	29. .0397	30. .78
31. .0093	32. .00093	33. .0827	34. .000095	35. .00936
36. .0051				

The ability to rapidly multiply and divide by powers of ten, simply by moving the decimal point the appropriate number of places to the right or the left is a skill that can be learned easily and will prove invaluable as a time saver in later operations. If you feel that you need more practice, make up lists of numerals with their decimal points in various places, and then practice multiplying and dividing them by moving the decimal point. An understanding of this skill is essential to understanding how division of decimals works.

DIVIDING DECIMAL FRACTIONS

When two numbers are to be divided, multiplying both of those numbers by the same quantity will not affect the quotient. Read that last sentence over again, to be certain that you understand its implications. Then look at the illustration of that statement below:

$$
3\overline{)15}^{\,5} \rightarrow \times 100 \rightarrow 300\overline{)1500}^{\,5}
$$

$$
\downarrow
$$

$$
\times 10
$$

$$
\downarrow
$$

$$
30\overline{)150}^{\,5}
$$

Across, both divisor and dividend are multiplied by 100. Down, both divisor and dividend are multiplied by 10. Whether the division is done as stated (15 divided by 3) or multiplied, the quotient is still 5.

You may try this trick with any division you like, multiplying divisor and dividend by any number you like, until you convince yourself that the process is a mathematically sound one—that is, it will always work.

The first thing you need to know in order to divide decimals is that **you never divide by** a decimal. Yes, that last sentence does sound contradictory, but nevertheless, it is true. In order to divide by a decimal, **it is first necessary to change that decimal to an integer** (whole number). Carefully study the examples below, and you will see that the process is, in reality, much easier to actually do than it is to describe.

Examples

1. $.6\overline{)1.8}$

   To change .6 to an integer, the decimal point must be moved to the right one place.

   $6\overline{)1.8.}$

   Moving that decimal point is the same as multiplying by 10. In order to avoid changing the problem, the decimal in the dividend must also be multiplied by 10.

   $6\overline{)18.}$

   The decimal point in the quotient is then placed directly above the one in the dividend.

   $6\overline{)18}^{\,3}$

   Dividing, we get a quotient of 3.

2. $.09\overline{)\,.081}$

To change .09 to an integer, multiply by 100. (That means moving the decimal point 2 places to the right.)

$\underline{.09.}\overline{)\,.08.1}$

Of course if one decimal point is moved 2 places, so must the other one be moved.

$09.\overline{)\,08.1} = 9\overline{)\,8.1}$

The zeroes to the left of the first figures are just excess baggage, and meaningless.

$9\overline{)\,8.1}$

Place the decimal point in the quotient directly above the one in the dividend.

$9\overline{)\,8.1}^{.9}$

Divide, and get a quotient of .9.

3. $1.2\overline{)\,144}$

Change 1.2 to an integer by moving the decimal point 1 place to the right.

$1\underline{.2.}\overline{)\,144.0}$

In order to move the decimal point in 144 one place to the right, a 0 must be inserted as a placeholder.

$12\overline{)\,1440\underline{.}}$

Place the decimal point in the quotient above the decimal point in the dividend.

$12\overline{)\,1440}^{120.}$

Divide and get a quotient of 120.

Notice that once the divisor has been made into an integer by moving the decimal point and the dividend's decimal point has been moved and the point placed in the quotient, the actual division process is identical to division with integers. Thus, when you have reached this point, you may then find the solution by whatever form of division you happen to normally use.

4. $3.2\overline{)\,7.6}$

Most divisions do not come out evenly. This one begins as have all the others—by moving decimal points:

$3\underline{.2.}\overline{)\,7.6\underline{.}}$

Points in the divisor and dividend are moved one place each.

$32\overline{)\,76\underline{.}}$

The decimal point in the quotient is then placed.

$32\overline{)\,76.}^{2.}$
$\underline{64}$
12

Dividing, we find that there are 2 32s in 76. We do not write remainders or fractional remainders when dividing decimals, however, so we must divide farther.

$32\overline{)\,76.0}^{2.3}$
$\underline{64}\downarrow$
$12\,0$
$\underline{9\,6}$
$2\,4$

To accomplish that, a zero is placed after the decimal point ...

$$\begin{array}{r} 2.37 \\ 32\overline{)76.00} \\ \underline{64} \\ 120 \\ \underline{96} \downarrow \\ 240 \\ \underline{224} \\ 16 \end{array}$$

... and, when needed, another zero ...

$$\begin{array}{r} 2.375 \\ 32\overline{)76.000} \\ \underline{64} \\ 120 \\ \underline{96} \\ 240 \\ \underline{224} \downarrow \\ 160 \\ \underline{160} \end{array}$$

... and when needed, still another zero. If this third zero had not yielded an exact ending, we would have rounded the answer back to the nearest hundredth.

Test Yourself

Directions: Divide. Where necessary, round the quotients to the nearest hundredth.

1. $.2\overline{)6.70}$ 2. $1.3\overline{).48}$ 3. $2.5\overline{)64}$ 4. $.16\overline{)35.4}$

5. $8\overline{)5.72}$ 6. $.9\overline{)56.3}$ 7. $4.1\overline{)57}$ 8. $.06\overline{)3.4}$

9. $.5\overline{)690}$ 10. $.031\overline{)5.97}$ 11. $.42\overline{)6.82}$ 12. $4.7\overline{).685}$

13. $.64\overline{)325}$ 14. $4.9\overline{).627}$ 15. $8.7\overline{)43.82}$ 16. $2.9\overline{)72.51}$

17. $6.7\overline{).3872}$ 18. $.53\overline{)4.691}$ 19. $.61\overline{)5.674}$ 21. $.83\overline{)37.46}$

21. $.009\overline{).5241}$ 22. $.37\overline{)65.71}$ 23. $.48\overline{)732.8}$ 24. $.036\overline{)4751}$

Answers

1. 33.5	**2.** .37	**3.** 25.6	**4.** 221.25	**5.** .715
6. 62.56	**7.** 13.90	**8.** 56.67	**9.** 1380	**10.** 192.58
11. 16.24	**12.** .15	**13.** 507.81	**14.** .13	**15.** 5.04
16. 25.00	**17.** .06	**18.** 8.85	**19.** 9.30	20. 45.13
20. 58.23	**22.** 177.59	**23.** 1526.67	**24.** 131,972.22	

EXERCISES: ADDING AND SUBTRACTING DECIMAL FRACTIONS

1. 5.3 + 16.48 + .792

2. 17.3 + 84 + .09

3. 11.6 + .75 + 2.34 + .357

4. 86 +.27 + 5.49 + 18.1 + .003

5. 62.79 + .381 + 5.1 + 87.4

6. .17 + .321 + .469 + .008 + 2

7. 5.94 + 6.38 + 47.3 + .29

8. .6531 + 2.4 + .087 + 1.2 + 9

9. 4.63 + .27 + 5.91 + .34

10. 17.2 + 126 + .28 + .953 + 15 + .04

11. 38.4 - 16.2

12. 56.9 - 24.7

13. 82.31 - 64.53

14. 12.34 - 7.89

15. 511.31 - 98.520

16. 76.3 - 48.95

17. 572.3 - 385.7

18. 621.4 - 8.374

19. 38.42 - 9.502

20. 1.732 -.9

21. 84.71 - 79.86

22. 54.82 - 9.634

23. 47.36 - 18

24. 75.81 -.3756

25. 2 -.8461

ANSWERS

1.	22.572	**2.**	101.39	**3.**	15.047	**4.**	109.863
5.	155.671	**6.**	2.968	**7.**	59.91	**8.**	13.3401
9.	11.15	**10.**	159.473	**11.**	22.2	**12.**	32.2
13.	17.78	**14.**	4.45	**15.**	412.79	**16.**	27.35
17.	186.6	**18.**	613.026	**19.**	28.918	**20.**	.832
21.	4.85	**22.**	45.186	**23.**	29.36	**24.**	75.4344
25.	1.1539						

EXERCISES: MULTIPLYING AND DIVIDING DECIMAL FRACTIONS

Directions: Multiply. Be sure to position the decimal point correctly.

1.
$$43.8 \times 2.4$$

2.
$$5.7 \times .6$$

3.
$$.69 \times .08$$

4.
$$.58 \times .32$$

5.
$$6.5 \times 3.4$$

6.
$$.75 \times .006$$

7.
$$.27 \times .11$$

8.
$$55 \times 2.7$$

9. $.006 \times .0071 =$ _____

10. $.83 \times 5.611 =$ _____

11. $.34 \times .00003 =$ _____

12. $1.8 \times 7 =$ _____

Directions: Divide. Where necessary, round the quotient to the nearest hundredth.

13. $.17\overline{)34.02}$ **14.** $2.5\overline{)500}$ **15.** $.09\overline{)81}$ **16.** $.34\overline{)6.92}$

17. $4.5\overline{)61.3}$ **18.** $.004\overline{).16}$ **19.** $1.02\overline{).476}$ **20.** $3.8\overline{)97}$

21. $11\overline{)57.3}$ **22.** $4.9\overline{)7.70}$ **23.** $.52\overline{)8.16}$ **24.** $.0014\overline{)70}$

ANSWERS

1.	105.12	**2.**	3.42	**3.**	.0552	**4.**	.1856
5.	22.1	**6.**	.0045	**7.**	.0297	**8.**	148.5
9.	.0000426	**10.**	4.65713	**11.**	.0000102	**12.**	12.6
13.	200.12	**14.**	200	**15.**	900	**16.**	20.35
17.	13.62	**18.**	40	**19.**	.47	**20.**	25.53
21.	5.21	**22.**	1.57	**23.**	15.69	**24.**	50,000

SUMMING IT UP

- A decimal fraction is a number with a decimal point (.). As a fraction, the denominator is understood to be 10 or some power of 10.

- The number of digits after a decimal point determines which power of 10 the denominator is. If there is one digit, the denominator is understood to be 10; if there are two, the denominator is understood to be 100, and so on.

- When zeros are added after a decimal point, the value of the decimal fraction is unchanged.

- Decimal fractions are added in the same way that whole numbers are added, as long as the decimal points are kept in a vertical line, one under the other.

- Decimal fractions are subtracted in the same way that whole numbers are subtracted, as long as the decimal points are kept in a vertical line, one under the other.

- Decimal fractions are multiplied in the same way that whole numbers are multiplied. A decimal fraction can be multiplied by a power of 10 by moving the decimal point to the right as many places as are indicated by the power: if multiplied by 10, the decimal point moves one place to the right; if multiplied by 100, the decimal point is moved two places to the right, and so on.

Introduction
to Algebra

OVERVIEW

- **Constants and variables**
- **A question of balance**
- **Signed numbers**
- **Solving equations by equal additions**
- **Solving equations by equal subtractions**
- **Solving equations by multiplication and division**
- **Putting it all together**
- **Ratio and proportion**
- **Percents**
- **Some useful properties**
- **Exponents**
- **Monomials, binomials, and polynomials**
- **Word problems**
- **Probability**
- **Statistics**
- **Summing it up**

Algebra is a branch of mathematics that was specifically developed to help solve word problems, as well as to organize one's thinking in a number of mathematical situations. Many people unfamiliar with the operations of algebra are afraid of them. However, algebra is an extremely useful tool that, once mastered, can be an invaluable aid in many situations that you are likely to encounter in the real world, as well as those that you will find on the GED examination.

CONSTANTS AND VARIABLES

In order to understand algebra, we need to first look at the two major elements of algebra—constants and variables. Since constants are by far the easier to understand, let's examine them first.

chapter 6

Constants

Any real number is a constant. $3 = 3 = 3$. The value of a constant is, for lack of a better word, constant. In case you are unclear on this, examine the number 25. It means 25 now and will mean 25 later. It is, therefore, a constant.

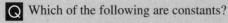

Q Which of the following are constants?

1. 36	**2.** 79	**3.** 251	**4.** y
5. $4n$	**6.** $\dfrac{4}{7}$	**7.** $2\dfrac{1}{15}$	**8.** .0792

A Each expression in the exercises above is a constant except y and $4n$. A constant may be a whole number, a fraction, a mixed numeral, a decimal; it may not, however, contain a letter.

As soon as you see a letter in an expression, you should recognize that you are not dealing with a constant. Rather, you have a variable.

Variables

A variable is a symbol that takes the place of a number. It is usually written as a lower case letter, $a, b, c, ...x, y, z$, although you will occasionally see capital letters, Greek letters, or even geometric symbols used to represent variables. Examine the following expression:

$$n = 4 + 7$$

In this mathematical sentence, the variable n has been used to represent a number. We can find out what number n is representing by combining the terms on the right hand side of the equal sign. Since $7 + 4 = 11$, the following is true:

$$n = 11$$

If that is clear to you, examine the following statement:

$$n = 19 - 4$$

What number does n represent this time?

By subtracting the 4 from the 19, we arrive at the conclusion that n represents 15. Notice that in two different mathematical sentences n has had two different values. The value of n has varied from expression to expression. n is therefore a variable.

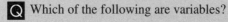

Q Which of the following are variables?

1. r	**2.** $3a$	**3.** $\dfrac{m}{7}$	**4.** 19
5. $\dfrac{9}{h}$	**6.** 97	**7.** D	**8.** Δ

A Each of the above is a variable, except for 19 and 97, which are constants. (If you're not sure, look back at the preceding section.)

As already noted, the meaning of a variable (that is, the number it represents) may change from mathematical sentence to mathematical sentence. *Within a single sentence,* however, *each time the same variable appears it must stand for the same number.* Look at the following sentence.

$$n + 3 = 11 - n$$

If the first $n = 4$, then so must the second. If we substitute 4 for each n in that sentence, we get the following:

$$4 + 3 = 11 - 4$$
$$7 = 7$$

This is a true statement. If, however, the second n had stood for anything but 4, the statement would have been false.

Combining Variables

Since variables represent numbers, they can be treated like numbers. That is to say, variables may be added, subtracted, multiplied, and divided. Let's look at addition first, and see if we can recognize a pattern.

Addition

Any time you see the term n it means that there is one n present. This seemingly obvious statement is rather critical to an understanding of how variables combine, so let's dwell on it for a moment. If there were 2 or more n's present, you would see them represented as $2n$, $3n$, $28n$, or whatever, meaning 2 n's, 3 n's, 28 n's, etc. If there were no n's being represented ($0n$), you'd see no symbol at all. Zero n's, after all, is the same as just plain 0. Consider:

$$0\ 2's = 0$$
$$0\ 5's = 0$$
$$0\ 35's = 0$$
$$0\ n's = 0$$

Just plain n, then, cannot mean 0 n's, does not mean 2 or more n's, and therefore must mean 1 n. However, the expression "$1n$" is rarely if ever written; mathematicians preferring to simply write "n."

A numeral written next to a variable is known as that variable's *coefficient*. In the expression $5n$, n's coefficient is 5. In the term $18x$, x's coefficient is 18. In the expression r, r's coefficient is ... What do you think r's coefficient is? Did you really think about it? It's 1.

Test Yourself

Directions: Name the coefficient of each variable.

1.	$3x$	**2.**	$7n$	**3.**	s	**4.**	$19b$
5.	w	**6.**	$4m$	**7.**	$\frac{2}{3}h$	**8.**	e

Answers

1. 3	**2.** 7	**3.** 1	**4.** 19	**5.** 1
6. 4	**7.** $\frac{2}{3}$	**8.** 1		

Did you get them all? How about the three 1s? Remember, if you do not see a numerical coefficient next to a variable, that variable's coefficient is 1.

It should be pointed out, in the interest of accuracy, that in any algebraic expression, all the members of a single term are, technically speaking, coefficients of each other. That is to say, in the term $5n$, 5 is the coefficient of n, but n is also the coefficient of 5. In the term, a, 1 is the coefficient of a and a is the coefficient of 1. Try to file this information somewhere in the back of your mind for later use. For the moment, however, we are going to be concerned exclusively with numerical coefficients of variables.

Only like variables may be combined into a single term by addition. Like variables are terms that have identical variable factors. For example, $2x$ and $5x$ are like variables. So are $5r$, $34r$, and $15r$. But, $2x$ and $3y$ are not like variables.

When adding like variables, only the numerical coefficients are added. This idea can be illustrated with the following examples:

$$2n + 3n = 5n$$
$$4x + 6x = 10x$$
$$2r + r = 3r \ \ \text{(do you know why?)}$$
$$w + w = 2w$$

Adding like variables may be likened to adding like pieces of fruit:

$$2 \text{ apples } + 3 \text{ apples } = 5 \text{ apples}$$
$$4 \text{ oranges } + 6 \text{ oranges } = 10 \text{ oranges}$$
$$2 \text{ peaches } + \text{ (a) peach } = 3 \text{ peaches}$$
$$\text{(a) grape } + \text{ (a) grape } = 2 \text{ grapes}$$

So far so good? Then try this one.

$$2x + 3y = \underline{\hspace{1cm}}$$

Think of:

$$2 \text{ oranges } + 3 \text{ grapes } = \underline{\hspace{1cm}}$$

Did you really think about that one? You might say that 2 oranges + 3 grapes = 5 pieces of fruit, but frankly, algebra would find that solution far too imprecise. Algebraically speaking:

$$2 \text{ oranges } + 3 \text{ grapes } = 2 \text{ oranges } + 3 \text{ grapes}$$

They cannot be further combined. Now bearing that in mind, you should recognize that:

$$2x + 3y = 2x + 3y$$

If the variables are not like, terms may not be combined into a single new term by adding. Only those terms containing identical variables may be combined into a single new term by addition.

Test Yourself

Directions: Complete the following.

1. $4n + 3n$
2. $5x + 2x$
3. $4w + w$

4. $r + r + r$
5. $m + n$
6. $\dfrac{1}{2}u + u$

7. $3v + 3w$
8. $\dfrac{1}{4}t + \dfrac{1}{2}t$
9. $\dfrac{z}{3} + \dfrac{z}{3}$

Answers

1. $7n$
2. $7x$
3. $5w$
4. $3r$
5. $m + n$

6. $1\dfrac{1}{2}u$
7. $3v + 3w$
8. $\dfrac{3}{4}t$
9.* $\dfrac{2z}{3}$

Subtraction

Subtraction of variables works in exactly the same fashion as addition, except, of course, that the numerical coefficient of one variable is subtracted from (rather than added to) the numerical coefficient of the other. All the other rules governing addition of variables apply.

$$4y - 2y = 2y$$
$$7m - 3m = 4m$$
$$s - s = 0$$
$$\text{and}$$
$$4p - 3q = 4p - 3q$$

Test Yourself

Directions: Solve each subtraction.

1. $5z - z$
2. $6r - 2r$
3. $9p - 3p$
4. $6x - 6x$
5. $8m - 4n$
6. $7c - 5c$
7. $u - u$
8. $9w - 4x$
9. $15v - 8v$

Answers

1. $4z$
2. $4r$
3. $6p$
4. 0
5. $8m - 4n$
6. $2c$
7. 0
8. $9w - 4x$
9. $7v$

* If problem 9 gave you any difficulty, bear in mind that you are combining two fractions with like denominators. The common denominator is 3. Then add the 2 numerators ($z + z$) and get $2z$.

If you are not sure about the reason for the answer to problem 1, remember that there is a 1 in front of the second z, even though you cannot see it there. As for problems 5 and 8, since the two variables are not like, they cannot be combined by subtraction.

A QUESTION OF BALANCE

You have probably, at some time or other, seen a twin pan scale. Such a scale is pictured below. Notice that the scale as shown is in balance.

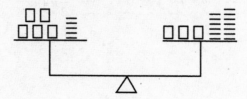

On the left arm of the scale you see 5 rolls of washers and 5 loose washers. We will assume that there is the same number of washers in each roll, and that the paper in which the washers are rolled is weightless. (While nothing can in fact be completely weightless, this fiction is often used for the sake of simplicity when solving problems in science. The actual weight of the paper would make very little difference anyway, when compared to that of the much heavier washers.) On the right arm, there are 3 rolls and 13 loose washers. Since the scale is in balance, the total weight on the left side must equal the total weight on the right.

What we wish to do is to determine how many washers there are in each roll. Do you have any idea of how we might do that? Because the scale is balanced, we may add the same amount of weight to each side or subtract the same amount of weight from each side without destroying that balance. How could we simplify the following picture?

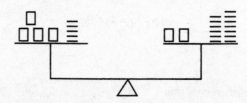

Here we've removed one roll from each side.

Is the scale still in balance?

Let's take the maximum number of rolls possible from each side while keeping the sides in balance.

Then the scale would look like this:

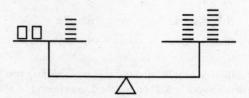

Why could we take no more than two additional rolls from each side? Think about it. If we took another roll from the left side, we would not be able to take another roll from the right side. There aren't any more rolls there.

Now, what is the maximum number of washers that we can remove from each side without losing the balance? How about this many?

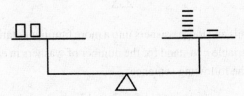

How many washers are in each roll? Can you tell now?

Since we took 5 loose washers from each side of the scale, we are left with 2 rolls on the left and 8 loose washers on the right. If 2 rolls balance 8 washers, then one roll would balance 4 washers, so there must be 4 washers in each roll.

Don't you think that the last balance picture was much easier to work with than the first one? And yet, the last balance picture is the result of having simplified the first one, one step at a time.

Now look at the balance picture below. Simplify it until you can see the solution.

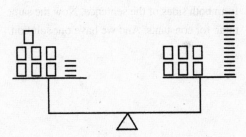

The simplest form for the balance picture above is obtained when 6 rolls and 4 loose washers are removed from each side. It looks like this:

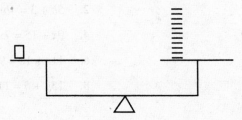

You can now see quite clearly how many washers are in a single roll. Can you see as clearly what this has to do with algebra?

Do you remember the original balance picture? Here it is again.

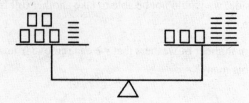

Now, let us translate the rolls and loose washers into a more familiar notation—that of variables and constants. If we use the variable n to stand for the number of washers in each roll, we can represent the balance picture with the following sentence:

$$5n + 5 = 3n + 13$$

Can you write the mathematical sentence for the balance picture above that had 7 rolls and 4 loose washers on the left?

If we let x be the number of washers in each roll this time, we get:

$$7x + 4 = 6x + 15$$

Mathematical sentences are often solved in the same manner as balance pictures. Look at these two solutions:

$$
\begin{array}{rr}
5n + 5 = & 3n + 13 \\
-3n \quad & -3n \\
\hline
2n + 5 = & 13 \\
-5 & -5 \\
\hline
2n \;\; = & 8 \\
n = 4 &
\end{array}
$$

First we remove as many variables as we can from both sides of the sentence. Now the same thing for constants. And we have our solution.

$$
\begin{array}{rr}
7x + 4 = & 6x + 15 \\
-6x \quad & -6x \\
\hline
x + 4 = & 15 \\
-4 & -4 \\
\hline
x \;\; = & 11
\end{array}
$$

Test Yourself

Directions: Use the above solutions as models as you try these.

1. $7a + 8 = 6a + 12$ 2. $5b + 3 = 4b + 19$

3. $4a + 11 = 3a + 20$ 4. $9x + 15 = 7x + 21$

5. $8k + 9 = 6k + 25$ 6. $11r + 10 = 9r + 24$

7. $7x + 8 = 4x + 20$ 8. $15n + 5 = 11n + 29$

Answers

1. 4	**2.** 16	**3.** 9	**4.** 3	**5.** 8
6. 7	**7.** 4	**8.** 6		

SIGNED NUMBERS

Before proceeding with an in-depth investigation into the workings of algebra, it is essential to look at the realm of numbers known as signed numbers (or integers). All numbers with one exception are either greater or less than zero. The exception is, of course, zero itself. Those numbers that are greater than zero are considered to be positive, and are denoted by a positive sign. For example, +6 is known as "positive six." Six is also known as positive 6. In fact, any numeral that is written without a sign in front of it is considered to be positive. Most of the numbers you have dealt with throughout your life have been positive numbers, as have all of those treated in this book until now.

Those numbers less than zero are called negative numbers. The number -6 is read "negative 6" and represents a distance of six away from zero on the left side of zero, if you are considering position on a number line:

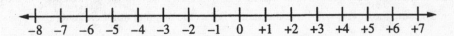

All numbers represented on a number line increase as you move from left to right and decrease as you move from right to left. So, for example, +6 is greater than +4, but -6 is less than -4. Notice the positions of these numbers on the number line and you'll readily see why.

Absolute Value

The absolute value of a number is the distance that that number is from zero. A glance at the number line above will show you that -2 is 2 spaces from 0. That means that the absolute value of negative two is 2. That would be written:

$$|{-2}| = 2$$

Now, +2 is also 2 spaces away from zero, so it too has an absolute value of 2, which leads to the following conclusion:

$$|{-2}| = |{+2}| = 2$$

Test Yourself

Directions: Find the absolute values.

1. $|{-7}| = $ ____   2. $|{-4}| = $ ____   3. $|{+8}| = $ ____
4. $|{+9}| = $ ____

Directions: Find the absolute values where $y = -6$:

5. $|{-y}| = $ ____   6. $|{+y}| = $ ____   7. $|y + 2| = $ ____

Directions: Find the absolute values where $m = 7$:

8. $|m - 3| = $ ____   9. $|m + {-2}| = $ ____   10. $|{+m} + {+3}| = $ ____

NOTE

A pair of vertical lines is read as "the absolute value of."

Answers

1. 7	**2.** 4	**3.** 8	**4.** 9	**5.** 6
6. 6	**7.** 4	**8.** 4	**9.** 9	**10.** 10

Combining Signed Numbers

Signed numbers can be combined by addition, subtraction, multiplication, and division, but the rules that govern these combinations are somewhat different from those that govern the combinations of numbers without signs. For that reason, we will study addition and subtraction separately. Multiplication and division follow the same rules and so will be considered together.

Adding Signed Numbers

When numbers with like signs are to be added together, their absolute values are added in the usual manner. The sum, however, will have the same sign as the original numbers had:

Example 1

To add

$$^{+}7 + {}^{+}6$$
$$\downarrow$$
$$|{}^{+}7| = 7;\ |{}^{+}6| = 6 \qquad \text{Find the absolute values.}$$
$$\downarrow$$
$$7 + 6 = 13 \qquad \text{Add the absolute values.}$$
$$\downarrow$$
$$^{+}7 + {}^{+}6 = {}^{+}13 \qquad \text{The sum takes the sign of the addends.}$$

Example 2

To add

$$^{-}5 + {}^{-}4$$
$$\downarrow$$
$$|{}^{-}5| = 5;\ |{}^{-}4| = 4 \qquad \text{Find the absolute values.}$$
$$\downarrow$$
$$5 + 4 = 9 \qquad \text{Add the absolute values.}$$
$$\downarrow$$
$$^{-}5 + {}^{-}4 = {}^{-}9 \qquad \text{The sum takes the sign of the addends.}$$

Test Yourself

Directions: Add the following.

1. $^{+}3 + {}^{+}7 =$ _____
2. $^{-}4 + {}^{-}8 =$ _____
3. $^{+}5 + {}^{+}8 =$ _____
4. $^{-}4 + {}^{-}6 =$ _____
5. $^{-}3 + {}^{-}12 =$ _____
6. $^{+}8 + {}^{+}9 =$ _____
7. $^{-}9 + {}^{-}12 =$ _____
8. $^{-}13 + {}^{-}4 =$ _____
9. $^{+}8 + {}^{+}12 =$ _____
10. $^{-}11 + {}^{-}10 =$ _____
11. $^{+}5 + {}^{+}21 =$ _____
12. $^{-}15 + {}^{-}13 =$ _____

Answers

1. $^+10$	**2.** $^-12$	**3.** $^+13$	**4.** $^-10$	**5.** $^-15$
6. $^+17$	**7.** $^-21$	**8.** $^-17$	**9.** $^+20$	**10.** $^-21$
11. $^+26$	**12.** $^-28$			

Adding signed numbers whose signs are different is accomplished by subtracting the absolute values of the two numbers. (That is correct. You have to subtract in order to add.) The sum (even though the operation performed is really subtraction, because it is the answer to an addition problem, it is still considered a sum) then takes the sign of the larger of the addends (that is, the number in the addition with the greater absolute value).

Example 3

$$^+7 + ^-5$$
$$\downarrow$$

$|^+7| = 7;\ |^-5| = 5$ Find the absolute values...
$$\downarrow$$

$7 - 5 = 2$ ...subtract them...
$$\downarrow$$

$^+7 + ^-5 = ^+2$ ...assign the sum the sign of the addend with the larger absolute value.

Example 4

$$^+6 + ^-11$$
$$\downarrow$$

$|^+6| = 6;\ |^-11| = 11$ Find the absolute values...
$$\downarrow$$

$11 - 6 = 5$ ...subtract them...
$$\downarrow$$

         ...assign the sum the sign of the addend with the larger absolute value.

$^+6 + ^-11 = ^-5$

Test Yourself

Directions: Add the following.

1. $^+3 + ^-4 = $ ____	**2.** $^-7 + ^+5 = $	**3.** $^-8 + ^+9 = $ ____
4. $^-6 + ^-4 = $ ____	**5.** $^+6 + ^-14 = $ ____	**6.** $^-8 + ^-5 = $ ____
7. $^+6 + ^-15 = $ ____	**8.** $^-9 + ^+12 = $ ____	**9.** $^+4 + ^+18 = $ ____
10. $^-17 + ^+8 = $ ____	**11.** $^-9 + ^+15 = $ ____	**12.** $^+15 + ^-24 = $ ____
13. $^-9 + ^+9 = $ ____	**14.** $^+31 + ^-27 = $ ____	**15.** $^-23 + ^+50 = $ ____

Answers

1. ⁻1		**2.** ⁻2		**3.** ⁺1		**4.** ⁻10		**5.** ⁻8	
6. ⁻13		**7.** ⁻9		**8.** ⁺3		**9.** ⁺22		**10.** ⁻9	
11. ⁺6		**12.** ⁻9		**13.** 0		**14.** ⁺4		**15.** ⁺27	

To add three or more signed numbers, first collect the numbers with like signs. Then add them together. Finally add the two resulting numbers with unlike signs together:

Example 5

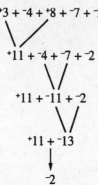

Combine the positive numbers...

Combine the negative numbers (two at a time)...

Combine the single resultant positive number with the single resultant negative.

Test Yourself

Directions: Add the following.

1. ⁺7 + ⁻5 + ⁺2 = _____   2. ⁻5 + ⁻9 + ⁺8 = _____

3. ⁺8 + ⁻4 + ⁺6 + ⁻12 = _____   4. ⁻7 + ⁺12 + ⁻8 + ⁻5 = _____

5. ⁻9 + ⁺5 + ⁻7 + ⁺8 = _____   6. ⁺6 + ⁻8 + ⁻7 + ⁺6 = _____

7. ⁺10 + ⁺7 + ⁻5 + ⁺8 + ⁻6 + ⁺11 + ⁻8 = _____

Answers

1. ⁺4		**2.** ⁻6		**3.** ⁻2		**4.** ⁻8		**5.** ⁻3	
6. ⁻3		**7.** ⁺17								

Subtracting Signed Numbers

Have you ever considered the meaning of a double negative? Think about this sentence: *I am not not going to the dentist.* What does it mean to you? If you are *not* not going to the dentist, then you must be going to the dentist. The same is true of a double negative when subtracting. Consider the following example:

Example 1

$^+12 - {}^-8 =$ _____ $- {}^-8$ is a double negative. It means $^+8$. Therefore we change the example:

$^+12 + 8$ Now we add 12 and 8 to get 20,...

$^+12 + 8 = {}^+20$ $\therefore {}^+12 - {}^-8 = {}^+20$

Notice that the key to effectively subtracting signed numbers is to first deal with the subtraction sign. That is to say, if a double negative is created, exchange both negatives for a single plus sign. After that, the subtraction proceeds exactly as a signed number addition would. Study the following examples, and you will see how each of the possible signed number subtractions works.

Example 2

$^-9 - {}^-4 =$ _____   Double negative becomes a plus.

$\downarrow$

$^-9 + 4$ Then add.

$\downarrow$

$^-9 + 4 = {}^-5$ $\therefore {}^-9 - {}^-4 = {}^-5$

Example 3

$^+7 - {}^+4 =$ _____   Since taking away a gain is the same as adding a loss,

$\downarrow$ exchange the $- +$ signs to make $+ -$.

$^+7 + {}^-4$ Then add.

$\downarrow$

$^+7 + {}^-4 = {}^+3$ $\therefore {}^+7 - {}^+4 = {}^+3$

Example 4

$^-11 - {}^+3 =$ _____   For the same reason as in example 3, exchange the $- +$ to make $+ -$.

$\downarrow$

$^-11 + {}^-3$ Then add.

$\downarrow$

$^-11 + {}^-3 = {}^-14$ $\therefore {}^-11 - {}^+3 = {}^-14$

Notice that in the third and fourth examples, by exchanging the + and - signs, we were able to change the signed number subtraction problem to a signed number addition problem. And, we have already seen from the first two examples that a double negative exchanged for a plus also creates an addition. It is therefore safe to conclude that any signed number subtraction can, by a careful manipulation of the two middle signs, be turned into a signed number addition, and then solved in the same way that any signed number addition is solved.

Test Yourself

Directions: Solve the following subtractions by first turning them into additions.

1. $^-8 - ^+6 = $ ___
2. $^+9 - ^+4 = $ ___
3. $^+12 - ^+5 = $ ___
4. $^-11 - ^+8 = $ ___
5. $^+7 - ^-4 = $ ___
6. $^-9 - ^-6 = $ ___
7. $^-13 - ^+6 = $ ___
8. $^+18 - ^-10 = $ ___
9. $^+21 - ^-8 = $ ___
10. $^+17 - ^+5 = $ ___
11. $^-16 - ^-4 = $ ___
12. $^-23 - ^-11 = $ ___
13. $^+15 - ^-8 = $ ___
14. $^-18 - ^-6 = $ ___
15. $^-20 - ^+5 = $ ___
16. $^+19 - ^-8 = $ ___
17. $^+16 - ^+7 = $ ___
18. $^-5 - ^-11 = $ ___
19. $^+4 - ^-12 = $ ___
20. $^-6 - ^+15 = $ ___

Answers

1. $^-14$
2. $^+5$
3. $^+7$
4. $^-19$
5. $^+11$
6. $^-3$
7. $^-19$
8. $^+28$
9. $^+29$
10. $^+12$
11. $^-12$
12. $^-12$
13. $^+23$
14. $^-12$
15. $^-25$
16. $^+11$
17. $^+9$
18. $^+6$
19. $^+16$
20. $^-21$

Multiplying and Dividing Signed Numbers

As noted earlier, the rules governing multiplication and division of signed numbers are identical, and so they will be treated at the same time. Multiplication of signed numbers is the same as any multiplication, as far as the computational part of the exercise is concerned. That is, 3 x 5 is 15, regardless of the sign of either of the numbers. It is in the determination of the sign of the product that the signs of the factors (numbers being multiplied) play a role. For determining the sign of the product or the quotient, there are exactly two rules:

If the signs are the same, the sign of the answer is positive.

If the signs are different, the sign of the answer is negative.

The following examples should serve to illustrate the application of these rules.

Example
$^+3 \times ^+4$
Signs are the same,
$\therefore ^+3 \times ^+4 = ^+12$

Example
$^+3 \times ^-4$
Signs are different,
$\therefore ^+3 \times ^-4 = ^-12$

Example
$^-3 \times ^+4$
Signs are different,
$\therefore ^-3 \times ^+4 = ^-12$

Example
$^-3 \times ^-4$
Signs are the same,
$\therefore ^-3 \times ^-4 = ^+12$

Since the same rules apply to division, we have not bothered to illustrate any divisions. Nevertheless, you will find both multiplications and divisions in the following exercises.

Test Yourself

> **Directions:** Solve the following.

1. $^+5 \times {}^+8 = $ _____
2. $^-6 \times {}^+7 = $ _____
3. $^-7 \times {}^-4 = $ _____
4. $^+6 \times {}^-7 = $ _____

5. $^-12 \div {}^+4 = $ _____
6. $^-6 \div {}^+3 = $ _____
7. $^-24 \div {}^-3 = $ _____
8. $^+7 \times {}^+4 = $ _____

9. $^+9 \div {}^+3 = $ _____
10. $^-3 \times {}^-8 = $ _____
11. $^+30 \div {}^-5 = $ _____
12. $^-5 \times {}^+8 = $ _____

13. $^+5 \times {}^+9 = $ _____
14. $^+16 \div {}^-2 = $ _____
15. $^-15 \div {}^-5 = $ _____
16. $^-10 \div {}^-2 = $ _____

17. $^+6 \times {}^-9 = $ _____
18. $^+8 \times {}^-8 = $ _____
19. $^-9 \times {}^+4 = $ _____
20. $^+18 \div {}^+9 = $ _____

21. $^+14 \div {}^-7 = $ _____
22. $^-20 \div {}^+4 = $ _____
23. $^+18 \div {}^-6 = $ _____
24. $^-5 \times {}^-8 = $ _____

Answers

1. $^+40$
2. $^-42$
3. $^+28$
4. $^-42$
5. $^-3$

6. $^-2$
7. $^+8$
8. $^+28$
9. $^+3$
10. $^+24$

11. $^-6$
12. $^-40$
13. $^+45$
14. $^-8$
15. $^+3$

16. $^+5$
17. $^-54$
18. $^-64$
19. $^-36$
20. $^+2$

21. $^-2$
22. $^-5$
23. $^-3$
24. $^+40$

SOLVING EQUATIONS BY EQUAL ADDITIONS

As noted earlier, an algebraic equation is like an equal-arm balance. As long as the same quantity is added to or subtracted from both sides of the balance, equilibrium is maintained. The same is true of an equation.

The main strategy employed when solving an algebraic equation is that of collecting like terms. That is to say, begin by getting all variables together on one side of the equal sign, and all constants together on the other side. The following examples explore the technique of adding the same quantity to each side of the equation in order to collect like terms.

Example

$$x - 5 = 14$$

$$\downarrow$$

$$\begin{array}{r} x - 5 = 14 \\ +5 \quad +5 \\ \hline \end{array}$$

$$\downarrow$$

$$x = 19$$

We don't want 5 on the same side of the equation as the x, so, since the 5 is combined with the x by subtraction, add 5 (the opposite of subtraction) to each side.

Example

$$^-n - 6 = 14 - 2n$$

$$\downarrow$$

$$\begin{array}{r} ^-n - 6 = 14 - 2n \\ +6 \quad\quad +6 \\ \hline \end{array}$$

$$\downarrow$$

$$\begin{array}{r} ^-n = 20 - 2n \\ +2n \quad\quad +2n \\ \hline n = 20 \end{array}$$

We add 6 to each side to move it from the left to the right side of the equation. Then we add $2n$ to each side to move all the n's to the left side of the equation.

When deciding the side of the equals sign to move the variables to (assuming that there are initially variables on both sides), ask yourself which side has the greater coefficient of the variable. $3x$, for example, has a greater coefficient than $2x$, and $2x$ has a greater coefficient than $-5x$. $-3x$ has a greater coefficient than $-9x$. Note the following:

Example

$$\begin{array}{r} 5 - 3w = {}^-2w \\ +3w \quad + 3w \\ \hline 5 \quad\quad = w \end{array}$$

$$w = 5$$

You will note that ^-2w has a greater coefficient than ^-3w. Therefore, it is desirable to collect the variables on the right side of the equation. You may then reverse the order of the solution if you like so that it reads $w = 5$, although there is no mathematical necessity to do so.

Test Yourself

Directions: Solve by equal additions.

1. $r - 7 = 15$
2. $x - 4 = 9$
3. $m - 12 = 81$
4. $^-2x = 7 - 3x$
5. $^-2y = {}^-9 - 3y$
6. $^-6 - 2y = {}^-3y - 5$
7. $^-4x - 6 = {}^-5x + 7$
8. $^-7r - 2 = {}^-6r - 5$
9. $^-4x - 9 = {}^-3x - 12$

Answers

1. 22
2. 13
3. 93
4. 7
5. $^-9$
6. 1
7. 13
8. 3
9. 3

SOLVING EQUATIONS BY EQUAL SUBTRACTIONS

Sometimes, the terms that you wish to move around in an equation will be connected by + signs. That means that those terms are combined by addition. (Remember, if you do not see any sign before a number, that means that there is a + sign there and that it is by convention not written.) To undo an addition, it is necessary to subtract. Study the examples to see how this works.

Example

$$5x + 11 = 6x + 3$$
$$\downarrow$$
$$5x + 11 = 6x + 3$$
$$\underline{-5x \qquad -5x}$$
$$11 = x + 3$$
$$\underline{-3 \qquad -3}$$
$$8 = x$$

First the variables are collected on the right side (by subtracting $5x$).

Then the constants are collected on the left (by subtracting 3).

or

$$x = 8$$

This last step is optional.

Example

$$7y + 5 = 6y + 4$$
$$\downarrow$$
$$7y + 5 = 6y + 4$$
$$\underline{-6y - 5 \quad -6y - 5}$$
$$y = {}^-1$$

Both $6y$ and 5 are subtracted from each side of the equation.

Test Yourself

Directions: Solve by subtracting the same quantities from both sides.

1. $y + 9 = 15$	**2.** $x + 8 = 5$	**3.** $y + 4 = {}^-3$
4. $5x + 7 = 6x + 4$	**5.** $5y + 3 = 4y + 6$	**6.** $7y - 9 = 8y + 5$
7. $11y + 4 = 10y + 6$	**8.** $5x + 2 = 4x + 8$	**9.** $3y + 5 = 2y + 7$
10. $4m + 8 = 3m - 2$	**11.** $7n - 4 = 8n$	**12.** $x + 6 = 2x + 6$

Answers

1. 6	**2.** $^-3$	**3.** $^-7$	**4.** 3	**5.** 3
6. $^-14$	**7.** 2	**8.** 6	**9.** 2	**10.** $^-10$
11. $^-4$	**12.** 0			

SOLVING EQUATIONS BY MULTIPLICATION AND DIVISION

Multiplication of a variable by a numerical coefficient is quite common in algebraic equations. To undo a multiplication, it is, of course, necessary to divide. Other equations indicate division of a variable by a constant or multiplication of a variable by a fractional coefficient. In either of those cases, the simplest way to get the variable by itself is to multiply by the reciprocal of that coefficient, so as to leave the variable with a coefficient of 1. Examine the examples below.

Example 1

$3a = 24$ The variable, a, is multiplied by 3 ...

... so divide both sides by 3 ...

$\dfrac{3a}{3} = \dfrac{24}{3}$ ... and find that $a = 8$.

$a = 8$

Example 2

$\dfrac{b}{4} = 9$ The variable, b, is divided by 4 ...

$4 \cdot \dfrac{b}{4} = 9 \cdot 4$ ... so multiply both sides by 4. Notice the dot used as the times sign. You'll get used to it.

$\dfrac{4}{1} \cdot \dfrac{b}{4} = 36$ $4 =$ the fraction $\dfrac{4}{1}$ ...

$\dfrac{4b}{4} = 36$ ... and that makes it easier to multiply.

$\dfrac{\cancel{4}b}{\cancel{4}} = 36$ The 4s cancel, since $\dfrac{4}{4} = 1$ ...

$b = 36$ ... and so, $b = 36$.

Example 3

$$\frac{1}{7}x = 5$$

The variable x is multiplied by $\frac{1}{7}$. You might divide both sides

$$\frac{7}{1} \cdot \frac{1}{7}x = 5 \cdot \frac{7}{1}$$

by $\frac{1}{7}$, but this is the same as multiplying both sides by 7, which

is $\frac{1}{7}$'s reciprocal.

$$\frac{7}{1} \cdot \frac{1}{7}x = \frac{5}{1} \cdot \frac{7}{1}$$

Multiply...

$$\frac{7}{7}x = \frac{35}{1}$$

...then simplify ...

$$x = 35$$

... and get $x = 35$.

Example 4

$$\frac{3}{4}x = 12$$

The variable, x, is multiplied by $\frac{3}{4}$.

$$\frac{4}{3} \cdot \frac{3}{4}x = 12 \cdot \frac{4}{3}$$

To get a coefficient of 1, multiply by $\frac{4}{3}$ ($\frac{3}{4}$'s reciprocal)...

$$\frac{4}{3} \cdot \frac{3}{4}x = \frac{12}{1} \cdot \frac{4}{3}$$

...then simplify ...

$$\frac{12}{12}x = \frac{48}{3}$$

... and find that $x = 16$.

$$x = 16$$

Actually, whether a variable is combined with a constant by multiplication or by division, you may consider that you are multiplying by that constant's reciprocal in order to get the variable alone (to have a coefficient of 1). In model example 1, you may consider the $3a$ to have been multiplied by $\frac{1}{3}$ as well as to have been divided by 3 (the two operations accomplish the same thing). In example 2, we multiplied by 4 because 4 is the reciprocal of $\frac{1}{4}$, and $\frac{b}{4}$ means $\frac{1}{4}b$.

Try your hand at the following equations for which a reciprocal multiplication is required.

Test Yourself

Directions: Solve for the variable.

1. $3x = 18$	**2.** $5x = 40$	**3.** $7n = 56$	**4.** $\frac{1}{2}n = 9$
5. $\frac{1}{3}r = 17$	**6.** $\frac{1}{8}y = 4$	**7.** $\frac{2}{3}m = 42$	**8.** $\frac{3}{4}y = 21$
9. $\frac{5}{6}x = 20$	**10.** $\frac{4}{5}w = 24$	**11.** $\frac{1}{3}m = 7$	**12.** $4v = 7$
13. $9u = 5$	**14.** $21m = 7$	**15.** $\frac{2}{3}l = \frac{3}{4}$	**16.** $\frac{1}{2}x = \frac{3}{4}$
17. $\frac{2}{3}x = \frac{7}{8}$	**18.** $6p = \frac{1}{7}$	**19.** $\frac{s}{5} = 9$	**20.** $\frac{t}{3} = 8$
21. $\frac{r}{5} = \frac{3}{8}$			

Answers

1. 6	**2.** 8	**3.** 8	**4.** 18	**5.** 51	**6.** 32
7. 63	**8.** 28	**9.** 24	**10.** 30	**11.** 21	**12.** $1\frac{3}{4}$
13. $\frac{5}{9}$	**14.** $\frac{1}{3}$	**15.** $1\frac{1}{8}$	**16.** $1\frac{1}{2}$	**17.** $1\frac{5}{16}$	**18.** $\frac{1}{42}$
19. 45	**20.** 24	**21.** $1\frac{7}{8}$			

PUTTING IT ALL TOGETHER

As you may have suspected, it is rather rare to find an algebraic equation that lends itself to being solved by just one operation alone. Most are solved by some combination of two or more operations. Nevertheless, since we have already looked at the procedures involved for undoing all of the operations by which variables and constants can be combined, being able to solve most equations is just a matter of being able to use the procedures we have already studied in the proper sequence. To assure that the proper sequence is followed, it is helpful to bear in mind three steps which, when followed, will lead to the solutions of most linear equations:

❶ Combine like terms on the same side of the equation.

❷ Collect terms.

❸ Multiply by the reciprocal of the variable's coefficient.

The application of these three steps is illustrated in the rather complex examples below.

Example

$$5x + 3 - 2x = 23 - x + 4$$

Inspecting the equation, we can see that there are two terms containing variables on the left side, and two terms that are constants on the right side. These like terms must be combined:

$$3x + 3 = 27 - x$$

Since the variable with the greater coefficient is on the left, we will collect variables on that side. Since 3 is added onto the left side, we subtract 3 from both sides:

$$
\begin{array}{rl}
3x + 3 & = 27 - x \\
\underline{- 3} & \underline{- 3} \\
3x & = 24 - x
\end{array}
$$

To move the variable **x** to the left side, we first notice how it is combined with the 24. Since it is combined by subtraction, we add **x** to both sides:

$$
\begin{array}{rl}
3x = & 24 - x \\
\underline{+x} & \underline{+ x} \\
4x = & 24
\end{array}
$$

What remains to be done is to undo the multiplication of x by 4. This can be done by multiplying by its reciprocal, or by dividing:

$$\left(\frac{1}{4}\right)(4x) = (24)\left(\frac{1}{4}\right) \qquad \text{or} \qquad \frac{4x}{4} = \frac{24}{4}$$

Either way, we finally determine that ...

$$x = 6$$

Notice in the previous example that the three steps were followed in the order stated before:

❶ Combine like terms on the same side.

❷ Collect terms.

❸ Multiply by the reciprocal of the variable's coefficient (or divide).

Not every one of these steps need be performed in every problem. Sometimes one or more steps cannot be done, simply because there are no terms with which to perform the step. Such a case can be seen in the following example.

Example

$$5x + 7 = 2x - 5$$

There are no like terms to combine, so we move right on to collecting like terms—first constants with constants ...

$$
\begin{array}{rcl}
5x + 7 & = & 2x - 5 \\
\underline{ - 7} & & \underline{ - 7} \\
5x & = & 2x - 12
\end{array}
$$

... and then variables with variables (although they did not need to be done in that order).

$$
\begin{array}{rcl}
5x & = & 2x - 12 \\
\underline{-2x} & & \underline{-2x} \\
3x & = & -12
\end{array}
$$

Finally, we multiply by the variable's coefficient's reciprocal, or we divide ...

$$\frac{3x}{3} = \frac{^-12}{3} \text{ or } \frac{1}{3}(3x) = \frac{1}{3}\left(^-12\right)$$

... and discover that:

$$x = {}^-4$$

In the next example, the need to collect terms does not exist:

Example

$$\frac{3}{4}x + \frac{7}{4}x = 12 - 9$$

Combine like terms on the same side of the equal sign ...

$$\frac{10}{4}x = 3$$

... Notice that there is no need to collect terms, so the next step is to multiply both sides by the reciprocal of $\frac{10}{4}$, which is $\frac{4}{10}$.

$$\left(\frac{4}{10}\right)\left(\frac{10}{4}x\right) = \left(\frac{4}{10}\right)\left(\frac{3}{1}\right)$$

The result is then expressed in lowest terms—in this case as a mixed numeral:

$$x = \frac{12}{10} = 1\frac{1}{5}$$

The following equations require a variety of operations to solve. Remember the order in which the operations should be performed, and you will have little difficulty with them. If necessary, refer to the rules and the model examples above.

Test Yourself

Directions: Solve for the variable.

1. $3x + 5 = 2x + 7$

2. $8y - 3 = 5y + 18$

3. $2x - 4 = 5x + 7$

4. $8y - 8 = 5y + 7$

5. $9z + 5 - 2z = 6 + 3z + 9$

6. $5 - z = 3z - 11$

7. $3 + 2n = 4n - 9$

8. $4 - 5m + 7 = 2m + 4m$

9. $a - 3 + a = 7 - 3a$

10. $9b - 6 = 6b + 9$

11. $\frac{b}{3} + 8 = 7$

12. $\frac{x}{2} - 5 = 6 - \frac{x}{2}$

13. $\frac{1}{4}y + \frac{1}{2} = \frac{2}{3}$

14. $11 - \frac{2}{3}x = \frac{1}{3}x - 6$

15. $\frac{3}{4}x - 12 = 15 - \frac{1}{2}x$

16. $\frac{5}{7}p - 4 = \frac{3}{7}p - 12$

17. $\frac{8r}{5} - 2 = \frac{3r}{10} + 8$

18. $\frac{1}{6}w + 11 = \frac{5}{6}w - 7$

19. $\frac{9q}{10} - \frac{3}{5}q = \frac{3}{4} - \frac{1}{8}$

20. $3m + 5 = \frac{3}{2}m - 6$

Answers

1. 2	2. 7	3. $\frac{-11}{3}$, or $^-3\frac{2}{3}$	4. 5
5. $\frac{5}{2}$ or $2\frac{1}{2}$	6. 4	7. 6	8. 1
9. 2	10. 5	11. $^-3$	12. 11
13. $\frac{2}{3}$	14. 17	15. $\frac{108}{5}$ or $21\frac{3}{5}$	16. $^-28$
17. $\frac{100}{13}$ or $7\frac{9}{13}$	18. 27	19. $\frac{25}{12}$ or $2\frac{1}{12}$	20. $\frac{-22}{3}$ or $^-7\frac{1}{3}$

RATIO AND PROPORTION

What Is a Ratio?

When we looked at the many different ways in which fractions are used (in Chapter 3), we noted that one of those uses was to represent a ratio. Ratio is a mathematical word that means comparison.

A ratio is a comparison.

There are two different notations commonly used to represent ratios. One of them is 3:4, and the other is $\frac{3}{4}$. Whether the colon or the fraction line is used, the ratio should read "3 is to 4 ..." Did you get the impression that something should have followed the symbols in quotation marks? Well something should have, and usually does, but before getting into that, let's consider a few common examples of ratios:

Example

There are 23 women and 17 men in an evening class. Find the ratio of men to women, women to men, and women to students in the class.

The ratio of men to women is 17:23 or $\frac{17}{23}$.

The ratio of women to men is 23:17 or $\frac{23}{17}$.

The ratio of women to students is 23:40 or $\frac{23}{40}$.

It is crucial to notice that in any ratio the order in which the terms or conditions are stated will determine what the ratio looks like. In model example 1, three ratios were formed from just two given numbers.

In fact, however, six ratios are possible from any two figures. In all of those ratios, the order of the terms is extremely significant. Examine the example, below, and you will see how the six ratios are formed.

Example

An order from a fast-food restaurant contains 5 hamburgers and 7 cheeseburgers. Form all possible ratios from this data.

Hamburgers to cheeseburgers:	5:7
Cheeseburgers to hamburgers:	7:5
Cheeseburgers to all burgers:	7:12
Hamburgers to all burgers:	5:12
All burgers to cheeseburgers:	12:7
All burgers to hamburgers:	12:5

... and hold the French fries.

Test Yourself

Directions: Now that you have seen how ratios are formed, see how you do at forming all possible ratios from the data given below.

1. 9 history professors and 2 mathematics professors:
   a) history to mathematics professors
   b) mathematics to history professors
   c) math professors to those of both subjects
   d) history professors to those of both subjects
   e) professors of both subjects to history professors
   f) professors of both subjects to math professors

2. 19 regular letters and 2 special delivery letters:
   a) regular to special delivery letters
   b) special delivery to regular letters
   c) regular to all letters
   d) special delivery to all letters
   e) all letters to regular letters
   f) all letters to special delivery letters

3. 7 spaghetti dinners and 3 fish dinners:
   a) fish dinners to spaghetti dinners
   b) spaghetti dinners to all dinners
   c) all dinners to fish dinners
   d) spaghetti dinners to fish dinners
   e) all dinners to spaghetti dinners
   f) fish dinners to all dinners

Answers

1. 9:2, 2:9, 2:11, 9:11, 11:9, 11:2
2. 19:2, 2:19, 19:21, 2:21, 21:19, 21:2
3. 3:7, 7:10, 10:3, 7:3, 10:7, 3:10

Test Yourself

Directions: Make 6 different ratios from each pair of numbers.

1. 4, 8 _____
2. 11, 5 _____
3. 6, 13 _____
4. 7, 12 _____
5. 9, 1 _____

Answers

1. 4:8, 8:4, 4:12, 8:12, 12:4, 12:8
2. 11:5, 5:11, 5:16, 11:16, 16:5, 16:11
3. 6:13, 13:6, 6:19, 13:19, 19:6, 19:13
4. 7:12, 12:7, 7:19, 12:19, 19:7, 19:12
5. 9:1, 1:9, 1:10, 9:10, 10:1, 10:9

Test Yourself

Directions: Express each ratio as a fraction in lowest terms.

1. 5:9 ____   2. 6:11 ____   3. 5:17 ____   4. 7:3 ____   5. 12:4
6. 8:4 ____   7. 9:12 ____   8. 24:36 ____   9. 18:15 ____   10. 6:9

Answers

1. $\dfrac{5}{9}$ 2. $\dfrac{6}{11}$ 3. $\dfrac{5}{17}$ 4. $\dfrac{7}{3}$ 5. $\dfrac{3}{1}$

6. $\dfrac{2}{1}$ 7. $\dfrac{3}{4}$ 8. $\dfrac{2}{3}$ 9. $\dfrac{6}{5}$ 10. $\dfrac{2}{3}$

What Is a Proportion?

Ratios in and of themselves can be handy for expressing a relationship between two quantities but are of little use in solving mathematical problems. Once ratios are placed into a proportion, however, they become very useful. *A proportion is an equation involving two ratios.*

6:9 = 2:3 is a proportion. It is read: "Six is to nine as two is to three." That means 6 has the same relationship to 9 as the relationship that 2 has to 3. In particular, 6 is $\frac{2}{3}$ of 9, and 2 is $\frac{2}{3}$ of 3. In a proportion, the two terms farthest from each other are known as the *extremes*. The two terms nearest each other are known as the *means:*

$$\overset{\textbf{extremes}}{\overset{\nearrow \quad \searrow}{3{:}4 = 15{:}20}}$$
$$\underset{\textbf{means}}{\underset{\searrow \quad \nearrow}{\phantom{3{:}4 = 15{:}20}}}$$

Can you recognize the means and the extremes in the following proportion?

$$\frac{1}{2} = \frac{4}{8}$$

The means are 2 and 4, while 1 and 8 are the extremes. If you are not sure of why that is the case, consider that the proportion could have been written as 1:2 = 4:8 (1 is to 2 as 4 is to 8). Using this format, the means and the extremes are obvious.

One rule governs operations involving proportions. That rule is:

The product of the means equals the product of the extremes.

To see how this rule works, consider the two previous proportions. In 3:4 = 15:20, multiply the extremes (3×20) and get 60. Then multiply the means (4×15) and get 60. Since 60 = 60, the product of the extremes equals the product of the means (and vice-versa).

In the proportion $\frac{1}{2} = \frac{4}{8}$, the product of the extremes (1×8) equals the product of the means (2×4). Since it is not customary to include colons in mathematical equations, but it is perfectly usual to find fractions, proportions are almost always seen in the fractional form. Since the fractional form is so usual, and since mathematicians are always looking for shortcuts to save time, the multiplication of means together and extremes together in fraction form is often thought of as "cross-multiplication," where the denominator of each ratio is multiplied by the numerator of the other:

$$\frac{1}{2} \overset{\times}{\underset{\times}{\times}} \frac{4}{8} \begin{matrix} = 8 \\ = 8 \end{matrix}$$

$$8 \ = \ 8$$

Note that cross-multiplication can only occur when there is a single fraction on either side of an equal sign. **That is the only instance in which cross-multiplication is permissible.**

Solving Proportions

Now that we have examined what ratios and proportions are and the rules that govern them, it is time to see how to put them to work. Below are proportions, each of which has one term missing. Find the missing term by cross-multiplying and then solving for the variable.

Test Yourself

Directions: Solve each proportion for the variable.

1. $\dfrac{x}{5} = \dfrac{8}{20}$
2. $\dfrac{x}{14} = \dfrac{21}{42}$
3. $\dfrac{y}{9} = \dfrac{4}{3}$
4. $\dfrac{5}{8} = \dfrac{x}{40}$

5. $\dfrac{5}{12} = \dfrac{n}{60}$
6. $\dfrac{24}{15} = \dfrac{56}{n}$
7. $\dfrac{35}{40} = \dfrac{14}{r}$
8. $\dfrac{6}{7} = \dfrac{z}{8}$

9. $\dfrac{5}{x} = \dfrac{4}{9}$
10. $\dfrac{3}{x} = \dfrac{7}{11}$
11. $\dfrac{4}{x} = \dfrac{32}{72}$
12. $\dfrac{5}{9} = \dfrac{x}{12}$

13. $\dfrac{y}{17} = \dfrac{8}{11}$
14. $\dfrac{6}{y} = \dfrac{8}{13}$
15. $\dfrac{3}{v} = \dfrac{8}{21}$
16. $\dfrac{2}{9} = \dfrac{g}{8}$

17. $\dfrac{15}{z} = \dfrac{6}{21}$
18. $\dfrac{4}{7} = \dfrac{9}{w}$
19. $\dfrac{3x}{5} = \dfrac{7}{9}$
20. $\dfrac{5}{4y} = \dfrac{3}{10}$

21. $\dfrac{7}{15} = \dfrac{8}{5y}$

Answers

1. 2
2. 7
3. 12
4. 25

5. 25
6. 35
7. 16
8. $\dfrac{48}{7}$ or $6\dfrac{6}{7}$

9. $\dfrac{45}{4}$ or $11\dfrac{1}{4}$
10. $\dfrac{33}{7}$ or $4\dfrac{5}{7}$
11. 9
12. $\dfrac{20}{3}$ or $6\dfrac{2}{3}$

13. $\dfrac{136}{11}$ or $12\dfrac{4}{11}$
14. $\dfrac{39}{4}$ or $9\dfrac{3}{4}$
15. $\dfrac{63}{8}$ or $7\dfrac{7}{8}$
16. $\dfrac{16}{9}$ or $1\dfrac{7}{9}$

17. $\dfrac{105}{2}$ or $52\dfrac{1}{2}$
18. $\dfrac{63}{4}$ or $15\dfrac{3}{4}$
19. $\dfrac{35}{27}$ or $1\dfrac{8}{27}$
20. $\dfrac{25}{6}$ or $4\dfrac{1}{6}$

21. $\dfrac{24}{7}$ or $3\dfrac{3}{7}$

PERCENTS

Many people have difficulty understanding percents and how they work. In fact, percents are ratios, but they can probably best be understood if they are thought of as fractions. Unlike common fractions (the kind with numerators and denominators) and decimal fractions, percents are not based on a unit of 1. That is to say, when dealing with percents, one whole is represented not by the numeral 1, but rather by 100. 100% = 1 whole. Any fraction, therefore, may be changed into a percent simply by establishing a proportion:

$$\frac{\text{numerator}}{\text{denominator}} = \frac{x}{100}$$

Solving for x will then give the percent.

Example

Express the fraction $\frac{3}{4}$ as a percent.

$$\frac{3}{4} = \frac{x}{100} \qquad \text{First a proportion is set up ...}$$
$$4x = 300 \qquad \text{... next we cross-multiply...}$$
$$\frac{4x}{4} = \frac{300}{4} \qquad \text{... divide each side by 4...}$$
$$x = 75\% \qquad \text{... and find the percent.}$$

To express any decimal as a percent, the task is even simpler. Since decimals are based upon the unit 1, and percents are based upon the unit 100, all that needs to be done is to multiply the decimal fraction by 100. (You may recall that that is accomplished by moving the decimal point two places to the right.)

Example

Express .35 as a percent.

$$.35 \times 100 = x$$
$$35 = x$$
$$x = 35\%$$

Below, you will find a table listing a few equivalent fractions, decimals, and percents. It is meant merely to demonstrate that any fraction can readily be expressed in any or all three forms. Note that all decimals have been rounded to the nearest hundredth, and all percents to the nearest whole number percent.

Fraction	Decimal	Percent
$\frac{10}{14}$	.71	71%
$\frac{2}{17}$	.12	12%
$\frac{1}{14}$	.07	7%
$\frac{4}{15}$	.27	27%
$\frac{3}{14}$	.21	21%
$\frac{4}{22}$	.18	18%
$\frac{6}{21}$	.29	29%
$\frac{2}{8}$	.25	25%
$\frac{9}{25}$	.36	36%
$\frac{8}{12}$	.67	67%
$\frac{8}{15}$	.53	53%
$\frac{9}{16}$	.56	56%
$\frac{4}{19}$	.21	21%
$\frac{8}{14}$	.57	57%
$\frac{2}{22}$	.09	9%

Test Yourself

Directions: Fill in the blank to write an equivalent fraction, decimal, or percent. Round all decimals to the nearest hundredth, and all percents to the nearest whole number percent.

	Fraction	Decimal	Percent
1.	$\frac{4}{7}$	——	57%
2.	$\frac{5}{16}$	.31	——
3.	——	.83	83%
4.	$\frac{2}{10}$	——	20%
5.	$\frac{4}{17}$	.24	——
6.	$\frac{8}{16}$	.5	——
7.	$\frac{4}{23}$	——	17%
8.	$\frac{9}{17}$	——	53%
9.	——	.25	25%
10.	$\frac{10}{23}$	——	43%
11.	——	.45	45%
12.	——	.67	67%
13.	$\frac{3}{10}$	.3	——
14.	$\frac{8}{12}$	——	67%
15.	——	.5	50%
16.	$\frac{9}{25}$	——	36%
17.	$\frac{5}{12}$	——	42%

	Fraction	Decimal	Percent
18.	$\frac{8}{25}$	___	___
19.	$\frac{5}{25}$	___	___
20.	$\frac{9}{14}$	___	___

Answers

1. .57	**2.** 31%	**3.** $\frac{83}{100}$	**4.** .2	**5.** 24%
6. 50%	**7.** .17	**8.** 53	**9.** $\frac{1}{4}$	**10.** .43
11. $\frac{9}{20}$	**12.** $\frac{67}{100}$	**13.** 30%	**14.** .67	**15.** $\frac{1}{2}$
16. .36	**17.** .42	**18.** .32, 32%	**19.** .2, 20%	**20.** .64, 64%

Test Yourself

Directions: Express each decimal as a percent.

1. .31	**2.** .12	**3.** .62	**4.** .71	**5.** .11	**6.** .6
7. .72	**8.** .2	**9.** .53	**10.** .46	**11.** .08	**12.** .41
13. .52	**14.** .57	**15.** .22	**16.** .63	**17.** .96	**18.** .25
19. .47	**20.** .36	**21.** .4	**22.** .77	**23.** .26	**24.** .45
25. .49	**26.** .72	**27.** .16	**28.** .42	**29.** .38	**30.** .35
31. .5	**32.** .92				

Answers

1. 31%	**2.** 12%	**3.** 62%	**4.** 71%	**5.** 11%
6. 60%	**7.** 72%	**8.** 20%	**9.** 53%	**10.** 46%
11. 8%	**12.** 41%	**13.** 52%	**14.** 57%	**15.** 22%
16. 63%	**17.** 96%	**18.** 25%	**19.** 47%	**20.** 36%
21. 40%	**22.** 77%	**23.** 26%	**24.** 45%	**25.** 49%
26. 72%	**27.** 16%	**28.** 42%	**29.** 38%	**30.** 35%
31. 50%	**32.** 92%			

Test Yourself

Directions: Express each fraction as a percent. Round where necessary.

1. $\dfrac{8}{21}$ 2. $\dfrac{9}{18}$ 3. $\dfrac{4}{6}$ 4. $\dfrac{1}{6}$

5. $\dfrac{5}{8}$ 6. $\dfrac{9}{17}$ 7. $\dfrac{4}{12}$ 8. $\dfrac{8}{12}$

9. $\dfrac{8}{9}$ 10. $\dfrac{3}{9}$ 11. $\dfrac{4}{14}$ 12. $\dfrac{10}{22}$

13. $\dfrac{8}{14}$ 14. $\dfrac{1}{8}$ 15. $\dfrac{9}{12}$ 16. $\dfrac{8}{24}$

17. $\dfrac{10}{11}$ 18. $\dfrac{4}{11}$ 19. $\dfrac{8}{8}$ 20. $\dfrac{2}{25}$

21. $\dfrac{4}{5}$ 22. $\dfrac{8}{13}$ 23. $\dfrac{9}{9}$ 24. $\dfrac{5}{15}$

25. $\dfrac{10}{24}$ 26. $\dfrac{11}{11}$ 27. $\dfrac{7}{23}$ 28. $\dfrac{5}{16}$

29. $\dfrac{12}{18}$ 30. $\dfrac{7}{14}$ 31. $\dfrac{2}{19}$ 32. $\dfrac{1}{13}$

Answers

1. 38%   2. 50%   3. 67%   4. 17%   5. 63%
6. 53%   7. 33%   8. 67%   9. 89%   10. 33%
11. 29%   12. 45%   13. 57%   14. 13%   15. 75%
16. 33%   17. 91%   18. 36%   19. 100%   20. 8%
21. 80%   22. 62%   23. 100%   24. 33%   25. 42%
26. 100%   27. 30%   28. 31%   29. 67%   30. 50%
31. 11%   32. 8%

Finding a Percent of a Number

You may recall that the words "times" and "of" are mathematically identical. Certainly you have heard expressions such as "Your tip should be 15% of the bill." Fifteen percent of the bill is found by multiplying. But how does one go about multiplying percents? Well, to be quite candid, one does not. Percents cannot be multiplied or divided, and it is quite rare to see them added or subtracted. How then, you might ask, can we figure out 15% of the bill, or is the waiter just not going to receive a tip? The answer lies in the close relationship between decimals and percents. In order to find fifteen percent of the bill, the 15% must first be converted to a decimal: .15. Then the multiplication can take place.

Example

Find 25% of 300.

25% = .25   First, change the percentage to a decimal.

$x = .25 \times 300$ Then multiply ...

$x = 75$ ... to get the answer.

Test Yourself

Directions: Find each amount.

1. 30% of 944 = ____	**2.** 44% of 5 = ____	**3.** 77% of 214 = ____
4. 28% of 504 = ____	**5.** 38% of 925 = ____	**6.** 20% of 148 = ____
7. 79% of 895 = ____	**8.** 23% of 993 = ____	**9.** 57% of 553 = ____
10. 3% of 450 = ____	**11.** 10% of 115 = ____	**12.** 17% of 329 = ____
13. 16% of 500 = ____	**14.** 5% of 641 = ____	**15.** 31% of 9 = ____
16. 70% of 141 = ____	**17.** 95% of 738 = ____	**18.** 21% of 625 = ____
19. 26% of 757 = ____	**20.** 1% of 981 = ____	**21.** 71% of 32 = ____
22. 82% of 979 = ____	**23.** 53% of 89 = ____	**24.** 41% of 824 = ____
25. 70% of 447 = ____	**26.** 7% of 450 = ____	**27.** 93% of 779 = ____
28. 82% of 567 = ____	**29.** 16% of 844 = ____	**30.** 47% of 457 = ____
31. 1% of 149 = ____	**32.** 27% of 525 = ____	**33.** 81% of 380 = ____
34. 43% of 592 = ____	**35.** 54% of 567 = ____	**36.** 38% of 941 = ____
37. 37% of 10 = ____	**38.** 71% of 117 = ____	**39.** 5% of 949 = ____
40. 92% of 474 = ____	**41.** 3% of 986 = ____	**42.** 91% of 469 = ____
43. 29% of 257 = ____	**44.** 22% of 709 = ____	**45.** 97% of 356 = ____
46. 100% of 753 = ____	**47.** 73% of 1 = ____	**48.** 76% of 651 = ____
49. 85% of 464 = ____	**50.** 82% of 249 = ____	

Answers

1. 283.2	**2.** 2.2	**3.** 164.78	**4.** 141.12	**5.** 351.5
6. 29.6	**7.** 707.05	**8.** 228.39	**9.** 315.21	**10.** 13.5
11. 11.5	**12.** 55.93	**13.** 80	**14.** 32.05	**15.** 2.79
16. 98.7	**17.** 701.1	**18.** 131.25	**19.** 196.82	**20.** 9.81
21. 22.72	**22.** 802.78	**23.** 47.17	**24.** 337.84	**25.** 312.9
26. 31.5	**27.** 724.47	**28.** 464.94	**29.** 135.04	**30.** 214.79
31. 1.49	**32.** 141.75	**33.** 307.8	**34.** 254.56	**35.** 306.18
36. 357.58	**37.** 3.7	**38.** 83.07	**39.** 47.45	**40.** 436.08
41. 29.58	**42.** 426.79	**43.** 74.53	**44.** 155.98	**45.** 345.32
46. 753	**47.** .73	**48.** 494.76	**49.** 394.4	**50.** 204.18

Finding a percent of a number is useful for figuring discounts on items that are on sale, tips paid as a percentage of the bill at a restaurant, sales tax, grades on examinations, and interest paid on a loan, or on money in a savings account. Salespeople often earn commissions as a percentage of what they sell, as do travel agents, ticket agents, and insurance agents. In short, there are many situations in which it is necessary to compute a percentage of a certain number—often a number that expresses a quantity of money. We will look further into some of these applications in the section dealing with word problems involving percents later in this chapter.

SOME USEFUL PROPERTIES

The Closure Property

When numbers are combined, they follow certain laws or rules. For example, we know that adding two whole positive numbers will always result in a sum that is a positive whole number. This leads to the somewhat useful, if technical, statement: The set of counting numbers is **closed** for addition. Stated more simply: When two counting numbers are added, a counting number will be the result.

Is the set of counting numbers closed for subtraction? Think about it. Select two counting numbers and subtract one from the other; say 8 take away 2. Does a counting number result? Will a counting number always result? How about if we take 8 away from 2? Well, 2 - 8 = -6. A negative number is not a counting number, therefore the set of counting numbers is not closed for subtraction. Note that you might have tried subtracting hundreds of pairs of counting numbers with the first number of each pair being larger than the second, and always have gotten a counting number as the difference. Just a single case where that does not occur, however, suffices to prove that the set of counting numbers is not closed for subtraction. It is an interesting feature of both mathematics and science that an exceedingly large number of positive results in an experiment never suffices to prove that something will always work. Just a single negative result, however, is adequate to prove that something will not always work.

Is the set of counting numbers closed for multiplication? Try multiplying two counting numbers and getting a product that is anything other than a counting number. No matter how many pairs of numbers you choose, you will get a product that is a counting number. Mathematicians are, therefore, pretty sure that the set of counting numbers is closed for multiplication. If, on the other hand, you multiply two counting numbers sometime and find that you get a negative number as a result, you may have just disproved this hypothesis. (In all candor, though, it is much more likely that you made a mistake.)

Divide one counting number by another, and the odds are very much in favor of the result being a fraction (otherwise known as a rational number). 4 divided by 2 is a counting number, 2. However, 2 divided by 4 equals $\frac{1}{2}$, and that is not a counting number. What statement may therefore be made about closure with respect to the set of counting numbers for division?

Realms of Numbers

It is sometimes useful to talk about numbers as belonging to certain realms. Each time a new condition is added to a realm of numbers, a new realm is formed that includes all the members of the previous realm, and a whole new group as well. We have just examined one property of the first realm of numbers that is usually studied, counting numbers. **Counting numbers** derive their name from the fact that when you count objects you start with the first object you see, and call it "1." Then you go on from there with 2, 3, and so on forever. Notice that the first counting number is 1.

The next realm of numbers is obtained by adding zero to the realm of counting numbers. This is the realm known as **whole numbers**. It is an infinite realm, as is the set of counting numbers, in that there is no highest number. Whole numbers start at zero and go on forever. Counting numbers also go on forever, but they start at 1.*

The next realm of numbers includes an infinite addition to the already infinite set of whole numbers. It includes all of the negatives of the counting numbers, that is, negative whole numbers. This new realm is known as the realm of **integers**. Positive integers consist of 1, 2, 3, ..., whereas negative integers consist of -1, -2, -3, ... (The series of three dots is used to indicate that the series of numbers continues in the same fashion forever.)

The next expansion consists of including any number that can be expressed as a fraction. Since any integer can be expressed as a fraction $\left(2 = \dfrac{2}{1}, \text{ or } \dfrac{4}{2} \right)$, all the numbers from previous realms are included in this new one, which is known as **rational numbers**. Often, you will see fractions referred to as rational numbers, but whole numbers (negative as well as positive) are rational numbers also. Rational numbers also include those numbers which can be expressed as repeating decimals. For example, .6666 ... is a rational number, since it is equal to the fraction $\dfrac{2}{3}$.

Irrational numbers are numbers which cannot be expressed as fractions. The set of irrational numbers is not a realm, since it does not include any of the numbers already discussed. $\sqrt{2}$ ** is an irrational number, as is the number π (pi). Mathematicians have calculated the values of these numbers to many decimal places, and there is no repeating pattern. Hence they are referred to as irrational.

When the set of irrational numbers is combined with the realm of rational numbers, we get the most important realm—that known as **real numbers**. The realm of real numbers encompasses all fractions, positive numbers, negative numbers, zero, and irrational numbers. To summarize, then, the realms of numbers are:

Counting numbers	(1, 2, 3, ...)
Whole numbers	(Counting numbers + 0)
Integers	(Whole numbers + the negatives of the whole numbers)
Rational numbers	(Integers + fractions)
Real numbers	(Rational numbers + irrational numbers)

---

* Counting numbers are positive whole numbers, i.e., 1, 2, 3, ...

** The square root of two: A number which, when multiplied by itself equals 2. $\sqrt{9}$ is a rational number, since 3 x 3 = 9.

The Commutative Property

Have you ever tried to add 8 + 6 and ended up adding 6 + 8 instead? Did it make a difference? Of course not. That is because **the order in which you add two numbers does not affect the sum**. That fact is referred to by mathematicians as the **Commutative Property of Addition**. The commutative property deals with two numbers, and might be shown symbolically as $a + b = b + a$.

There is also a commutative property for multiplication. It states that **the order in which two numbers are multiplied together does not affect the product**. Stated symbolically, $a \times b = b \times a$. Consider that $9 \times 5 = 45$, and that $5 \times 9 = 45$. That is the commutative property for multiplication in action. Try applying the commutative property to subtraction or to division, and convince yourself that it does not work.

The Associative Property

All arithmetic operations are **binary**. That is a fancy way of saying that they involve only two numbers at a time. If you are skeptical, try adding 3 + 4 + 5. If you think that you just added 3 + 4 + 5, you are just fooling yourself. Nobody, and for that matter no computer, is capable of adding three numbers together at the same time. Try it again, only this time do it out loud. Did you do it? You probably said 3 + 4 = 7; + 5 = 12. Notice the ";" that separates the two steps. What you actually did was find the 7, and then combine 7 and 5 to get 12. There were two other options, however. Without changing the order of the numbers, you might have added 4 + 5 to get 9, and then combined 9 with 3 to get 12. The other possibility was to combine the 3 and 5 to make 8, and then add 4 to get 12.

Whichever way you combined the numbers, what you were doing was selectively grouping the numbers in order to facilitate adding them. Grouping numbers together for addition is called associating them. The fact that **the way in which numbers are grouped for addition does not affect the sum** is known as the **Associative Property for Addition**. Using the previous addition problem as an illustration, the associative property would look like this:

$$(3 + 4) + 5 = 3 + (4 + 5)$$

In general terms, the associative property for addition may be illustrated as:

$$(a + b) + c = a + (b + c)$$

If you think about it, you will recognize that the associative property may also be applied to multiplication. In that case, you would discover that $(2 \times 3) \times 4 = 2 \times (3 \times 4)$, or, generally stated, $(a \times b) \times c = a \times (b \times c)$.

The Distributive Property

While the three properties discussed above are useful in computational work, they are, for the most part, intuitive. The odds are that even if you had not seen it stated here and given a fancy name, you still would have known that 3 × 4 gives the same result as 4 × 3. The Distributive Property of Multiplication over Addition is a different story. It is a property that is usually seen only in algebraic applications. The **Distributive Property states that if the sum of two numbers is multiplied by a third, the result is the same as if each of the numbers was first multiplied by the third, and their products added together**. It may be illustrated as follows:

$$a(b + c) = ab + ac$$

Consider the meaning if we substitute the numbers 2, 3, and 4 for *a*, *b*, and *c*, respectively. We would then get:

$$2(3+4) = 2\cdot3 + 2\cdot4$$
$$2(7) = 6+8$$
$$14 = 14$$

And there you have it!! You will find a need to apply this property repeatedly in algebraic equations, so it is a good idea to get some practice with it. The following exercises should help you to do just that.

Test Yourself

> **Directions:** Write an equivalent expression that is the result of applying the distributive property to the expression already written.

1. $5(n + 3) = 5$ _____ + _____
2. $6(r + 8) =$ _____ $r +$ _____
3. $4(7 - b) =$ _____ - _____ b
4. $a(7 + c) =$ _____
5. $9(r + w) =$ _____
6. $a(x + r) =$ _____
7. $c(l + m) =$ _____
8. $r(v - h) =$ _____
9. $w(2p - 8) =$ _____
10. $v(3r - 2x) =$ _____

Answers

1. $5n + 15$
2. $6r + 48$
3. $28 - 4b$
4. $7a + ac$
5. $9r + 9w$
6. $ax + ar$
7. $cl + cm$
8. $rv - rh$
9. $2pw - 8w$
10. $3rv - 2vx$

EXPONENTS

Exponents are little numerals written above and to the right of other numerals. They are read as **powers**. For example, x^3 is read "*x* to the third power," and 5^2 is read "5 to the second power."

Exponents are used to represent a number being multiplied by itself a certain number of times.

x^3 means $x \cdot x \cdot x$
n^4 means $n \cdot n \cdot n \cdot n$
2^3 means $2 \cdot 2 \cdot 2 = 8$
3^4 means $3 \cdot 3 \cdot 3 \cdot 3 = 81$

When adding or subtracting numbers that have exponents, only like variables raised to the identical power may be combined. Examine the examples below, and you will see what is meant by the last statement:

$x^3 + x^3 = 2x^3$ $2y^2 + 5y^2 = 7y^2$
$9m^4 + 5m^4 = 14m^4$ $6r^5 - 2r^5 = 4r^5$
$8w^3 - 5w^3 = 3w^3$ $10ab^4 - 5ab^4 = 5ab^4$

but...

$x^3 + x^2 = x^3 + x^2$ and $5z^5 - 2z^2 = 5z^5 - 2z^2$

Since the variables in each of the last two expressions differ with respect to their exponents, they cannot be further combined by addition or subtraction.

Combining Variables with Exponents by Multiplication and Division

If two variables have the same base, that is, the same letter, they may be combined by multiplication. To combine variables with exponents, it is only necessary to add their exponents together:

$$x^2 \cdot x^3 = x^{2+3} = x^5 \qquad\qquad y^4 \cdot y^3 = y^{4+3} = y^7$$

See if you can figure these two out:

$$x \cdot x = x^2 \qquad\qquad n \cdot n \cdot n = n^3$$

In case you have not guessed, a variable with no exponent showing has, in fact, an exponent of 1. Hence $x \cdot x$ is $x^1 \cdot x^1 = x^{1+1} = x^2$ $n \cdot n \cdot n = n^{1+1+1} = n^3$.

What do you suppose is the way to divide numbers with exponents? Since exponents were added in multiplication, it would seem to be reasonable to suppose that they will be subtracted during division. Not only is that supposition reasonable, t is also accurate:

$$\frac{x^5}{x^2} = x^{5-2} = x^3$$

$$n^8 \div n^5 = n^{8-5} = n^3$$

Now, since we have seen how exponentiated numbers may be combined by multiplication and division, let us take a little space to justify the rules that we have just examined. First of all, consider a multiplication:

$$2^3 \cdot 2^4$$

$$2^3 = 2 \cdot 2 \cdot 2 = 8$$

$$2^4 = 2 \cdot 2 \cdot 2 \cdot 2 = 16$$

Then, $2^3 \cdot 2^4 = 8 \cdot 16$, which $= 128$.

Now, by the rules just discussed,

$$2^3 \cdot 2^4 = 2^{3+4} = 2^7$$

$$2^7 = 2 \cdot 2 \cdot 2 \cdot 2 \cdot 2 \cdot 2 \cdot 2 = 128$$

... and there you have an illustration of why the rule works for multiplication. You may wish to work out an illustration for division for yourself. As a bit of a helping hand, however, consider 2^{10} divided by 2^6. Note that 2^{10} is worth 1024, while 2^6 is worth 64.

Test Yourself

Directions: Evaluate each of the following.

1. 3^2
2. 2^3
3. 5^2
4. 2^5
5. 3^4
6) 4^3
7. 2^{11}

Directions: Express each in simplest form.

8. $a^5 + a^4$
9. $2a^3 + a^4$
10. $3a^2 + a^2$
11. $4b^5 - 3b^3$

12. $f^3 \cdot f^2$
13. $m^5 \cdot m^8$
14. $\dfrac{c^5}{c^3}$
15. $r^9 + r^9$

16. $2j^2 \cdot 3j^3$
17. $2s^4 \cdot 3s^5$
18. $5x^9 \cdot 2y^3$
19. $13d^4 - 9d^4$

20. $\dfrac{6w^8}{3w^6}$
21. $\dfrac{12z^9}{4z^8}$
22. $\dfrac{9k^5}{3k^3}$
23. $5g^7t^4 \cdot 6g^2s^3$

Answers

1. 9
2. 8
3. 25
4. 32
5. 81
6. 64
7. 2048
8. $a^5 + a^4$
9. $2a^3 + a^4$
10. $4a^2$
11. $4b^5 - 3b^3$
12. f^5
13. m^{13}
14. c^2
15. $2r^9$
16. $6j^5$
17. $6s^9$
18. $10x^9y^3$
19. $4d^4$
20. $2w^2$
21. $3z$
22. $3k^2$
23. $30g^9s^3t^4$

The Zero Power

Any number, except zero, raised to the zero power has a value of 1. That is to say, $3^0 = 1$, $17^0 = 1$, $335^0 = 1$, and $n^0 = 1$. In order to justify this rather unusual fact, consider the following:

$$n^x \div n^x = n^{x - x} = n^0 \qquad \text{(since anything minus itself is zero.)}$$

$$\text{But } \frac{n^x}{n^x} = 1 \qquad \text{(since anything divided by itself} = 1.)$$

If any number divided by itself equals that number to the zero power, and any number divided by itself equals 1, then it follows that any number raised to the zero power equals 1. The exception to the preceding explanation is zero raised to the zero power since the result of division by zero is undefined.

Squares and Square Roots

Special names are reserved for situations that occur frequently. In mathematics there are special names for a number raised to the second power and a number raised to the third power. A number raised to the second power is called its **square**. That is because the area of a square is found by multiplying a side of the figure by itself (s^2). A cube's volume is found by multiplying one side of the figure by itself and by itself again ($V = s^3$). For that reason a number raised to the third power is known as that number **cubed**. We will not give any special attention to cubes at this time, but squares play a rather important role in GED mathematics.

Consider the first ten squares of positive integers. To find them, all that is necessary is to start with the first counting number, and square (raise to the second power) it and each of its consecutive successors. $1^2 = 1$; $2^2 = 4$; $3^2 = 9$; $4^2 = 16$; .. $10^2 = 100$. You can fill in the blanks represented by the dots. The numbers just considered are known as **perfect squares**. A perfect square is the result of multiplying any integer by itself.

It should be pointed out that there are two ways to make any perfect square. A perfect square is 25, being the product of $5 \cdot 5$. But 25 is also the product of (-5)(-5). Both 5 and -5 are considered to be **square roots** of 25. A square root is the integer which, when multiplied by itself, yields a square. Not all numbers have integers as square roots. In fact, as you may have noticed, of the first 100 natural numbers, only 10 of them are perfect squares (1, 4, 9, 16, 25, 36, 49, 64, 81, and 100).

Each of these perfect squares, technically, has two square roots, one positive and one negative. For example, the square root of 64 can be either +8 or –8. In some algebraic problems, when taking a square root, you must consider each of the two possible square roots. Most of the time in geometry, however, you will only need to consider the positive square root, since geometry deals with lengths and lengths cannot be negative. For this reason, the positive square root of a number is called the primary square root. Unless otherwise specified, this is the number we will mean when we refer to the square root of a number.

Since the square of 2 is 4, the primary square root of 4 is 2.

Since the square of 3 is 9, the primary square root of 9 is 3.

Since the square of 4 is 16, the primary square root of 16 is 4.

Now let's continue, but using the appropriate symbols. $\sqrt{}$ is called a radical sign. It means "the square root of" whatever number is under it:

$$\text{Since } 5^2 \text{ is } 25, \sqrt{25} = 5.$$
$$\text{Since } 6^2 \text{ is } 36, \sqrt{36} = 6.$$
$$\text{Since } 7^2 \text{ is } 49, \sqrt{49} = 7.$$
$$\text{Since } 8^2 \text{ is } 64, \sqrt{64} = 8...$$

Obviously, most numbers do not have perfect (i.e. whole number) square roots. There is no whole number which, when multiplied by itself will give a product of 3. The same is true for 2, 5, 7, and, in fact, most numbers. It has therefore become customary to express the square root of a number in as simple terms as possible, by removing as great as perfect square from beneath a radical sign as possible. Consider, for example, the radical $\sqrt{20}$. 20 contains a perfect square, 4, so that $\sqrt{20} = \sqrt{4 \cdot 5}$. But, since the square root of 4 is 2, $\sqrt{20} = 2\sqrt{5}$. The square root of 4, which is multiplied by the square root of 5, remains in radical form. By the way, $\sqrt{2}$ is one of those irrational numbers that we've mentioned before. Even though we can not exactly calculate its value, it comes in handy from time to time.

Examine the following examples.

Example 1	Example 2	Example 3
$\sqrt{160} = \sqrt{16 \cdot 10}$	$\sqrt{90} = \sqrt{9 \cdot 10}$	$\sqrt{500} = \sqrt{100 \cdot 5}$
$= 4\sqrt{10}$	$= 3\sqrt{10}$	$= 10\sqrt{5}$

The trick, as you can see in the models, is to extract a perfect square from the number under the radical sign. It is not essential that you get the largest perfect square each time you simplify a radical, as long as you eventually get the radical down to its simplest form. Example 4 illustrates two different approaches to the same factorization.

Example 4

$\sqrt{320} = \sqrt{64 \cdot 5}$ $= 8\sqrt{5}$ (That is the most efficient approach.)

$\sqrt{320} = \sqrt{4 \cdot 80}$ $= 2\sqrt{80}$ Here only a 4 was recognized . . .

$2\sqrt{80} = 2\sqrt{4 \cdot 20}$ $= 4\sqrt{20}$ . . . then another 4. The extracted 2 is multiplied by the 2 that was already outside (the radical's coefficient) . . .

$4\sqrt{20} = 4\sqrt{4 \cdot 5}$ $= 8\sqrt{5}$ . . . and so one more time. Note that the final answer obtained this way is identical to the one obtained in the first line of the problem.

What if there were a variable under the radical sign? Examine examples 5, 6, and 7. You'll see that the process used to simplify them is almost identical to what we did with constants.

Example 5	Example 6	Example 7
$\sqrt{7n^2} = \sqrt{7 \cdot n^2} = \|n\|\sqrt{7}$	$\sqrt{25b^3} = \sqrt{25 \cdot b^3} = 5b\sqrt{b}$	$\sqrt{36m^4} = \sqrt{36 \cdot m^4} = 6m^2$

Test Yourself

Directions: Express the following radicals in simplest form. Assume that the variables represent positive numbers.

1. $\sqrt{36}$	2. $\sqrt{49}$	3. $\sqrt{121}$	4. $\sqrt{25}$
5. $\sqrt{81}$	6. $\sqrt{169}$	7. $\sqrt{144}$	8. $\sqrt{18}$
9. $\sqrt{50}$	10. $\sqrt{27}$	11. $\sqrt{98}$	12. $\sqrt{72}$
13. $\sqrt{75}$	14. $\sqrt{80}$	15. $\sqrt{x^2}$	16. $\sqrt{y^2}$
17. $\sqrt{m^6}$	18. $\sqrt{2x^2}$	19. $\sqrt{3q^2}$	20. $\sqrt{4n^2}$
21. $\sqrt{y^4}$	22. $\sqrt{16p^2}$	23. $\sqrt{8l^2}$	24. $\sqrt{24y}$
25. $\sqrt{81r}$	26. $\sqrt{144c}$	27. $\sqrt{162r^2}$	28. $\sqrt{16m^3}$

Answers

1. 6		**2.** 7		**3.** 11		**4.** 5		**5.** 9	
6. 13		**7.** 12		**8.** $3\sqrt{2}$		**9.** $5\sqrt{2}$		**10.** $3\sqrt{3}$	
11. $7\sqrt{2}$		**12.** $6\sqrt{2}$		**13.** $5\sqrt{3}$		**14.** $4\sqrt{5}$		**15.** $|x|$	
16. $|y|$		**17.** m^3		**18.** $|x|\sqrt{2}$		**19.** $|q|\sqrt{3}$		**20.** $2|n|$	
21. y^2		**22.** $4|p|$		**23.** $2|t|\sqrt{2}$		**24.** $2\sqrt{6y}$		**25.** $9\sqrt{r}$	
26. $12\sqrt{c}$		**27.** $9|r|\sqrt{2}$		**28.** $4m\sqrt{m}$						

MONOMIALS, BINOMIALS, AND POLYNOMIALS

When two or more numbers are multiplied together in an algebraic expression, they are considered to be a single term. $3x$ is a single term, as is $4y^2$, or $7a^2b$. Single terms are known as **monomials** (from the prefix "mono" meaning one, and "nomial" meaning number). That is to say, monomials are considered to be single numbers. As an extreme case, $12a \cdot 6b \cdot 9a \cdot 11c$ is a monomial, and hence a single number. You may evaluate it if you know the values of a, b, and c, but that is not the point.

When two terms are separated by a plus or minus sign they are considered to be two numbers—hence $4a + b$ is a **binomial**. Any expression with more than two terms, regardless of how many terms there happen to be, is known as a polynomial (the prefix "poly-" meaning many). $4x^2 + 3x + 2$ and $7z^3 + 5r^2 + 6m + 3r + 8$ are both examples of polynomials.

Polynomials, binomials, and monomials may all be added together, but, as is the case with all expressions containing variables, to get any terms to actually combine, the variables in the terms must be identical with regard to variable name **and** power:

$4x^2 + 3x^2$ can be combined to form $7x^2$.

$4x^2 + 3x$ can be combined no further.

$5ax^3 + 5bx^3$ can be combined no further.

$4ax^6 + 8ax^6$ can be combined to form $12ax^6$.

The binomials $3x + 2$ and $5x - 8$ can be combined as follows:

$$\begin{array}{r} 3x + 2 \\ 5x - 8 \\ \hline 8x - 6 \end{array}$$

In the case of binomials where all terms have different variables, a polynomial will result from combining them:

Adding the binomials $3x + 8$ and $4x^2 + y$ results in $4x^2 + 3x + y + 8$.

Adding the binomials $2x^2 + 7x$ and $3x - 5$ results in $2x^2 + 10x - 5$.

Multiplying Binomials

Binomials are multiplied together in the same way that two-digit numbers are multiplied. The problem usually is that most people do not think of them as working that way. Suppose, for example, you wished to multiply $(x + 2)(x - 3)$. You might set it up as follows:

$$\begin{array}{r} x - 3 \\ (x + 2) \\ \hline 2x \ -6 \\ \hline x^2 - 3x \\ \hline x^2 - x - 6 \end{array}$$

First multiply $2(\bar{\ }3)$ and $2(x)$.

Then multiply $x(\bar{\ }3)$ and $x(x)$.

Finally add, combining like terms.

Simple as this procedure is, some people delight in doing such a binomial multiplication mentally. They use a method known as **FOIL**, an acronym that stands for First, Outer, Inner, Last. Consider once more the multiplication $(x + 2)(x - 3)$. Multiply together the

First,	Outer,	Inner,	and	Last	terms.
$x \cdot x$	$x \cdot \bar{\ }3$	$2 \cdot x$		$2 \cdot \bar{\ }3$	
x^2 +	$\bar{\ }3x$	$2x$		$\bar{\ }6$	

Next, combine like terms, which gives

$$x^2 \qquad\qquad \bar{\ }x \qquad\qquad \bar{\ }6$$

or: $x^2 - x - 6$

Test Yourself

Directions: Combine the following expressions as far as possible.

1. $5x + 7x$
2. $3x^2 + 5x$
3. $(2x + 5) + (6x - 3)$
4. $(8y - 6) + (6y + 9)$
5. $(3y - 8) + (3y + 8)$
6. $(5k - 6) + (6 - 5k)$
7. $(4b + 3)(2b + 6)$
8. $(x + 3)(x - 3)$
9. $(x - 3)(x - 3)$
10. $(x + 3)(x + 3)$
11. $(m - 4)(m + 8)$
12. $(r + 7)(r + 5)$
13. $(7r - 3) - (2r + 5)$
14. $(8g + 7) - (5g + 3)$
15. $(9m - 6) - (3m - 8)$
16. $(12q - 5)(m + 3)$
17. $(5b - 9) + (3b - 9)$
18. $(5b - 9) - (3b - 9)$
19. $(6d + 7)(2d + 3)$
20. $(s - 4)(s + 6)$
21. $\left(b + \dfrac{1}{2}\right)\left(b + \dfrac{1}{4}\right)$
22. $(n + 2)(n - 2)$
23. $\left(u + \dfrac{1}{4}\right)\left(u - \dfrac{1}{4}\right)$
24. $(w + 11) - (w + z)$
25. $(r - 8) + (7 - z)$
26. $(t + 5)(8 + j)$
27. $(z - 5)(2z + 9)$
28. $(h + 3)(3h + 4)$
29. $(t - 3)(4 + g)$
30. $(x + y)(x - y)$

Answers

1.	$12x$	2.	$3x^2 + 5x$	3.	$8x + 2$
4.	$14y + 3$	5.	$6y$	6.	0
7.	$8b^2 + 30b + 18$	8.	$x^2 - 9$	9.	$x^2 - 6x + 9$
10.	$x^2 + 6x + 9$	11.	$m^2 + 4m - 32$	12.	$r^2 + 12r + 35$
13.	$5r - 8$	14.	$3g + 4$	15.	$6m + 2$
16.	$12mq - 5m + 36q - 15$	17.	$8b - 18$	18.	$2b$
19.	$12d^2 + 32d + 21$	20.	$s^2 + 2s - 24$	21.	$b^2 + \dfrac{3}{4}b + \dfrac{1}{8}$
22.	$n^2 - 4$	23.	$u^2 - \dfrac{1}{16}$	24.	$11 - z$
25.	$r - z - 1$	26.	$8t + tj + 5j + 40$	27.	$2z^2 - z - 45$
28.	$3h^2 + 13h + 12$	29.	$4t + gt - 3g - 12$	30.	$x^2 - y^2$

Factoring Binomials

Do you recall the distributive property of multiplication? If you do not, this would be an excellent time to refresh your memory by reviewing that section earlier in this chapter. The reason a knowledge of the distributive property is so essential to the factoring of binomials is that the latter may be thought of as applying the distributive property backwards.

Factors are numbers which are multiplied together. One way of factoring was discussed in the section dealing with finding least common denominators in Chapter 3. Factoring a binomial consists of finding a number that is a factor of both terms in the binomial (2 and 3 are two of the factors of 6) and removing it. Examine the examples below.

Example 1

Factor $2b + 24$.

In order to factor a binomial, first think:

"What factor is common to $2b$ and to 24?"

The factors of $2b$ are 2 and b. The factors of 24 are 2, 3, 4, 6, 8, 12. What do they have in common?

Since 2 is common to both, that is what will be "factored out."

$$2b + 24 = 2(b + 12)$$

As a final check, apply the distributive property to the factored form, $2(b + 12)$, and you will see that the result is $2b + 24$.

Example 2

Factor $x^2 - 5x$.

Both terms contain x as a factor, therefore:

$$x^2 - 5x = x(x - 5)$$

Example 3

Factor $3w^2 + 9w$

w is a factor of each term. So is 3. You may factor out each of these one at a time, or both simultaneously. Below it is done one at a time, beginning with the w.

$$3w^2 + 9w = w(3w + 9)$$
$$= 3w(w + 3)$$

Bear in mind when working the exercises below that it is not essential to factor out all possible factors in a single step. Certainly, there is nothing wrong with doing so, but if you are more comfortable with two or three steps, then do it that way.

Test Yourself

Directions: Factor completely.

1. $5x + 25$	**2.** $3y + 18$	**3.** $4r + 24$	**4.** $8z + 32$
5. $4x^2 - 12y$	**6.** $7b^2 - 49a$	**7.** $t^2 + 6t$	**8.** $h^2 + 5h$
9. $p^2 + 3p$	**10.** $k^2 + dk$	**11.** $v^2 + 2v$	**12.** $s^2 + s$
13. $5n^2 + n$	**14.** $6y^2 + y$	**15.** $vw - 8w$	**16.** $rb + ab^2$
17. $3d^2 - 6d$	**18.** $12m^2 - 18m$	**19.** $11f^3 + 66f$	**20.** $9t^2 - 6t$
21. $14t^2 + 21t$	**22.** $15r^3 - 10r^2$	**23.** $6m^3 + 6m^2$	**24.** $8k^4 - 12k^3$

Answers

1. $5(x + 5)$	**2.** $3(y + 6)$	**3.** $4(r + 6)$	**4.** $8(z + 4)$
5. $4(x^2 - 3y)$	**6.** $7(b^2 - 7a)$	**7.** $t(t + 6)$	**8.** $h(h + 5)$
9. $p(p + 3)$	**10.** $k(k + d)$	**11.** $v(v + 2)$	**12.** $s(s + 1)$
13. $n(5n + 1)$	**14.** $y(6y + 1)$	**15.** $w(v - 8)$	**16.** $b(r + ab)$
17. $3d(d - 2)$	**18.** $6m(2m - 3)$	**19.** $11f(f^2 + 6)$	**20.** $3t(3t - 2)$
21. $7t(2t + 3)$	**22.** $5r^2(3r - 2)$	**23.** $6m^2(m + 1)$	**24.** $4k^3(2k - 3)$

Factoring the Difference of Two Squares

Examine the binomial $b^2 - g^2$. Notice that each of the two terms is a perfect square. (Any number or variable raised to the second power is known as a perfect square, as well as numbers such as 1, 4, 9, etc.) Note also that the two terms are separated by a minus sign. Hence, the entire binomial is known as the difference (subtraction) of two squares. A difference of two squares can be factored by taking the square root of each term and then multiplying both their sum and difference together:

$$b^2 - g^2 = (b + g)(b - g)$$

Study that factorization carefully. To prove that it is valid, FOIL or multiply the two factors $(b + g)(b - g)$ together:

$$
\begin{array}{r}
b + g \\
\underline{(b - g)} \\
- bg - g^2 \\
\underline{b^2 + bg} \\
b^2 \qquad - g^2 = b^2 - g^2
\end{array}
$$

Take a look at this one:

Example 1

$x^2 - 4$ Notice it is a difference of two squares:

$$x^2 = x \cdot x \qquad 4 = 2 \cdot 2$$

$= (x + 2)(x - 2)$ Multiply the difference of the roots by their sum.

Example 2

$3x^2 - 27$ At first, this binomial does not appear to be the difference of two squares. The only hint at that possibility, in fact, is the minus sign.

$3(x^2 - 9)$ We can, however, factor a 3 out of both terms. Having done so, you might now notice that both x^2 and 9 are perfect squares ($x \cdot x$ and $3 \cdot 3$).

$3(x + 3)(x - 3)$ Factoring further, here is what we find.

In general, when given an expression to factor, begin by factoring out the largest common factor, if possible, and then see if the resulting term in parentheses can be factored further.

Test Yourself

Directions: Factor completely.

1. $w^2 - 25$
2. $x^2 - 16$
3. $y^2 - 49$
4. $r^2 - 100$
5. $a^2 - b^2$
6. $u^2 - m^2$
7. $v^2 - 4c^2$
8. $q^2 - 9a^2$
9. $k^2 - 36d^2$
10. $25b^2 - f^2$
11. $100h^2 - t^2$
12. $81z^2 - w^2$
13. $64m^2 - n^2$
14. $64p^2 - 9q^2$
15. $121s^2 - 81t^2$
16. $32b^2 - 128c^2$
17. $18x^2 - 2$
18. $3x^2 - 75$
19. $98x^2 - 72$
20. $6x^2 - 24$
21. $7y^2 - 63$
22. $200g^2 - 98h^2$
23. $72r^2 - 32w^2$
24. $400w^2 - 256x^2$

Answers

1. $(w + 5)(w - 5)$ 2. $(x + 4)(x - 4)$ 3. $(y + 7)(y - 7)$

4. $(r + 10)(r - 10)$ 5. $(a + b)(a - b)$ 6. $(u + m)(u - m)$

7. $(v + 2c)(v - 2c)$ 8. $(q + 3a)(q - 3a)$ 9. $(k + 6d)(k - 6d)$

10. $(5b + f)(5b - f)$ 11. $(10h + t)(10h - t)$ 12. $(9z + w)(9z - w)$

13. $(8m + n)(8m - n)$ 14. $(8p + 3q)(8p - 3q)$ 15. $(11s + 9t)(11s - 9t)$

16. $32(b + 2c)(b - 2c)$ 17. $2(3x + 1)(3x - 1)$ 18. $3(x + 5)(x - 5)$

19. $2(7x + 6)(7x - 6)$ 20. $6(x + 2)(x - 2)$ 21. $7(y + 3)(y - 3)$

22. $2(10g + 7h)(10g - 7h)$ 23. $8(3r + 2w)(3r - 2w)$ 24. $16(5w + 4x)(5w - 4x)$

Note: $x^2 + y^2$ or any binomial of that form is *not* the difference of two squares and *cannot* be factored.

Factoring Quadratic Expressions

A three-place polynomial that contains a variable raised to the second power is called a **quadratic expression**. Quadratic expressions take the form:

$$ax^2 + bx + c$$

In the form above, a and b stand for numerical coefficients of the x^2 and the x terms respectively, while c is a constant. A factorization of a quadratic expression is always going to take the form of two binomials multiplied together. The FOIL form must be used to insure that the factors multiply together to form the correct three term quadratic polynomial[†]. The form will be as follows:

(variable * constant) (variable * constant)

Each * above represents either a "+" or a "-" sign. To determine which signs belong, examine the signs in the polynomial. There are the following possibilities.

1. $ax^2 + bx + c$ Both signs will be +.

2. $ax^2 - bx + c$ Both signs will be -.

3. $ax^2 + bx - c$ ⎫
4. $ax^2 - bx - c$ ⎭   One sign will be + and one will be -.

If we follow the First, Outer, Inner, Last scheme of FOIL, then the value of ax^2 and of c are determined exclusively by multiplication of the Fs and the Ls respectively. Since c's sign is determined by multiplication, two signs that are the same will give a + and two signs that are different will give a - as the constant's sign. The sign of the middle term (bx) is determined by adding together the O and the I terms. Examine the examples below, and you will see how this works.

[†]Of course, in cases involving the difference of two squares, the middle term will disappear, as we have already seen.

Example 1

Factor $x^2 + 2x + 1$ Note that both signs are positive.

$(\quad+\quad)(\quad+\quad)$ Next look at the first term. What times what makes x^2?

$(x+\quad)(x+\quad)$ Obviously $x \cdot x$ does. Now look at the last term of the polynomial. We need two numbers that multiply together to make 1 ...

(Look at the middle term of the polynomial.) ... and add together to make 2. Therefore,

$(x + 1)(x + 1)$ FOIL this solution and see what you get.

Example 2

Factor $x^2 - 3x + 2$ Note that the sign of the c term is positive. That means that both signs in the

$(\quad-\quad)(\quad-\quad)$ factors must be the same, either both positive, or both negative. The sign of the middle term tells us both signs must be negative.

What multiplies together to make x^2?

$(x-\quad)(x-\quad)$ Now, from the c term, you need two numbers that multiply together to make 2, and the b term tells you that they must also add together to make ⁻3, hence:

$(x - 2)(x - 1)$ FOIL the solution and see what you get.

Example 3

Factor $x^2 + 2x - 3$ Note that the last sign is negative. That means that the two factors contain

$(x+\quad)(x-\quad)$ different signs.

Two numbers that multiply together to make ⁻3 but add together to make ⁺2 are ⁺3 and ⁻1.

$(x + 3)(x - 1)$ FOIL the solution and see what happens.

Example 4

Factor $x^2 - 3x - 4$ Last term is negative, so the signs are different:

$(x+\quad)(x-\quad)$ ⁻4 can be obtained by multiplying ⁻2 · 2, ⁻4 · 1, or ⁻1 · 4. Which of those pairs add together to make ⁻3?

$(x + 1)(x - 4)$ FOIL and see for yourself.

Refer to the examples as many times as you have to as you proceed through the exercises below.

Test Yourself

Directions: Factor completely.

1. $x^2 + 7x + 6$
2. $x^2 + 5x + 4$
3. $x^2 + 5x + 6$
4. $x^2 + 7x + 10$
5. $x^2 - 7x + 10$
6. $x^2 - 8x + 15$
7. $x^2 - 8x + 12$
8. $b^2 - 12b + 20$
9. $y^2 - 7y - 8$
10. $m^2 + 9m - 10$
11. $t^2 + 11t - 12$
12. $z^2 - z - 12$
13. $w^2 + 7w - 18$
14. $x^2 - 19x + 18$
15. $s^2 + 10s + 21$
16. $k^2 - 11k + 28$
17. $v^2 - 10v - 24$
18. $q^2 + 8q - 65$

Answers

1. $(x + 6)(x + 1)$
2. $(x + 4)(x + 1)$
3. $(x + 3)(x + 2)$
4. $(x + 5)(x + 2)$
5. $(x - 5)(x - 2)$
6. $(x - 3)(x - 5)$
7. $(x - 2)(x - 6)$
8. $(b - 2)(b - 10)$
9. $(y - 8)(y + 1)$
10. $(m + 10)(m - 1)$
11. $(t + 12)(t - 1)$
12. $(z - 4)(z + 3)$
13. $(w + 9)(w - 2)$
14. $(x - 18)(x - 1)$
15. $(s + 7)(s + 3)$
16. $(k - 7)(k - 4)$
17. $(v - 12)(v + 2)$
18. $(q + 13)(q - 5)$

Solving Quadratic Equations by Factoring

A quadratic equation is any equation that can be expressed in the form $ax^2 + bx + c = 0$. That, in fact, is known as the **standard form** for a quadratic equation. While not all quadratic equations are capable of being solved by factoring, all of those on the GED test can be. For that reason, we will concern ourselves here exclusively with factorable quadratics.

The examples below include all the steps needed for solution of any quadratic equation that is factorable.

Example 1

Solve for x: $x^2 + 7x - 60 = 0$

First, the polynomial must be in standard form. This one already is. Next, factor the polynomial:

$$(x + 12)(x - 5) = 0$$

Now, for two factors multiplied together to equal zero, either one or both of those factors must equal zero. That is because multiplication by zero is the only way to get a product of zero. Let us therefore set each factor equal to zero:

$$(x + 12) = 0 \qquad (x - 5) = 0$$

Then solve each for x.

$$x + 12 = 0 \qquad x - 5 = 0$$
$$x = -12 \qquad x = 5$$

The **roots** of the equation are -12 and/or 5.

This may be checked by substituting either $^-12$ or 5 into the original equation:

$$x^2 + 7x - 60 = 0 \qquad x^2 + 7x - 60 = 0$$
$$\left(^-12\right)^2 + 7\left(^-12\right) - 60 = 0 \qquad (5)^2 + 7 \cdot 5 - 60 = 0$$
$$144 - 84 - 60 = 0 \qquad 25 + 35 - 60 = 0$$
$$60 - 60 = 0 \qquad 60 - 60 = 0$$
$$0 = 0 \qquad 0 = 0$$

Note that, while linear equations have only one solution, quadratic equations typically have two!

Example 2

Solve for x: $x^2 - 7x = ^-12$

$$x^2 - 7x + 12 = 0 \quad \text{First put into standard form.}$$
$$(x - 4)(x - 3) = 0 \quad \text{Then factor.}$$
$$x - 4 = 0 \qquad x - 3 = 0 \quad \text{Set each factor equal to zero.}$$
$$x = 4 \qquad x = 3 \quad \text{Find the two roots.}$$

You may refer to the two examples if you need to while solving the following equations.

Test Yourself

Directions: Solve each equation for both values of x.

1. $x^2 + 7x + 6 = 0$
2. $x^2 + 5x = ^-4$
3. $x^2 + 5x = ^-6$
4. $7x + 10 = ^-x^2$
5. $x^2 - 7x + 10 = 0$
6. $x^2 = 8x - 15$
7. $x^2 = 8x - 12$
8. $x^2 - 12x + 20 = 0$
9. $x^2 = 8 + 7x$
10. $x^2 + 9x = 10$
11. $12 = x^2 + 11x$
12. $x^2 = 12 + x$
13. $18 = 7x + x^2$
14. $x^2 - 19x + 18 = 0$
15. $10x = -x^2 - 21$
16. $x^2 + 28 = 11x$
17. $x^2 + 10x + 24 = 0$
18. $x^2 + 8x = 65$

Answers

1. $^-6, ^-1$
2. $^-4, ^-1$
3. $^-3, ^-2$
4. $^-2, ^-5$
5. $2, 5$
6. $5, 3$
7. $2, 6$
8. $2, 10$
9. $8, ^-1$
10. $1, ^-10$
11. $1, ^-12$
12. $4, ^-3$
13. $2, ^-9$
14. $1, 18$
15. $^-7, ^-3$
16. $4, 7$
17. $^-4, ^-6$
18. $5, ^-13$

WORD PROBLEMS

Preparing to Solve Word Problems

Algebra, as a branch of mathematics, is most useful for the solution of word problems. It is for that reason that we have avoided the introduction of any type of word problem until we had progressed to the stage where enough algebra had been learned to make it possible to view word problems from an algebraic standpoint. A word problem is a mathematical situation requiring a mathematical solution, but stated in verbal terms. That is to say, it is a mathematical problem presented in English sentences. The key to solving word problems successfully is being able to translate the English sentences into a mathematical sentence. A mathematical sentence is another name for an algebraic equation.

Changing English sentences into mathematical sentences can be viewed as similar to changing from one language to another. Some words translate directly from one language to the other, while certain phrases are idiomatic and have to be translated as groups of words. An example of a word that translates directly is "is," or "the result is." "Exceeds" means "+." "Is more than" also means "+." "The sum is" is yet another way of indicating that things must be added together.

Below, you will find a list of English phrases commonly found in word problems.

Test Yourself

> **Directions:** Translate each into the appropriate mathematical symbol or group of symbols.

1. 5 more than the number n _____

2. n more than the number 7 _____

3. The product of n and 3 _____

4. The difference of x and 5 _____

5. The difference of 5 and x _____

6. 8 increased by y _____

7. Twice y _____

8. Twice x increased by 6 _____

9. Twice 7 decreased by m _____

10. A number r diminished by 5 _____

11. 3 more than half of p _____

12. Half the sum of x and 7 _____

13. Twice the difference of y and 5 _____

14. The quotient of 23 and a _____

15. The quotient of m and 16 _____

16. The number that exceeds 5 by k _____

17. The number that exceeds the sum of t and 3 by 11 _____

18. 5 less than the product of 9 and u _____

19. The sum of t and 4 diminished by 5 _____

20. One-fourth the difference of h and 9 _____

Answers

1. $n + 5$ or $5 + n$ 2. $7 + n$ 3. $3n$

4. $x - 5$ 5. $5 - x$ 6. $8 + y$

7. $2y$ 8. $2x + 6$ 9. $2 \cdot 7 - m$ or $14 - m$

10. $r - 5$ 11. $\dfrac{1}{2}p + 3$ or $\dfrac{p}{2} + 3$ 12. $\dfrac{1}{2}(x + 7)$ or $\dfrac{x + 7}{2}$

13. $2(y - 5)$ 14. $\dfrac{23}{a}$ 15. $\dfrac{m}{16}$

16. $5 + k$ 17. $(t + 3) + 11$ or $t + 14$ 18. $9u - 5$

19. $(t + 4) - 5$ or $t - 1$ 20. $\dfrac{1}{4}(h - 9)$ or $\dfrac{h - 9}{4}$

Number Problems

The last section dealt with mathematical phrases. A sentence, however, usually consists of more than one phrase, together with connecting words. So, a mathematical sentence will always consist of more than one phrase and connectives. A mathematical sentence will always express a complete thought, just as an English sentence expresses a complete thought. Mathematical sentences are easier to construct than English sentences, however, because the main verb is always "is," expressed as an equal sign.* Let us examine a few examples and see how this whole business works.

Example 1

A certain number when added to twice itself is 39. Find the number.

First, we make a "let statement:"

$$\text{Let } x = \text{the number.}$$

Then we start to translate phrase by phrase:

A certain number ...	That means x.
... when added ...	That means $+$.
... to twice itself ...	That means $2x$.
... is ...	That means $=$.
... 39.	That means 39.

In other words, the problem translates to the sentence:

$$x + 2x = 39$$

Solving the equation, we find

$$3x = 39$$
$$x = 13$$

Therefore, the number is 13.

* At least this is true of equations. Inequalities are a different story.

Example 2

3 more than half a certain number is 22. Find the number.

First the let statement:

Let x = the number

Then the translation:

3 *more than* ...	means + 3. Of course, we have yet to get to the spot to where we add the 3.
... *half a certain number* ...	means either $\frac{1}{2}x$ or $\frac{x}{2}$. Both mean the same thing. This is the expression that we will be adding the 3 onto.)
... *is 22.*	That means = 22.

The equation, therefore, is:

$$\frac{1}{2}x + 3 = 22$$

Solving that equation we find that:

$$\frac{1}{2}x = 19$$
$$x = 38$$

Therefore the number is 38.

Example 3

When 5 is added to a number, the result is 9 less than twice the number increased by 7. Find the number.

As complicated as this problem may seem, starting with the let statement, we can solve it by following the usual steps.

Let x = the number

Then translate the sentence bit by bit:

When 5 is added to a number ...	That means $x + 5$.
... the result is ...	That means =.
... 9 less than ...	That means 9 is subtracted from something.
... twice the number ...	$2x$ is twice the number.
... increased by 7.	That means 7 is added to $2x$. It also means that $2x + 7$ is what we must take 9 from.

The equation, then, is:

$$x + 5 = 2x + 7 - 9$$

Solving that equation, we find:

$$x + 5 = 2x - 2$$
$$7 = x$$

Therefore, the number is 7.

Example 4

Twice the sum of 3 and a number is 1 less than three times the number. Find the number.

Solution: Let x = the number

Twice the ...	We'll have to multiply something by 2.
... sum of 3 and a number ...	The sum of 3 and x is what we must multiply by 2. That's $2(3 + x)$.
... is ...	=
... 1 less than ...	We'll have to -1 from something.
... three times the number.	$3x$ is what we -1 from.

The equation must be:

$$2(3 + x) = 3x - 1$$

Solving, we find:

$$2(3 + x) = 3x - 1$$
$$6 + 2x = 3x - 1$$
$$\underline{+\quad 1 - 2x = -2x + 1}$$
$$7 \qquad = x$$

The number must be 7.

Notice that, in this example, we wrote the multiplication "twice the sum of a number and 3" as $2(3 + x)$. Writing it this way indicates that the entire sum $3 + x$ is to be multiplied by 2. A common mistake in solving such problems is to incorrectly write $2 \times 3 + x$, instead of $2(3 + x)$. Since $2 \times 3 + x = 6 + x$, and $2(3 + x) = 6 + 2x$, these two expressions are indeed different. Be careful, and use parentheses when necessary to correctly set up the conditions required by the problem.

Test Yourself

Directions: Find the number described in each of the following word problems.

1. A certain number added to 5 gives a result of 16. What is it?

2. When a certain number is added to 11, the result is 35. Find it.

3. When a certain number is taken from 47, 29 results. Find the number.

4. Forty-three is 7 less than twice a number. What is the number?

5. When three times a number is added to 17, 68 results. What is the number?

6. Three times a number increased by 4 is the same as four times that number decreased by 11. Find the number.

7. When twice a number is added to itself, the result is 7 less than four times the number. What is the number?

8. When 9 is added to a number, the result is the same as when twice the number is diminished by 6. Find the number.

9. When a number is increased by 11, the result is the same as when twice the number is decreased by 4. Find the number.

10. Five more than half a certain number is 32. Find the number.

11. Nine more than $\dfrac{1}{4}$ a certain number is 14. What is the number?

12. When half a certain number is increased by 18, the result is the same as when the number is increased by 6. Find the number.

13. Twice the sum of 5 and a number is 3 more than three times the number. What is the number?

14. Three times the sum of a number and 8 is 6 more than 5 times the number. Find the number.

15. Half the sum of a number and 32 is 2 less than twice the number. What is the number?

16. When the product of 3 and a number is increased by 7, the result is 5 less than the product of the number and 5. Find the number.

17. Twice the difference of 31 and a number exceeds twice the number by 30. Find the number.

18. The difference between a number and 29 is 14 less than one-fourth the number. Find the number.

19. Half a number increased by 40 is the same as four times the number decreased by 16. Find the number.

20. Twice the product of 3 and a number exceeds the sum of the number and 80 by 10. Find the number.

Answers

1. $\begin{aligned} x + 5 &= 16 \\ x &= 11 \end{aligned}$ **2.** $\begin{aligned} 11 + x &= 35 \\ x &= 24 \end{aligned}$ **3.** $\begin{aligned} 47 - x &= 29 \\ x &= 18 \end{aligned}$

4. $\begin{aligned} 43 &= 2x - 7 \\ 50 &= 2x \\ x &= 25 \end{aligned}$ **5.** $\begin{aligned} 3x + 17 &= 68 \\ 3x &= 51 \\ x &= 17 \end{aligned}$ **6.** $\begin{aligned} 3x + 4 &= 4x - 11 \\ 15 &= x \end{aligned}$

7. $\begin{aligned} 2x + x &= 4x - 7 \\ 3x &= 4x - 7 \\ 7 &= x \end{aligned}$ **8.** $\begin{aligned} x + 9 &= 2x - 6 \\ 15 &= x \end{aligned}$ **9.** $\begin{aligned} x + 11 &= 2x - 4 \\ 15 &= x \end{aligned}$

10. $\begin{aligned} \tfrac{1}{2}x + 5 &= 32 \\ \tfrac{1}{2}x &= 27 \\ x &= 54 \end{aligned}$ **11.** $\begin{aligned} \tfrac{1}{4}x + 9 &= 14 \\ \tfrac{1}{4}x &= 5 \\ x &= 20 \end{aligned}$ **12.** $\begin{aligned} \tfrac{1}{2}x + 18 &= x + 6 \\ 12 &= \tfrac{1}{2}x \\ 24 &= x \end{aligned}$

13. $2(5 + x) = 3x + 3$
$10 + 2x = 3x + 3$
$7 = x$

14. $3(x + 8) = 5x + 6$
$3x + 24 = 5x + 6$
$18 = 2x$
$9 = x$

15. $\frac{1}{2}(x + 32) = 2x - 2$
$1(x + 32) = 4x - 4$
$x + 32 = 4x - 4$
$36 = 3x$
$12 = x$

16. $3x + 7 = 5x - 5$
$12 = 2x$
$6 = x$

17. $2(31 - x) = 2x + 30$
$62 - 2x = 2x + 30$
$32 = 4x$
$8 = x$

18. $x - 29 = \frac{1}{4}x - 14$
$\frac{3}{4}x = 15$
$x = 20$

19. $\frac{1}{2}x + 40 = 4x - 16$
$x + 80 = 8x - 32$
$112 = 7x$
$16 = x$

20. $2(3x) = (x + 80) + 10$
$6x = x + 90$
$5x = 90$
$x = 18$

Consecutive Integer Problems

Consecutive integer problems differ from the number problems of the last section in two ways. First, you are seeking to find more than a single answer. Secondly, it is necessary to write a multiple let statement. When we talk about consecutive integers, we mean whole numbers that follow one after the other in counting sequence. For example, 3, 4, and 5 are consecutive integers. So are -6, -5, and -4, or -2, -1, 0, and 1.

Examine any of the sets of consecutive integers above, and you will discover a pattern in the way each integer in any set is related to the others. To get from 3 to 4, for example, you must add 1. To get from 4 to 5, you must add another 1. If we wished to write each of the integers in the first set based upon the first element of that set, we might call them 3, 3 + 1, and 3 + 2. Examine the last sentence carefully, because it forms the basis for all consecutive integer equations. The last set of integers in the first paragraph might be represented as -2, -2 + 1, -2 + 2, and -2 + 3. Check the value of each of these expressions, and you will find that they equal -2, -1, 0, and 1 respectively.

Now, consider the problem of finding four consecutive integers. We are given some information about them, but we do not know the values of the integers. We only know that they are consecutive. Obviously a variable is needed. The following let statement defines the four consecutive integers that we are looking for:

Let $x =$ the first   number
" $x + 1 =$ the second   "
" $x + 2 =$ the third   "
" $x + 3 =$ the fourth   "

And so, we have taken care of representing each of the four consecutive integers.

What if what we are seeking is a set of consecutive even integers? How are even integers related to one another? Consider the first three consecutive positive even integers: 2, 4, 6. How does one get from one to another? Hopefully, you recognized the fact that each is two more than the one that came before it. Three consecutive even integers may therefore be represented as x, $x + 2$, and $x + 4$. The let statement should then be obvious:

$$
\begin{aligned}
\text{Let } x &= \text{the first} & \text{even integer} \\
" \ x + 2 &= \text{the second} & " & \quad " \\
" \ x + 4 &= \text{the third} & " & \quad "
\end{aligned}
$$

Now for an odd question: What if we are looking for a set of consecutive odd integers? How are odd integers related to one another? Consider the first three positive odd integers, 1, 3, and 5. Are you surprised at your discovery? Odd integers are related to one another in exactly the same way as even ones, that is, they are two apart. The let statement for three consecutive odd integers would look identical to that for three consecutive even integers as shown above, except that the word "even" would be replaced by the word "odd."

Now, examine the following examples and you will get an idea of how to deal with consecutive integer problems.

Example 1

Find three consecutive integers whose sum is 72.

$$
\begin{aligned}
\text{Let } x &= \text{the first} \\
" \ x + 1 &= \text{the second} \\
" \ x + 2 &= \text{the third}
\end{aligned}
$$

Next, we set up the equation:

$$
\begin{aligned}
x + x + 1 + x + 2 &= 72 & \text{Sum means add the integers together.} \\
3x + 3 &= 72 & \text{Combine like terms.} \\
3x &= 69 & \text{Collect terms.} \\
x &= 23 & \text{Divide both sides by 3.}
\end{aligned}
$$

Answer: 23, 24, 25   Refer to let - statement.

Example 2

Find three consecutive even integers such that the sum of the first and third equals 40.

$$
\begin{aligned}
\text{Let } x &= \text{the first} \\
" \ x + 2 &= \text{the second} \\
" \ x + 4 &= \text{the third}
\end{aligned}
$$

$$
\begin{aligned}
x + x + 4 &= 40 & \text{Set up the equation.} \\
2x + 4 &= 40 & \text{Combine like terms.} \\
2x &= 36 & \text{Collect terms.} \\
x &= 18 & \text{Divide both sides by 2.}
\end{aligned}
$$

Answer: 18, 20, 22

Example 3

Find three consecutive odd integers such that the sum of the second and third exceeds the first by 35.

$$\text{Let } x \quad = \text{the first}$$
$$\text{" } x + 2 = \text{the second}$$
$$\text{" } x + 4 = \text{the third}$$

$x + 2 + x + 4 = x + 35$	Set up the equation.
$2x + 6 = x + 35$	Combine like terms.
$x = 29$	Collect terms.
Answer: 29, 31, 33	Refer to let statement.

Test Yourself

Directions: Solve the following.

1. Find three consecutive integers whose sum is 21.

2. Find three consecutive integers whose sum is 57.

3. Find three consecutive integers whose sum is 96.

4. Find four consecutive integers whose sum is 66.

5. Find four consecutive integers whose sum is 106.

6. Find five consecutive integers whose sum is 140.

7. Find three consecutive even integers whose sum is 66.

8. Find three consecutive even integers whose sum is 42.

9. Find three consecutive odd integers whose sum is 51.

10. Find four consecutive odd integers whose sum is 112.

11. Find three consecutive integers such that twice the first exceeds the third by 3.

12. Find three consecutive integers such that twice the sum of the first and third exceeds the second by 18.

13. Find three consecutive integers such that the first increased by twice the second exceeds the third by 24.

14. Find three consecutive integers such that one-third the sum of the first two is 93.

15. Find three consecutive odd integers such that twice the sum of the first and second is 144.

16. Find four consecutive integers such that three times the sum of the second and fourth is 24.

17. Find three consecutive integers such that the difference between twice the second and the first is 22.

18. Find three consecutive even integers such that $\frac{1}{2}$ the sum of the first and second is 37.

19. Find three consecutive odd integers such that 480 is 4 times the sum of the first two.

20. Find three consecutive odd integers such that four times the sum of the second and third exceed three times the sum of the first and second by 28.

Answers

1. Let $x = $ first integer
$$x + 1 = \text{second}$$
$$x + 2 = \text{third}$$

$$x + x + 1 + x + 2 = 21$$
$$3x + 3 = 21$$
$$3x = 18$$
$$x = 6$$
$$x + 1 = 7$$
$$x + 2 = 8$$

2. Let $x = $ first integer
$$x + 1 = \text{second}$$
$$x + 2 = \text{third}$$

$$x + x + 1 + x + 2 = 57$$
$$3x + 3 = 57$$
$$3x = 54$$
$$x = 18$$
$$x + 1 = 19$$
$$x + 2 = 20$$

3. Let $x = $ first integer
$$x + 1 = \text{second}$$
$$x + 2 = \text{third}$$

$$x + x + 1 + x + 2 = 96$$
$$3x + 3 = 96$$
$$3x = 93$$
$$x = 31$$
$$x + 1 = 32$$
$$x + 2 = 33$$

4. Let $x = $ first integer
$$x + 1 = \text{second}$$
$$x + 2 = \text{third}$$
$$x + 3 = \text{fourth}$$

$$x + x + 1 + x + 2 + x + 3 = 66$$
$$4x + 6 = 66$$
$$4x = 60$$
$$x = 15$$
$$x + 1 = 16$$
$$x + 2 = 17$$
$$x + 3 = 18$$

5. Let $x = $ first integer
$$x + 1 = \text{second}$$
$$x + 2 = \text{third}$$
$$x + 3 = \text{fourth}$$

$$x + x + 1 + x + 2 + x + 3 = 106$$
$$4x + 6 = 106$$
$$4x = 100$$
$$x = 25$$
$$x + 1 = 26$$
$$x + 2 = 27$$
$$x + 3 = 28$$

6. Let $x = $ first integer
$$x + 1 = \text{second}$$
$$x + 2 = \text{third}$$
$$x + 3 = \text{fourth}; \; x + 4 = \text{fifth}$$

$$x + x + 1 + x + 2 + x + 3 + x + 4 = 140$$
$$5x + 10 = 140$$
$$5x = 130$$
$$x = 26$$
$$x + 1 = 27$$
$$x + 2 = 28$$
$$x + 3 = 29$$
$$x + 4 = 30$$

7. Let x = first integer
$x + 2$ = second
$x + 4$ = third

$x + x + 2 + x + 4 = 66$
$3x + 6 = 66$
$3x = 60$
$x = 20$
$x + 2 = 22$
$x + 4 = 24$

8. Let x = first integer
$x + 2$ = second
$x + 4$ = third

$x + x + 2 + x + 4 = 42$
$3x + 6 = 42$
$3x = 36$
$x = 12$
$x + 2 = 14$
$x + 4 = 16$

9. Let x = first integer
$x + 2$ = second
$x + 4$ = third

$x + x + 2 + x + 4 = 51$
$3x + 6 = 51$
$3x = 45$
$x = 15$
$x + 2 = 17$
$x + 4 = 19$

10. Let x = first integer
$x + 2$ = second
$x + 4$ = third
$x + 6$ = fourth

$x + x + 2 + x + 4 + x + 6 = 112$
$4x + 12 = 112$
$4x = 100$
$x = 25$
$x + 2 = 27$
$x + 4 = 29$
$x + 6 = 31$

11. Let x = first integer
$x + 1$ = second
$x + 2$ = third

$2x = x + 2 + 3$
$2x = x + 5$
$x = 5$
$x + 1 = 6$
$x + 2 = 7$

12. Let x = first integer
$x + 1$ = second
$x + 2$ = third

$2(x + x + 2) = x + 1 + 18$
$2(2x + 2) = x + 19$
$4x + 4 = x + 19$
$3x = 15$
$x = 5$
$x + 1 = 6$
$x + 2 = 7$

13. Let x = first integer
$x + 1$ = second
$x + 2$ = third

$x + 2(x + 1) = x + 2 + 24$
$x + 2x + 2 = x + 26$
$3x + 2 = x + 26$
$2x = 24$
$x = 12$
$x + 1 = 13$
$x + 2 = 14$

14. Let x = first integer
$x + 1$ = second
$x + 2$ = third

$\frac{1}{3}(x + x + 1) = 93$
$\frac{1}{3}(2x + 1) = 93$
(Triple both sides)
$2x + 1 = 279$
$2x = 278$
$x = 139$
$x + 1 = 140$
$x + 2 = 141$

15. Let x = first integer
$x + 2$ = second
$x + 4$ = third

$2(x + x + 2) = 144$
$2(2x + 2) = 144$
$4x + 4 = 144$
$4x = 140$
$x = 35$
$x + 2 = 37$
$x + 4 = 39$

16. Let x = first integer
$x + 1$ = second
$x + 2$ = third
$x + 3$ = fourth

$3(x + 1 + x + 3) = 24$
$3(2x + 4) = 24$
$6x + 12 = 24$
$6x = 12$
$x = 2$
$x + 1 = 3$
$x + 2 = 4$
$x + 3 = 5$

17. Let x = first integer
$x + 1$ = second
$x + 2$ = third

$2(x + 1) - x = 22$
$2x + 2 - x = 22$
$x + 2 = 22$
$x = 20$
$x + 1 = 21$
$x + 2 = 22$

18. Let x = first integer
$x + 2$ = second
$x + 4$ = third

$\frac{1}{2}(x + x + 2) = 37$
(Double both sides)
$x + x + 2 = 74$
$2x + 2 = 74$
$2x = 72$
$x = 36$
$x + 2 = 38$
$x + 4 = 40$

19. Let x = first integer
$x + 2$ = second
$x + 4$ = third

$$4(x + x + 2) = 480$$
$$4(2x + 2) = 480$$
$$8x + 8 = 480$$
$$8x = 472$$
$$x = 59$$
$$x + 2 = 61$$
$$x + 4 = 63$$

20. Let x = first integer
$x + 2$ = second
$x + 4$ = third

$$4(x + 2 + x + 4) = 3(x + x + 2) + 28$$
$$4(2x + 6) = 3(2x + 2) + 28$$
$$8x + 24 = 6x + 6 + 28$$
$$8x + 24 = 6x + 34$$
$$2x = 10$$
$$x = 5$$
$$x + 2 = 7$$
$$x + 4 = 9$$

Problems Involving Ratios and Proportions

Ratios can be extremely useful in all sorts of problem-solving situations. Examine the examples below, and you will see how this is done. If you need to be refreshed as to how to set up ratios or how to solve proportions, refer to the section on ratio and proportion earlier in this chapter.

They are particularly useful for solving proportion problems. In proportion problems, we are told the rate at which two quantities vary and are asked to find the value of one of them, given the value of the other.

Example 1

Henry got 12 hits in the first 20 ball games of the year. If he were to continue hitting at the same pace, how many hits would he get in a 75-game season?

Let h = number of hits in the season   1. First write the let statement.

$$\frac{\text{hits}}{\text{games}} = \frac{\text{hits}}{\text{games}}$$ 2. Then describe the proportion.

$$\frac{12}{20} = \frac{h}{75}$$

$$20h = 12(75)$$ 3. Establish the proportion.

$$20h = 900$$ 4. Solve.

$$h = 45$$ Henry would get 45 hits.

Example 2

Karen's stock portfolio cost her $600. It paid her an income of $90 per year. If she had invested a total of $1400 at the same rate, what would her yearly income have been?

Let i = yearly income	1. Write the let statement
(from her new investment)	
$\dfrac{\text{old investment}}{\text{old income}} = \dfrac{\text{new investment}}{\text{new income}}$	2. Describe the proportion.
$\dfrac{600}{90} = \dfrac{1400}{i}$	3. Establish the proportion.
$600i = 90(1400)$	4. Solve.
$600i = 126,000$	
$i = 210$	Her income would have been $210.

Example 3

A 6-foot-tall person casts a 9-foot-long shadow at the same time that a nearby flagpole casts a 75-foot shadow. How tall is the flagpole?

Let f = the height of the flagpole	1. Write the let statement.
$\dfrac{\text{person's height}}{\text{person's shadow}} = \dfrac{\text{flagpole's height}}{\text{flagpole's shadow}}$	2. Describe the proportion.
$\dfrac{6}{9} = \dfrac{f}{75}$	3. Establish the proportion.
$9f = 6(75)$	4. Solve.
$9f = 450$	
$f = 50$	The flagpole is 50 feet tall.

The following word problems are all solvable by use of ratio and proportion. Examine each problem carefully before deciding the ratios that you will establish. Always make sure that both ratios are in the same order (that is, if one ratio is height to weight, then the second must also be height to weight—not weight to height).

Test Yourself

Directions: Solve by using ratios and proportions.

1. Six raffle tickets cost $1.85. How much would 67 raffle tickets cost?

2. Five people consume 3 pounds of food at a party. How much food should be ordered for a party of 37?

3. Three records cost $12.96. At that price, how much should 7 records cost?

4. It costs the government 8 million dollars to build 5 supersonic fighter planes. At the same unit cost, how much should the bill for 65 such planes come to?

5. A park with a perimeter of 150 meters is adequate for 24 children's recreational needs. Based upon the same perimeter to child ratio, how many children could be serviced by a park of 525 meters perimeter?

6. Twenty-five acres of forestland supply a certain paper company with all of its needs for wood pulp for a period of 7 weeks. How many acres are required by the company for a year's supply of pulp?

7. Four washes require 68 gallons of water. How many gallons of water are needed to do 11 washes?

8. At a ballpark, 23,000 people consumed 630 pounds of frankfurters. At that same rate of consumption, how many frankfurters would have been needed for a crowd of 49,000?

9. The fire department raised $518 at a dance that attracted 67 persons. At that same rate, how much money would have been raised if 200 persons had attended?

10. A chicken farmer got 526 eggs per day from 712 chickens. How many chickens would he have needed to get 1000 eggs per day, if the new chickens produced at the same rate as the old ones?

11. A truck traveled 165 miles in 3 hours. Traveling at the same average rate of speed, how long would it take the truck to go 715 miles?

12. If 6 pounds of peanuts sell for $11.45, to the nearest penny, how much should a 19 pound bag of peanuts cost (assuming the same price per pound)?

13. Seven feet of lead pipe weigh 25 pounds. How much will 40 feet of the same pipe weigh?

14. Five cups of flour were required to bake 24 cupcakes. Using the same recipe, how much flour would be needed for 136 cupcakes?

15. A bus travels 39 miles on 8 gallons of diesel fuel. How much fuel would it use to travel from New York City to Miami, Florida—a distance of about 1200 miles?

16. Forty-five volts of electricity cause an electric motor to turn at 1100 rpm. If the speed varies directly with the voltage, how fast would the motor turn when 75 volts were applied?

Answers

1. Let c = cost of 67 tickets
$$\frac{6}{1.85} = \frac{67}{c}$$
$$6c = 123.95$$
$$c = \$20.66$$

2. Let f = food for party
$$\frac{5}{3} = \frac{37}{f}$$
$$5f = 111$$
$$f = 22\frac{1}{5} \text{ pounds}$$

3. Let r = cost of 7 records
$$\frac{3}{12.96} = \frac{7}{r}$$
$$3r = 90.72$$
$$r = \$30.24$$

4. Let b = bill for 65 planes
$$\frac{8}{5} = \frac{b}{65}$$
$$5b = 520$$
$$b = \$104 \text{ million}$$

5. Let n = number of children
$$\frac{150}{24} = \frac{525}{n}$$
$$150n = 12,600$$
$$n = 84 \text{ children}$$

6. Let a = acres needed
$$\frac{25}{7} = \frac{a}{52} *$$
$$7a = 1300$$
$$a = 185\frac{5}{7} \text{ acres}$$

7. Let g = gallons needed
$$\frac{4}{68} = \frac{11}{g}$$
$$4g = 748$$
$$g = 187 \text{ gallons}$$

8. Let f = number of pounds of frankfurters
$$\frac{630}{23,000} = \frac{f}{49,000} **$$
$$2300f = 63(49,000)$$
$$2300f = 3,087,000$$
$$23f = 30,870$$
$$f = 1342\frac{4}{23} \text{ pounds}$$

9. Let m = money
$$\frac{518}{67} = \frac{m}{200}$$
$$67m = 103,600$$
$$m = \$1546.27$$

10. Let c = number of chickens
$$\frac{526}{712} = \frac{1000}{c}$$
$$526c = 712,000$$
$$c = 1354 \text{ chickens } ***$$

* 52 weeks = 1 year.

** Simplifying the fraction lets us work with smaller numbers.

*** Although the answer is actually $1353\frac{161}{263}, \frac{161}{263}$ of a chicken cannot lay that 1000th egg. Therefore, we must round our answer up to the nearest whole number.

11. Let t = time

$$\frac{165}{3} = \frac{715}{t}$$
$$165t = 2145$$
$$t = 13 \text{ hours}$$

12. Let a = amount (or cost)

$$\frac{6}{11.45} = \frac{19}{a}$$
$$6a = 19(11.45)$$
$$6a = 217.55$$
$$a = \$36.26$$

13. Let w = weight

$$\frac{7}{25} = \frac{40}{w}$$
$$7w = 1000$$
$$w = 142\frac{6}{7} \text{ pounds}$$

14. Let c = cups of flour

$$\frac{5}{24} = \frac{c}{136}$$
$$24c = 680$$
$$c = 28\frac{1}{3} \text{ cups}$$

15. Let x = amount of fuel

$$\frac{39}{8} = \frac{1200}{x}$$
$$39x = 9600$$
$$x = 246\frac{2}{13} \text{ gallons}$$

16. Let s = speed

$$\frac{45}{1100} = \frac{75}{s}$$
$$45s = 82,500$$
$$s = 1833\frac{1}{3} \text{ rpm}$$

It should be pointed out that each ratio in any of the solutions could have been inverted, as long as the other ratio was also inverted. Each answer would still have been the same. The choices of variable names were, of course, purely arbitrary.

Solving Motion Problems

There are many different ways to get from one place to another. It is only natural then, that there should be many different types of problems about getting from one place to another. These problems do not take into account missing the train, or forgetting your passport or traveler's checks.

All motion problems deal with three quantities. The first is distance. Usually, distance is expressed in miles, kilometers, meters, yards, or feet. The second quantity is time, usually expressed in seconds, minutes, or hours. The third and final quantity is rate of travel. This is not quite as cut and dried a quantity as the previous two. That is because there is no ruler nor clock with which rate can be measured. Rate, in fact, is a comparison between the distance traveled, and the amount of time that it took to travel that distance.

$$\text{Rate} = \frac{\text{distance traveled}}{\text{traveling time}}$$

The units in which rates are expressed are derived from the formula above. Here are some, but by no means all, units of rate:

$$\frac{\text{miles}}{\text{hour}} \qquad \frac{\text{miles}}{\text{minute}} \qquad \frac{\text{feet}}{\text{second}} \qquad \frac{\text{meters}}{\text{hour}} \qquad \frac{\text{kilometers}}{\text{hour}}$$

In reading these rates, the fraction-line is usually replaced by the word "per," and therefore the rates above would read "miles per hour," "miles per minute," "feet per second," "meters per hour," and "kilometers per hour." As you can see, a rate may be made up of any unit of distance **per** any unit of time.

Learning to solve motion problems is mainly a matter of "getting the hang of it." Any motion problem can be broken down in such a way as to make it fit into the following formula: Distance = rate × time. This formula is usually abbreviated as follows:

$$d = rt$$

This formula tells us that, to find the actual distance traveled, the average rate of travel must be multiplied by the actual time used in making the trip. Any straightforward motion problem can be solved by simply substituting the known quantities into the formula and then solving the equation that results for the unknown quantity.

Following are three straightforward motion problems and their solutions. You will notice that a box has been used to organize the data from each problem. Using this box is a good habit. While it is not really necessary for the problems in this section, as you reach more difficult problems, you'll discover it to be essential. It never hurts to develop good work habits early.

Example 1

Geoffrey drives for 4 hours at 45 mph. How far does he travel?

Remember: $d = rt$

4 hours is time. 45 mph is rate, so:

d	r	t
d	45	4

$$d = (45)(4) \quad \text{(substitute)}$$
$$d = 180 \quad \text{(multiply)}$$

Geoffrey drove 180 miles.

Example 2

Carol drove 320 miles in 4 hours. What was her average speed?

320 miles is distance. 4 hours is time. Speed is just another word for rate, so:

d	r	t
320	r	4

$$d = rt$$
$$320 = (r)(4) \quad \text{(substitute)}$$
$$320 = 4r \quad \text{(multiply)}$$
$$80 = r \quad \text{(divide both sides by 4)}$$

Having found the value of r, you must now look back at the problem. Notice that the distance was given in miles and the time in hours. That means that rate must be expressed as a relation of those two units. Carol's speed, therefore, was 80 mph. (Tsk, tsk! Speeding again.)

Example 3

David flew from New York to Des Moines (in an airplane). He covered the 1200 miles at an average rate of 400 mph. How long did the trip take?

Notice that this time we have been given distance and rate. It is time that we must find.

d	r	t
1200	400	4

$$d = rt$$
$$1200 = 400t \qquad \text{(substitute)}$$
$$3 = t \qquad \text{(divide both sides by 400)}$$

Here again, we must look back at the problem to find the unit in which to express time. Since rate was expressed in miles per hour, time will be in hours. The trip, therefore, took 3 hours.

Below are some problems to try on your own. The starred ones (*) are trickier than the rest.

Test Yourself

Directions: Solve the following.

1. Erica can run 60 meters in 10 seconds. What is her average rate?

2. How fast must a train go in order to cover 264 kilometers in 3 hours?

3. Jason rows for 3 hours at an average rate of 6 mph. How far does he get?

4. Alessandra drives at 55 mph for 6 hours. How far does she travel?

5. The tooth fairy flies from pillow to pillow covering 3 kilometers per minute. How long does it take her to go 120 kilometers?

6. Dylan runs at an average speed of 12 yards per second. How long will it take him to run 120 yards?

*7. A plane flies 6 miles in 2 minutes. How far will it go in 3 hours?

*8. A snail travels 7 feet in 7 minutes. At that snail's pace, how long will it take for it to travel 1 mile?

Answers

Problems 1–6 are straightforward. Each of them may be solved by substituting the appropriate values into the "distance formula."

1.

d	r	t
60	r	10

$$d = rt$$
$$60 = (r)(10) \qquad \text{(substitute)}$$
$$60 = 10r \qquad \text{(multiply)}$$
$$6 = r \qquad \text{(divide)}$$
$$r = 6 \text{ meters per second} \qquad \text{(include units)}$$

2.

d	r	t
264	r	3

$$d = rt$$
$$264 = (r)(3)$$
$$264 = 3r$$
$$88 = r$$
$$r = 88 \text{ kilometers per hour}$$

3.

d	r	t
d	6	3

$d = rt$

$d = (6)(3)$ (substitute)

$d = 18$ (multiply)

$d = 18$ miles (include units)

4.

d	r	t
d	55	6

$d = rt$

$d = (55)(6)$

$d = 330$

$d = 330$ miles

5.

d	r	t
120	3	t

$d = rt$

$120 = 3t$ (substitute)

$40 = t$ (divide)

$t = 40$ minutes (include units)

6.

d	r	t
120	12	t

$d = rt$

$120 = 12t$

$10 = t$

$t = 10$ seconds

7. The time for flying 2 miles was given in minutes. The information asked for, however, is expressed in hours. Now it just is not possible to deal with two different units of time within the same problem. Did you know what to do about that? Well, 3 hours just happens to be 180 minutes. (60 minutes per hour × 3 hours = 180 minutes.)

Six miles in 2 minutes is not a very useful rate. A rate must be per unit of time. If the plane flies 6 miles in 2 minutes, how far does it fly in 1 minute?

d	r	t
d	3	180

$d = rt$

$d = (3)(180)$ (substitute)

$d = 540$ (multiply)

$d = 540$ miles (include units)

8. This time, feet and miles don't mix. Do you know how many feet there are in a mile? There are 5280, and if you did not know that, this is as good a time as any to incorporate that fact into your general knowledge. The rate, 7 feet in 7 minutes, must also be interpreted. If the snail goes at the pace mentioned, how far will it go in 1 minute? Now look at the solution below.

d	r	t
5280	1	t

$d = rt$

$5280 = 1t$ (substitute)

$t = 5280$ minutes, or 88 hours, or 3 days,16 hours.

Some motion problems are not quite as easy to substitute into the "distance formula" as those on the previous page. They take a good deal of analysis in order to figure out just exactly how to hook the equation that will provide the answer you are looking for. Each of the examples that follow will show you how to deal with a different type of motion problem. Keep in mind, though, that no matter how mixed up each problem appears to be, the data provided must in some way fit into the formula. Remember, the more experience you get, the better you'll become at solving any kind of motion problem.

Example 4

Two trains leave the same station at the same time and travel in opposite directions. The first train averages 60 mph, while the second averages 45 mph. How long will it take for them to be 315 miles apart?

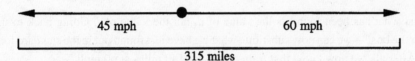

Often a diagram is helpful in visualizing a problem. The one above shows this one. Each arrow represents the direction and rate of a train. The dot on the line represents the station that they depart from. The bracket shows the distance being traveled.

Now to organize the data:

	d	r	t
Train #1	d	60	t
Train #2	$315 - d$	45	t

Notice that we do not know either the distance that each train traveled, nor the amount of time that it took. If together the two trains traveled 315 miles, one train traveled part of that distance, and the other traveled the remainder. (Remainder is a key mathematical word, indicating subtraction.) If we call d the distance that the first train traveled, then the second train must have traveled the remainder, or $315 - d$. Since both trains must travel for the same amount of time, we can call each time t.

If you are comfortable solving systems of equations in two variables, you need go no further. Two equations can be derived just from the data in the box:

$$d = 60t$$
$$315 - d = 45t$$

For the rest of us, however, an alternative solution may suggest itself.

Since the times of the two trains are the same, we can make an equation relating the two times:

$$t^1 = t^2$$

All that the above equation says is "the time of train #1 equals the time of train #2." (Mathematicians love symbols.) Now, to make use of this fact, we must rearrange the terms of the "distance formula" by solving it for time.

$$\text{If } d = rt, \text{ then } t = \frac{d}{r}$$

If that does not quite make sense to you, then consider this. To solve the equation $d = rt$ for t, each side must be divided by r. Now, we can substitute values from the box above.

$$t_1 = t_2$$
$$\frac{d_1}{r_1} = \frac{d_2}{r_2}$$

$$\frac{d}{60} = \frac{315 - d}{45} \qquad \text{(substitute)}$$
$$45d = 18,900 - 60d \qquad \text{(cross multiply)}$$
$$105d = 18,900 \qquad \text{(add } 60d \text{ to each side)}$$
$$d = 180 \text{ miles} \qquad \text{(divide)}$$

Notice that what has been found is the value of d, not the value of t. Going back to the box—remember the box?—we can now substitute what we have just found out in the standard formula for train #1. After all, we now know that train #1 travels 180 miles at 60 mph.

$$d = rt$$
$$180 = 60t$$
$$3 = t$$
$$t = 3 \text{ hours}$$

Now let's try a different approach to the same problem:

Two trains leave the same station at the same time and travel in opposite directions. The first train averages 60 mph, while the second averages 45 mph. How long will it take for them to be 315 miles apart?

Since the two trains are traveling in opposite directions, they are moving apart at a speed equal to their combined rates of travel. That is to say, if you imagine yourself to be on one train, the other train is moving away from you at 105 mph (the 60 you're going away from it plus the 45 it's going away from you). It is just as if you were standing still, and the other train is doing all the moving:

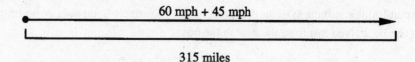

60 mph + 45 mph

315 miles

Taking advantage of this fact,

d	r	t
315	105	t

therefore:

$$d = rt$$
$$315 = 105t \qquad \text{(substitute)}$$
$$3 = t \qquad \text{(divide)}$$
$$t = 3 \text{ hours} \qquad \text{(include units)}$$

So, you see, there is more than one way to approach the same problem.

The problems below are similar to the model that was just worked. You may try to do them in any of the outlined ways, or maybe you'll come up with a method of your own. Before trying to invent your own way, however, see if one of the ways described might fit your style. It will make things a lot easier if it does.

Test Yourself

Directions: Solve the following.

1. Two ships leave the same harbor at the same time. One steams due south at 35 mph and the other due north at 30 mph. How far apart will they be after 5 hours?

2. A plane takes off and flies due west at 350 mph. At the same time, a second plane leaves the same airport and flies due east at 400 mph. How long after they take off will the two planes be 1500 miles apart?

3. Two cars start from Chicago at the same time and travel in opposite directions—one averaging 88 km/hr and the other 76 km/hr. How far apart will they be after $4\frac{1}{2}$ hours?

4. An aircraft carrier steams south from Norfolk at 30 km/hr. At the same time, a destroyer leaves Norfolk heading north at 50 km/hr. After how many hours will the ships be 360 kilometers apart?

5. Centerville and Middletown are 450 kilometers apart. A car leaves Centerville for Middletown traveling at an average speed of 60 km/hr at the same time that another car, traveling 75 km/hr leaves Middletown for Centerville. How long will the first car have been travelling when the two cars meet?

6. A car leaves South Grapefruit at 9 a.m. and travels north at 35 mph. An hour later, a second car leaves from the same town and travels in the same direction as the first. It travels at 60 mph. How long will it be until the second car overtakes the first?

7. An airliner leaves Kennedy airport and travels northward at 600 mph. Ninety minutes later, a supersonic plane leaves Kennedy traveling in the same direction as the first airliner. How far will the airliner have traveled before it is overtaken, if the second plane travels at 850 mph?

8. A train leaves Grand Central Station and travels toward Albany at an average speed of 80 km/hr. A second train leaves Grand Central on the same track two hours later. If the second train overtakes the first in four hours after the second train departed, what was the speed of the second train?

9. A ship leaves Baltimore sailing toward France, traveling at an average speed of 28 mph. Two days later to the minute, an airplane leaves Baltimore travelling in the same direction as the ship and flying at 630 mph. How far from Baltimore will the plane be over the ship?

10. A crop-dusting plane averaging 130 km/hr leaves an airport and is airborne for 3 hours before returning to the same airport. What is the maximum distance from the airport that the plane might have flown?

11. A car drives from 9 a.m. until noon at 40 mph. From noon until it stops it goes 55 mph. Altogether, 230 miles are covered. How long does the car drive at each speed?

12. Geoffrey jogged a certain distance at 8 mph. He then walked back at $2\frac{1}{2}$ miles per hour. Altogether, he covered 13 miles. He was gone from his home for 3 hours. How long did he jog? How long did he walk?

13. A boat travels downstream at a speed of 24 km/hr. A round trip covers 96 kilometers and requires 5 hours. What is the average speed of the upstream portion of the journey?

*14. A man started walking at 2 mph, while a woman 2 miles behind him began walking at the same time at a rate of 4 mph, and in the same direction. Just then, the man's dog left him and ran toward the woman. Upon reaching her, it instantly turned around and ran back toward the man. And so, the dog continued to run back and forth between them, at a constant rate of 5 mph, until the woman finally overtook the man. How far did the dog run?

Answers

1.
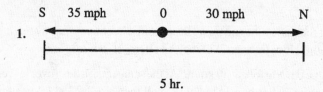

5 hr.

Pretend one ship stays put and the other travels away from it at 65 mph.

$d = r \cdot t$

$d = 65 \cdot 5$

$d = 325$ miles

2.
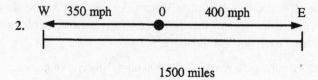

1500 miles

As in question 1, pretend that only one plane moves. Its speed would be 750 mph.

$d = r \cdot t$

$1500 = 750t$

$t = 2$ hours

---

* This is a classic problem for those who enjoy doing some really heavy thinking.

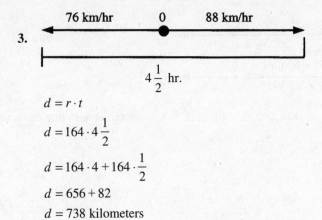

3.

$d = r \cdot t$

$d = 164 \cdot 4\frac{1}{2}$

$d = 164 \cdot 4 + 164 \cdot \frac{1}{2}$

$d = 656 + 82$

$d = 738$ kilometers

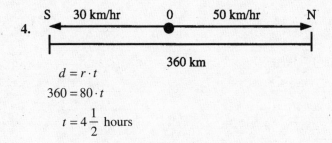

4.

$d = r \cdot t$

$360 = 80 \cdot t$

$t = 4\frac{1}{2}$ hours

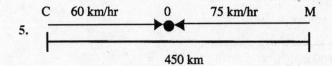

5.

If both cars leave at the same time, they must both travel for the same time. They close the gap between them at a speed totaling their combined individual speeds.

$d = r \cdot t$

$450 = 135t$

$t = 3\frac{1}{3}$ hours or 3 hours, 20 minutes

6.

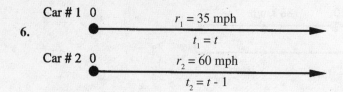

The second car travels for an hour less than the first. Both cars travel the same distance, hence

$$d_1 = d_2$$
$$\therefore r_1 \cdot t_1 = r_2 \cdot t_2$$
$$35t = 60(t-1)$$
$$35t = 60t - 60$$
$$25t = 60$$
$$t = 2\frac{2}{5} \text{ hours}$$

Second car travels $t - 1$, or $1\frac{2}{5}$ hours.

7.

#1 0 $r_1 = 600$ mph
    $t_1 = t$

#2 0
    $r_2 = 850$ mph

    $t_2 = t - 1\frac{1}{2}$

The distances traveled are identical.

$$d_1 = d_2$$
$$\therefore r_1 \cdot t_1 = r_2 \cdot t_2$$
$$600t = 850\left(t - 1\frac{1}{2}\right)$$
$$600t = 850t - 1275$$
$$250t = 1275$$
$$t = 5\frac{1}{10} \text{ or } 5.1 \text{ hours}$$

Then, to find distance,

$$d = rt$$
$$d = 600 \cdot 5.1$$
$$d = 3060 \text{ miles}$$

8.

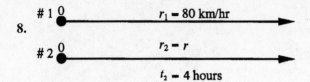

If the second train overtakes the first in 4 hours, then the first train must have traveled 4 hours + the 2 hour head start, or a total of 6 hours.

$$d_1 = r_1 \cdot t$$
$$d_1 = 80 \cdot 6$$
$$d_1 = 480 \text{ kilometers}$$

Since the second train traveled the same distance as the first, $d_1 = d_2$:

$$d_2 = r \cdot t_2$$
$$480 = r \cdot 4$$
$$r = 120 \text{ km/hr}$$

9.

	d	$=$	r	$\cdot$	t
ship	$28t$		28		t
plane	$630(t-48)$		630		$t - 48$

Once more, the distances are the same, $\therefore$,

$$28t = 630(t-48)$$
$$28t = 630t - 630 \cdot 48$$
$$28t = 630t - 30,240$$
$$602t = 30,240$$
$$t = 50.23 \text{ hours}$$

Since $\quad d = 28t$ (see chart),

$$d = 28 \cdot 50.23$$
$$d = 1,406.44 \text{ miles}$$

10. If the plane is airborne for 3 hours at 130 km/hr, then

$$d = r \cdot t$$
$$d = 130 \cdot 3$$
$$d = 390 \text{ kilometers}$$

... but that is total distance. The farthest away the plane could have gone is $\frac{1}{2}$ that distance, or 195 kilometers.

11. Before noon, the car drove for 3 hours, since it is 3 hours from 9 a.m. to noon. That means that in the morning the car drove

$$3 \cdot 40 = 120 \text{ miles}$$

That left 230 - 120, or 110 miles, to be covered in the afternoon:

$$d = r \cdot t$$
$$110 = 55 \cdot t$$
$$t = 2 \text{ hours}$$

12.

	d	=	r	$\cdot$	t
jog	$8t$		8		t
walk	$2\frac{1}{2}(3-t)$		$2\frac{1}{2}$		$3-t$

$$8t + 2\frac{1}{2}(3-t) = 13$$

$$8t + 7\frac{1}{2} - 2\frac{1}{2}t = 13$$

$$5\frac{1}{2}t = 5\frac{1}{2}$$

$$t = 1 \text{ hour jogging}$$

$$3 - t = 2 \text{ hours walking}$$

13.

	d	=	r	$\cdot$	t
downstream	48		24		t
upstream	48		$\dfrac{48}{5-t}$		$5-t$

If a round trip was 96 kilometers, one way must be $\dfrac{96}{2}$, or 48 kilometers. Time downstream is found by:

$$48 = 24 \cdot t$$

$$t = 2 \text{ hours}$$

That means it took $5 - 2 = 3$ hours to go upstream.

$$d = r \cdot t$$

$$48 = r \cdot 3$$

$$r = 16 \text{ km/hr}$$

14. Start out by ignoring the dog entirely. Considering the man and woman only, it is as if he is standing still, and she is walking toward him at 2 mph. She will then close the two mile gap in one hour.

$(d = rt; 2 = 2t; t = 1)$

Since the dog is running at 5 mph, it will continue to run for the 1 hour necessary for the woman to catch the man.

Dog's $d = 5t$; $d = 5 \cdot 1$; $d = 5$.

The dog will run 5 miles.

Problems Involving Percents

Problems involving percents generally fall into one of three categories. The first is the type of problem where it is necessary to take a percent of something in order to find a discounted price or the interest on a loan, the tax on a sale, etc. The second type of problem involving percents will tell you the cost of something both before and after a discount has been taken, and then ask you to determine what the percent discount was. The third and final type of percentage problem will tell you the amount of discount that was taken on an item, as well as the discounted price, and then require that you determine what the original price was before the discount was taken.

Percentage problems have a wide range of applications, most of them being in the world of business and finance. Examine the examples below carefully. If you are not certain of how to find a percent of a number, refer to "Finding a Percent of a Number" earlier in this chapter.

Example 1

The list price on a sport jacket is $84. It is being sold at a 23% discount. What is the sale price?

There are two ways in which to solve this type of problem. Both take into account the fact that there is an original price, a percent discount, and a reduction of the original price by the amount of the discount. In other words, both multiplication and subtraction are involved. The difference in the two approaches is the order in which those two operations are executed.

Solution 1:

First, figure the amount of the discount. In this case, it will be 23% of $84.

$$23\% \text{ of } \$84 = .23 \times 84 = \$19.32$$

Next, the discount must be subtracted from the list price:

$$\$84 - \$19.32 = \$64.68$$

Notice that the multiplication was done first, followed by the subtraction.

Solution 2:

This is a somewhat more interesting solution in that it is useful for other types of problems as well. It entails doing the subtraction first, and it illustrates a formula that may be used in many different discount situations. The formula is as follows:

$$S = L(1.00 - d)$$

In the formula, S stands for the sale price, L stands for the list price, and d stands for the amount of discount expressed as a decimal.

First consider the logic of the solution without the formula: If the discount is to be 23%, then the amount being paid for the jacket is 100% - 23%, or 77% of the list price. To find 77% of $84, multiply .77 × 84, and get $64.68.

Now, examine the formula, and you'll see that the 1.00 - d is responsible for subtracting the percent of the discount from 100%, in order to find the percent actually paid. That percent is then multiplied by the list price to get the sale price. Now see how the formula works for the same problem:

$$S = L(1.00 - d)$$
$$S = 84(1.00 - .23)$$
$$S = 84(.77)$$
$$S = \$64.68$$

Example 2

A dress on sale costs $30. Before it was marked down, it sold for $50. Find the percent of the discount.

Remember, a percent may be expressed as a ratio with a denominator of 100. We can therefore establish a proportion of the sale price to the original price versus the unknown percent to 100:

$$\frac{30}{50} = \frac{x}{100}$$
$$50x = 3000$$
$$x = 60$$

That means that 60% is the part of the original price that $30 is. If you are paying 60% of the original price, then the discount must have been 100% - 60%, or 40%.

This problem might also be solved by using the formula:

$$S = L(1.00 - d)$$

Substituting the two quantities that we know, we get:

$$30 = 50(1.00 - d)$$
$$30 = 50 - 50d$$
$$50d = 20$$
$$d = \frac{20}{50} = .40 = 40\%$$

Example 3

After a discount of 35%, a suit was selling for $130. What was its original selling price?

Here, we can once more use the formula that relates sale price, discount, and list price:

$$S = L(1.00 - d)$$

Since we know two of the three quantities involved, we should be able to substitute them into the formula and get the third.

$$130 = L(1.00 - .35)$$
$$130 = L(.65)$$
$$130 = .65L$$
$$L = \$200$$

Test Yourself

> **Directions:** Solve the following discount problems.

1. Alessandra wishes to buy a portable tape recorder that lists for $89.00. She sees it advertised as selling for 30% off. How much should she expect to pay for it?

2. Jason saw an ad for a dirt bike on sale. The ad offered a discount of 27% from the $342 price. How much money does he need?

3. A $620 stereo was on sale at Sam Baddy's for 45% off. What was the sale price?

4. Karen saw a $69 dress that had been reduced in price by 29%. Just as she was about to buy it, it was marked for final clearance and its sale price was reduced by 30%. How much did Karen pay for the dress?

5. Dylan's Computer Store deducted an additional 17% off the sale price of a Peach II Computer bought during the month of May. This was on top of the 26% discount that was already being offered. If a Peach II lists for $1195, how much did it cost to buy one at Dylan's during May?

6. A car that listed for $8000 was selling for $6000. What was the percent of the discount?

7. A dining room set was on sale for $245. It had been reduced from $300. What was the percent of the price reduction?

8. A $70,000 house had been reduced for quick sale. The sale price was $53,500. What was the percent of the price reduction?

9. James bought a $575 guitar for $310. What percent did he save?

10. A telephone answering machine was on sale for $204. It had been discounted by 37%. What was the answering machine's list price?

11. Boozer's Liquor Store advertised a bottle of Lafitte Rothschild 1968 vintage wine at $54 instead of the usual $81. By what percent had the bottle been discounted?

12. A rare stamp that normally sells for $240 was being advertised at $210 by Sticky Tongues, Inc. What percent had been taken off the stamp's normal price?

13. A set of stainless flatware was on sale at The Knife and Spoon for $36, instead of at its nationally advertised price of $45. By what percent had the set been discounted?

14. After an 18% discount, a sofa sold for $403. What was the sofa's list price?

15. A set of drums sold for $320 after it had been discounted by 20%. What was the list price of the drums?

16. Rosalie got a $33\frac{1}{3}\%$ discount on a set of encyclopedias. She paid $480 for the set. How much would she have had to pay if she had not received the discount?

17. After a 35% discount, a television set sells for $130. What was the set's list price?

18. A $56 radio was on sale for 30% less. After Melissa bought the radio and paid 8% sales tax on it, how much change did she get from a $50 bill?

19. Dale got a 32% discount on a $43 blouse. She paid 7% sales tax. How much money did she have to give the cashier?

Answers

1.
$S = L(1.00 - d)$
$S = 89(.70)$
$S = \$62.30$

2.
$S = 342(.73)$
$S = \$249.66$

3.
$S = 620(.55)$
$S = \$341$

4.
$S = 69(.71)$
$S = \$48.99 \ldots$

But that is only the initial sale price. To figure the clearance price:

$S = 48.99(.70)$
$S = \$34.29$

5.
$S = 1195(.74)$
$S = \$884.30 \ldots$

before the special May discount. To figure the final price:

$S = 884.30(.83)$
$S = \$733.97$

6.
$S = L(1.00 - d)$
$6000 = 8000(1.00 - d)$
$6000 = 8000 - 8000d$
$8000d = 2000$
$d = \dfrac{2000}{8000} = \dfrac{1}{4} = .25$
$d = 25\%$

7.
$245 = 300(1.00 - d)$
$245 = 300 - 300d$
$300d = 55$
$d = .1833 = 18\dfrac{1}{3}\%$

8.
$53,500 = 70,000(1.00 - d)$
$53,500 = 70,000 - 70,000d$
$70,000d = 16,500$
$d = .2357 = 23.6\%$

9.
$310 = 575(1.00 - d)$
$310 = 575 - 575d$
$575d = 265$
$d = .4608 = 46.1\%$

10.
$S = L(1.00 - d)$
$204 = L(1.00 - .37)$
$204 = L(0.63)$
$.63L = 204$
$L = \$323.81$

11.
$54 = 81(1.00 - d)$
$54 = 81 - 81d$
$81d = 27$
$d = \dfrac{1}{3} = .333 = 33\dfrac{1}{3}\%$

12.
$210 = 240(1.00 - d)$
$210 = 240 - 240d$
$240d = 30$
$d = \dfrac{30}{240} = \dfrac{1}{8} = .125 = 12\dfrac{1}{2}\%$

13.
$36 = 45(1.00 - d)$
$36 = 45 - 45d$
$45d = 9$
$d = \dfrac{1}{5} = .20 = 20\%$

14.
$S = L(1.00 - d)$
$403 = L(1.00 - .18)$
$403 = L(.82)$
$.82L = 403$
$L = \$491.46$

15.
$320 = L(1.00 - .20)$
$320 = L(.80)$
$.80L = 320$
$L = \$400$

16.
$480 = L(1.00 - .333)$
$480 = L(.667)$
$.667L = 480$
$L = \$719.64$

17.
$130 = L(1.00 - .35)$
$130 = L(.65)$
$.65L = 130$
$L = \$200$

18. First figure the sale price:

$S = 56(1.00 - .30)$

$S = 56(.70)$

$S = \$39.20$

Then take 8% of the sale price ...

$.08(39.20) = t$

$t = 3.136 = \$3.14$

... and add it to the sale price:

$\$39.20 + 3.14 = \$42.34$

Her change is:

$\$50 - 42.34 = \$7.66$

19.

$S = L(1.00 - d)$

$S = 43(1.00 - .32)$

$S = 43(.68)$

$S = \$29.24$

$t = (29.24)(.07)$

$t = \$2.05$

Total $= 29.24 + 2.05 = \$31.29$

Interest Problems

Another type of percent problem deals with simple interest. Interest may be figured on money in a savings account, on stocks and bonds, on a mortgage, or on a loan. Remember, though, simple interest assumes that interest is being paid on the amount that was originally in the account (the principal). It does not take into account interest earned on interest.

The standard formula for figuring simple interest is:

$I = prt$ I stands for interest.

p stands for principal (upon which the interest is paid).

r is the rate at which the interest is paid (a percent).

t is time (usually in years).

Example 1

Andrew put $5000 into a savings certificate at 15% annual interest. What was the total amount of money in his account at the end of a year?

$I = prt$

$I = 5000 \times .15 \times 1$

$I = 750$ We have found that Andrew will receive $750 interest.

$5000 + 750 = \$5750$ Add the interest to the principal to find the total amount in his account

Example 2

Mr. Appleby invested part of $18,000 at 10% and the rest at 20%. His total income was $2400. How much did he invest at each rate?

Let p = the amount invested at 10%.

$18000 - p$ = the amount invested at 20%.

Since no time period was stated, we may assume that one year is the time involved.

The interest earned by the amount at 10% will be $.10(p)$.

The interest earned at 20% will be $.20(18,000 - p)$.

Since we know that the total interest earned was $2400, we write an equation that adds the interest earned at each rate together to get the total interest:

$$.10p + .20(18,000 - p) = 2400$$
$$.10p + 3600 - .20p = 2400$$
$$-.10p = -1200$$
$$.10p = 1200$$
$$p = 12,000 \qquad \text{That means \$12,000 was invested at 10\%.}$$
$$18,000 - p = 6,000 \qquad \text{So \$6,000 was invested at 20\%.}$$

When you work the practice tests later in this book, you will notice that some of the problems may present interest problems like the above, but, instead of asking you for the solution, will just ask you for the equation that *could be solved* to give you the solution. In such a problem, the answer would be the equation $.10p + .20(18,000 - p) = 2400$. As you will see, you do not need to solve the equation to answer such a problem. Problems like this are most likely to occur on the non-calculator section.

Test Yourself

Directions: Solve the following interest problems.

1. Find the annual interest on $5000 if the rate of interest is:

   a) 5%

   b) $12\frac{1}{2}\%$

   c) 13%

   d) $7\frac{1}{2}\%$

   e) 19%

2. Ms. Bernbach invested $12,000. Part of it was invested at 12% and part at $15\frac{1}{2}\%$. Suppose that p represents the amount invested at 12%. Represent:

   a) the amount invested at $15\frac{1}{2}\%$

   b) the interest on the amount invested at $15\frac{1}{2}\%$

   c) the interest on the amount invested at 12%

3. Sue Tree invested x dollars at 9%, and $900 more than that at 12%. The total interest from both investments was $1200. How much did she invest at each rate?

4. F. Rugal placed $10,000 into two accounts. The first paid 14% interest, and the second, 16%. After a year, Mr. Rugal collected $1520 interest. How much money was invested in each account?

5. Scott invested a sum of money at 9% and a second sum at 18%. The second amount was $450 less than the first. He received $162 annually from the investments. How much was placed at each rate?

6. Lynn bought 12% municipal bonds and 15% company bonds. The amount invested in the company bonds was $600 greater than that in the municipals. If her total annual income was $390, how much was invested in each type?

7. Rose Frescas put a sum of money into a money-market certificate yielding 9% interest. She put twice as much into a 12% bond. Her annual income from her investments was $870. How much was invested at each rate?

8. After receiving a large inheritance, Lucy Stiff invested $50,000. Part of it was invested at 8% and part at 14%. She received $5890 at the end of a year. Find the amount that she invested at each rate.

9. Andre invested $20,000 in two parts. The first part was put into an investment yielding 10%, while the second part went into 13% bonds. The total annual return was $2150. How much was invested at each rate?

10. Joyce bought 12% bonds and 15% debentures. She invested a total of $6000. If the annual interest that she received on both investments was the same, find the amount she invested at each rate.

11. David had two small bank accounts totaling $500. One of them paid 5% interest, while the other paid $6\frac{1}{4}$%. After a year, he received interest from both accounts totaling $30. How much was in each account?

Answers

1. a) $5000 \times .05 = $250

   b) $5000 \times .125 = $625

   c) $5000 \times .13 = $650

   d) $5000 \times .075 = $375

   e) $5000 \times .19 = $950

2. a) $12,000 - p$

   b) $.155(12,000 - p)$

   c) $.12p$

3. $$.09x + .12(x + 900) = 1200$$
   $$.09x + .12x + 108 = 1200$$
   $$.21x = 1092$$
   $$x = $5200 \text{ at } 9\%$$
   $$x + 900 = $6100 \text{ at } 12\%$$

4. $$.14p + .16(10,000 - p) = 1520$$
   $$.14p + 1600 - .16p = 1520$$
   $$-.02p = -80$$
   $$p = $4000 \text{ at } 14\%$$
   $$10,000 - p = $6000 \text{ at } 16\%$$

5.
$$.09p + .18(p - 450) = 162$$
$$.09p + .18p - 81 = 162$$
$$.27p = 243$$
$$p = \$900 \text{ at } 9\%$$
$$p - 450 = \$450 \text{ at } 18\%$$

6.
$$.12p + .15(p + 600) = 390$$
$$.12p + .15p + 90 = 390$$
$$.27p = 300$$
$$p = \$1111.11$$
$$\text{in municipals}$$
$$p + 600 = \$1711.11$$
$$\text{in companies}$$

7.
$$.09p + .12(2p) = 870$$
$$.09 + .24p = 870$$
$$.33p = 870$$
$$p = \$2636.36 \text{ at } 9\%$$
$$2p = \$5272.72 \text{ at } 12\%$$

8.
$$.08p + .14(50,000 - p) = 5890$$
$$8p + 14(50,000 - p) = 589,000$$

(Multiplying through by 100 gets rid of the decimals and makes computation easier.)

$$8p + 700,000 - 14p = 589,000$$
$$^-6p = ^-111,000$$
$$p = \$18,500 \text{ at } 8\%$$
$$50,000 - p = \$31,500 \text{ at } 14\%$$

9.
$$.10p + .13(20,000 - p) = 2150$$
$$.10p + 2600 - .13p = 2150$$
$$^-.03p = ^-450$$
$$p = \$15,000 \text{ at } 10\%$$
$$20,000 - p = \$5000 \text{ at } 13\%$$

10.
$$.12p = .15(6000 - p)$$
$$.12p = 900 - .15p$$
$$.27p = 900$$
$$p = \$3333.33 \text{ at } 12\%$$
$$6000 - p = \$2666.67 \text{ at } 15\%$$

11.
$$.05p + .0625(500 - p) = 30$$
$$.05p + 31.25 - .0625p = 30$$
$$^-.0125p = ^-1.25$$
$$p = \$100 \text{ at } 5\%$$
$$500 - p = \$400 \text{ at } 6\frac{1}{4}\%$$

Coin and Stamp Problems

Coin and stamp problems are useful in laying the foundation for another class of word problems—mixture problems. The main thing to bear in mind is that two quantities are being dealt with: the number of coins or stamps, and the value of each. A diagram is most helpful when it comes to organizing the data for each problem. Remember that the number of coins multiplied by the value of each coin gives the total value of the coins. In order to eliminate the need for decimal multiplication and division, it is helpful to express the value of each coin in cents, i.e. a nickel which is .05 dollars would normally have its value expressed as 5 cents. $3.20 would be written as 320 cents.

Since it is necessary to use only one variable in an equation, it is helpful to express the quantity of each type of coin in terms of the coin upon which the other numbers in the problem are based. Examine the **Example**s, and you'll see what we mean.

Example 1

Jason had a collection of dimes, nickels, and quarters in a jar. He had twice as many nickels as dimes, and four more quarters than nickels. Altogether, he had $3.80. How many coins of each type did he have?

First of all, notice that the number of quarters is expressed in terms of the number of nickels, but the number of nickels is expressed in terms of the number of dimes. Dimes, then, are the coins upon which the situation is based.

Let x = the number of dimes

$2x$ = the number of nickels

$2x + 4 =$ the number of quarters

	Number of coins	Value of each	Total value
dimes	x	10	$10x$
nickels	$2x$	5	$5(2x)$
quarters	$2x + 4$	25	$25(2x + 4)$

Notice that a dime is worth 10 cents, a nickel, 5 cents, and a quarter, 25 cents.

Notice too that total value is determined by multiplying the number of each type of coin by the value of each.

The equation is formed primarily from the last column of the diagram. In essence, it states that if you add the total value of dimes to the total value of nickels and the total value of quarters, you end up with the total amount of money in the jar:

$$10x + 5(2x) + 25(2x + 4) = 380 \qquad \text{(Note that \$3.80 is 380 cents.)}$$
$$10x + 10x + 50x + 100 = 380$$
$$70x + 100 = 380$$
$$70x = 280$$
$$x = 4 \qquad\qquad \text{Now refer back to the let statement.}$$
$$x = 4 \text{ dimes}$$
$$2x = 8 \text{ nickels}$$
$$2x + 4 = 12 \text{ quarters}$$

It never hurts to check the answer, so:

$$4 \text{ dimes} = \quad .40$$
$$8 \text{ nickels} = \quad .40$$
$$12 \text{ quarters} = \quad \underline{3.00}$$
$$\$3.80 \qquad \text{which is what Jason started with.}$$

Example 2

The local high school held a carnival and charged 75¢ admission for adults and 30¢ for children under 12. Twice as many children as adults were admitted, with the total receipts coming to $135. How many tickets of each type were sold?

Let x = the number of adult tickets sold

$2x$ = the number of children's tickets

	# of tickets	Unit value	Total value
adult	x	75	$75x$
children	$2x$	30	$30(2x)$

Once more, the total value of the children's tickets combined with the total value of the adult's tickets gives the total receipts:

$$75x + 30(2x) = 13,500 \qquad (\$135 = 13,500\text{¢})$$
$$75x + 60x = 13,500$$
$$135x = 13,500$$
$$x = 100$$

There were 100 adult and 200 children's tickets.

Check

$$100 \times .75 = \$75.00$$
$$200 \times .30 = \underline{60.00}$$
$$\$135.00$$

Test Yourself

Directions: Solve the following coin and stamp problems.

1. Represent the value of each of the following in cents.

   a) 5 nickels
   b) 6 quarters
   c) 9 dimes
   d) 3 dimes and 7 nickels
   e) 5 quarters and 7 dimes
   f) 3 quarters, 5 dimes, 9 nickels
   g) 6 dollars
   h) 12 dollars
   i) $5.79
   j) $6.75

2. Represent the value of each of the following in cents.

   a) x nickels, $3x$ dimes
   b) y quarters, $4y$ nickels
   c) 5 quarters and $9n$ nickels
   d) r nickels and $(r + 3)$ dimes
   e) $(x + 7)$ quarters and $(2x - 3)$ dimes
   f) $5z$ dimes, $2z$ nickels, and $(3 - z)$ quarters

Directions: Make a let statement and a diagram to organize the data for each. Then solve.

3. Anthony has a can full of coins. There are six times as many quarters as nickels and four more dimes than quarters. Altogether, the can contains $11.15. How many coins of each type are in the can?

4. A vending machine operator is counting the change taken in from a single candy machine. He finds that he has 43 quarters, twice as many nickels, and 4 more than three times as many dimes as nickels. How much money was in the machine?

5. A vending machine contains $4.20 in dimes and nickels. Altogether there are 61 coins. How many coins of each type were in the machine?

6. Mary buys 20¢ regular stamps and 35¢ airmail stamps at the post office. Altogether she spends $6.20 and receives 22 stamps. How many of each denomination did she buy?

7. Marge paid a grocery bill of $9.20 with quarters, dimes, and nickels. The number of nickels was 6 less than the number of dimes, and there were 10 more dimes than quarters. How many of each type of coin were there?

8. A post office clerk sold 75 stamps for $14.40. If some of the stamps were 15¢ ones, and the others were 24¢ stamps, how many of each kind did she sell?

9. Henry has $8.20 in dimes and quarters. He has 16 more quarters than he has dimes. How many coins of each type does he have?

10. Karen put $11.50 into her savings account. There were 15 more dimes than nickels and 48 fewer quarters than nickels. How many coins of each type did Karen deposit?

11. Alan collected $7.15 for the volunteer fire department. There were 46 coins altogether, some of which were quarters, and some dimes. How many coins of each type did he collect?

12. Mema took $6.30 to the grocery store in nickels, dimes, and quarters. She had three more dimes than quarters, and three fewer nickels than dimes. How many coins of each type did she have?

Answers

1. a) 25¢   b) 150¢   c) 90¢   d) 65¢   e) 195¢

   f) 170¢   g) 600¢   h) 1200¢   i) 579¢   j) 675¢

2. a) $5x + 30x = 35x$¢

   b) $\cdot 25y + 20y = 45y$¢

   c) $(125 + 45n)$ ¢

   d) $5r + 10(r + 3) = 5r + 10r + 30 = 15r + 30$¢

   e) $25(x + 7) + 10(2x - 3) = 25x + 175 + 20x - 30 = (45x + 145)$ ¢

   f) $10(5z) + 5(2z) + 25(3 - z) = 50z + 10z + 75 - 25z = (35z + 75)$ ¢

3.   Let x =# of nickels

        $6x$ =# of quarters

     $6x + 4$ =# of dimes

Type of coin	# of each	Value of each	Total value
Nickels	x	5	$5x$
Dimes	$6x + 4$	10	$10(6x + 4)$
Quarters	$6x$	25	$25(6x)$

$$5x + 10(6x + 4) + 25(6x) = 1115$$
$$5x + 60x + 40 + 150x = 1115$$
$$215x = 1075$$
$$x = 5 \text{ nickels}$$
$$6x = 30 \text{ quarters}$$
$$6x + 4 = 34 \text{ dimes}$$

4.

Type of coin	# of each	Value of each	Total value
Nickels	2(43)	5	5[2(43)]
Dimes	3[2(43)] + 4	10	10{3[2(43)] + 4}
Quarters	43	25	25(43)

Let v = total value

$$v = 5[2(43)] + 10\{3[2(43)] + 4\} + 25(43)$$
$$v = 5(86) + 10[3(86) + 4] + 1075$$
$$v = 430 + 10(258 + 4) + 1075$$
$$v = 430 + 2620 + 1075$$
$$v = 4125 = \$41.25$$

5.   Let n = # of nickels

$61 - n$ = # of dimes

Type of coin	# of each	Value of each	Total value
Nickels	n	5	$5n$
Dimes	$61 - n$	10	$10(61 - n)$

$$5n + 10(61 - n) = 420$$
$$5n + 610 - 10n = 420$$
$$-5n = -190$$
$$n = 38 \text{ nickels}$$
$$61 - n = 23 \text{ dimes}$$

6.   Let a = # of 35¢ airmail stamps

$22 - a$ = # of 20¢ stamps

Type of stamp	# of each	Value of each	Total value
Regular	$22 - a$	20	$20(22 - a)$
Airmail	a	35	$35a$

$$20(22 - a) + 35a = 620$$
$$440 - 20a + 35a = 620$$
$$15a = 180$$
$$a = 12 \text{ airmail stamps}$$
$$22 - a = 10 \text{ regular stamps}$$

7.   Let x =# of quarters

$x + 10$ =# of dimes

$x + 4^*$ =# of nickels

Type of coin	# of each	Value of each	Total value
Nickels	$x + 4$	5	$5(x + 4)$
Dimes	$x + 10$	10	$10(x + 10)$
Quarters	x	25	$25x$

$$5(x + 4) + 10(x + 10) + 25x = 920$$
$$5x + 20 + 10x + 100 + 25x = 920$$
$$40x + 120 = 920$$
$$40x = 800$$
$$x = 20 \text{ quarters}$$
$$x + 4 = 24 \text{ nickels}$$
$$x + 10 = 30 \text{ dimes}$$

8.   Let x =# of 15¢ stamps

$75 - x$ =# of 24¢ stamps

Type of stamp	# of each	Value of each	Total value
15¢	x	15	$15x$
24¢	$75 - x$	24	$24(75 - x)$

$$15x + 24(75 - x) = 1440$$
$$15x + 1800 - 24x = 1440$$
$$^-9x + 1800 = 1440$$
$$^-9x = ^-360$$
$$x = 40 \text{ 15¢ stamps}$$
$$75 - x = 35 \text{ 24¢ stamps}$$

9. Let x =# of dimes

$x + 16$ =# of quarters

Type of coin	# of each	Value of each	Total value
Dimes	x	10	$10x$
Quarters	$x + 16$	25	$25(x + 16)$

*$[(x + 10) - 6] = x + 4$

$$10x + 25(x + 16) = 820$$
$$10x + 25 + 400 = 820$$
$$35x + 400 = 820$$
$$35x = 420$$
$$x = 12 \text{ dimes}$$
$$x + 16 = 28 \text{ quarters}$$

10. Let x =# of nickels

$x + 15$ =# of dimes

$x - 48$ =# of quarters

Type of coin	# of each	Value of each	Total value
Nickels	x	5	$5x$
Dimes	$x + 15$	10	$10(x + 15)$
Quarters	$x - 48$	25	$25(x - 48)$

$$5x + 10(x + 15) + 25(x - 48) = 1150$$
$$5x + 10x + 150 + 25x - 1200 = 1150$$
$$40x - 1050 = 1150$$
$$40x = 2200$$
$$x = 55 \text{ nickels}$$
$$x + 15 = 70 \text{ dimes}$$
$$x - 48 = 7 \text{ quarters}$$

11. Let x =# of quarters

$46 - x$ =# of dimes

Type of coin	# of each	Value of each	Total value
Dimes	$46 - x$	10	$10(46 - x)$
Quarters	x	25	$25x$

$$10(46 - x) + 25x = 715$$
$$460 - 10x + 25x = 715$$
$$15x = 255$$
$$x = 17 \text{ quarters}$$
$$46 - x = 29 \text{ dimes}$$

12.   Let $x = $ # of quarters

$x + 3 = $ # of dimes

$(x + 3) - 3 = x = $ # of nickels

Type of coin	# of each	Value of each	Total value
Nickels	x	5	$5x$
Dimes	$x + 3$	10	$10(x + 3)$
Quarters	x	25	$25x$

$$5x + 10(x + 3) + 25x = 630$$
$$5x + 10x + 30 + 25x = 630$$
$$40x + 30 = 630$$
$$40x = 600$$
$$x = 15 \text{ nickels}$$
$$x = 15 \text{ quarters}$$
$$x + 3 = 18 \text{ dimes}$$

Mixture Problems

At the beginning of the last section it was noted that coin and stamp problems serve as a lead-in to mixture problems. After all, very few people are likely to wonder how many quarters, dimes, and nickels they have in a jar or purse. All they are really concerned with is how much they can buy with the total. Nonetheless, coin and stamp problems serve to introduce the notion that the value of each item times the number of items is a useful way of calculating the total value of purchases. This approach is essential to a successful experience with mixture problems.

The following examples will serve to acquaint you with the way in which mixture problems are solved, as well as different types of mixture problems. Once more, a diagram is used to organize the data and to help you construct the equation. Let statements, as always, are essential.

Example 1

A supermarket manager wishes to mix raisins worth $1.29 per pound with almonds worth $2.20 per pound in order to make 35 pounds of a mixture worth $1.81 per pound. How many pounds of each must he use?

Let $x = $ the number of pounds of raisins

$35 - x = $ the number of pounds of almonds

Ingredients	# of pounds	Price/pound (in cents)	Total value
raisins	x	129	$129x$
almonds	$35 - x$	220	$220(35 - x)$
mixture	35	181	$181(35)$

Notice that since there are 35 pounds altogether, and we have arbitrarily decided that x would represent the number of pounds of raisins, then $35 - x$ (what is left when x pounds are removed from 35 pounds) represents the number of pounds of almonds.

The equation comes from the total value of the parts of the mixture being added together to find the total value of the mixture itself:

$$129x + 220(35 - x) = 181(35)$$
$$129x + 7700 - 220x = 6335$$
$$^-91x = ^-1365$$
$$x = 15 \quad \text{pounds of raisins}$$
$$35 - x = 20 \quad \text{pounds of almonds}$$

Example 2

How many kilograms of pretzels worth $1.05 per kilogram must be mixed with 45 kilograms of potato chips worth $1.45 per kilogram, in order to make a mixture which will sell for $1.15 per kilogram?

Let x = # of kilograms of pretzels

Ingredients	Amt. of each	Price (in ¢)	Total value
pretzels	x	105	$105x$
potato chips	45	145	145 (45)
mixture	$45 + x$	115	115 (45 + x)

Once more (as always), the equation is derived by combining the total value of the components and setting that equal to the value of the mixture:

$$105x + 145(45) = 115(45 + x)$$
$$105x + 6525 = 5175 + 115x$$
$$10x = 1350$$
$$x = 135 \text{ kilograms of pretzels}$$

Test Yourself

Directions: Solve the following mixture problems.

1. A certain type of nut is worth $3 per pound. What is the value of

   a) 6 pounds?   b) 2 pounds?   c) 5 pounds?   d) x pounds?   e) (5 - x) pounds?

2. How many pounds of $2.40 per pound coffee must be mixed together with $3.48 per pound coffee in order for the roaster to sell 360 pounds of the mixture for $2.76 per pound?

3. How many gallons of $1.20 gasoline must be mixed with $1.45 per gallon gasoline to get 500 gallons of gasoline that will sell for $1.35 per gallon?

4. A baker has cookies worth 90¢ per pound and cookies worth $1.35 per pound. How many pounds of each should he mix together in order to get 30 pounds of a mixture that will sell for $1.05 per pound?

5. How many tons of tea that sells for $345 per ton must be mixed with 15 tons of tea selling for $270 per ton in order to get a blend that can be wholesaled at $295 per ton?

6. How many kilograms of $4.20 per kilogram cattle fodder must be mixed with 310 kilograms of $2.40 per kilogram fodder in order to get a mixture worth $3.33 per kilogram?

7. A candy-maker wishes to make an assortment from $3.89 per pound filled chocolates and $4.38 per pound covered nuts. How many pounds of the filled chocolates must he combine with 13 pounds of the nuts in order to get an assortment that will be valued at $4.19? (Express your answer to the nearest pound.)

8. A landscape supply store owner mixes rye grass seed at $2.07 per ounce with Kentucky blue grass at $3.45 per ounce. He mixes them in the ratio of 4:7. How much should the mixture sell for?

9. A distiller combines 500 gallons of neutral grain spirits worth $3.59 per gallon with 1200 gallons of rye whiskey worth $7.38 per gallon. How much per gallon should he charge the bottler in order to recognize a 20% profit?

10. A canner of fruit cocktail combines pears, peaches, and pineapple in the ratio 5:3:1. If pears cost the canner $.49 per pound, peaches $.65 per pound, and pineapple $1.12 per pound, how much should the canner charge per pound of fruit cocktail in order to recognize a 35% profit?

11. Sesame seeds cost $.76 per ounce, cashews $5.13 per pound, and raisins, $1.68 per pound. A health food store makes a mix of the three by weight in the ratio 1:3:4. How much should the store charge for $\frac{1}{2}$ pound of the mixture if it is to make a 15% profit on each sale?

Answers

1. a) $18   b) $6   c) $15   d) $3x$ e) $15 - 3x$

2.   Let $x =$ # of pounds of $2.40 coffee
$360 - x =$ # of pounds of $3.48 coffee

Ingredients	Amt. of each	Price (¢)	Total value
$2.40 coffee	x	240	$240x$
$3.48 coffee	$360 - x$	348	$348(360 - x)$
Mixture	360	276	$360(276)$

$$240x + 348(360 - x) = 360(276)$$
$$240x + 125,280 - 348x = 99,360$$
$$^-108x + 125,280 = 99,360$$
$$^-108x = ^-25,920$$
$$x = 240 \text{ pounds of } \$2.40 \text{ coffee}$$
$$360 - x = 120 \text{ pounds of } \$3.48 \text{ coffee}$$

3.   Let $x =$ # of gallons of $1.20 gasoline
$500 - x =$ # of gallons of $1.45 gasoline

Ingredients	Amt. of each	Price (¢)	Total value
$1.20 gasoline	x	120	$120x$
$1.45 gasoline	$500 - x$	145	$145(500 - x)$
Mixture	500	135	$135(500)$

$$120x + 145(500 - x) = 135(500)$$
$$120x + 72,500 - 145x = 67,500$$
$$^-25x = ^-5000$$
$$x = 200 \text{ gallons @ } \$1.20$$
$$500 - x = 300 \text{ gallons @ } \$1.45$$

4.   Let $x =$ # of pounds of 90¢ cookies
$30 - x =$ # of pounds of $1.35 cookies

Ingredients	Amt. of each	Price (¢)	Total value
90¢ cookies	x	90	$90x$
$1.35 cookies	$30 - x$	135	$135(30 - x)$
Mixture	30	105	$30(105)$

$$90x + 135(30 - x) = 30(105)$$
$$90x + 4050 - 135x = 3150$$
$$-45x = -900$$
$$x = 20 \text{ pounds of } 90¢ \text{ cookies}$$
$$30 - x = 10 \text{ pounds of } \$1.35 \text{ cookies}$$

5.   Let $x = $# of \$345 per ton of tea

Then $x + 15 = $# of tons of mixture

Ingredients	Amt. of each	Price ($)	Total value
$345 per ton of tea	x	345	$345x$
$270 per ton of tea	15	270	15 (270)
Mixture	$x + 15$	295	295 $(x + 15)$

$$345x + 15(270) = 295(x + 15)$$
$$345x + 4050 = 295x + 4425$$
$$50x = 375$$
$$x = 7\frac{1}{2} \text{ tons @ } \$345$$

6.   Let $x = $# of kilograms of \$4.20 fodder

$x + 310 = $# of kilograms of mixture

Ingredients	Amt. of each	Price (¢)	Total value
$4.20 per kilogram of fodder	x	420	$420x$
$2.40 per kilogram of fodder	310	240	240 (310)
Mixture	$x + 310$	333	333 $(x + 310)$

$$420x + 240(310) = 333(x + 310)$$
$$420x + 74,400 = 333x + 103,230$$
$$87x = 28,830$$
$$x = 331.38 \text{ kilograms}$$

7. Let x =# of pounds of chocolate

$x + 13$ =# of pounds of assortment

Ingredients	Amt. of each	Price (in ¢)	Total value
chocolates	x	389	$389x$
nuts	13	438	$13(438)$
assortment	$x + 13$	419	$419(x + 13)$

$$389x + 13(438) = 419(x + 13)$$
$$389x + 5694 = 419x + 5447$$
$$^-30x = ^-247$$
$$x = 8.23$$
$$x = 8 \text{ pounds}$$

8. If the grass seeds are mixed in the ratio 4:7, there are a total of 11 parts to the mixture $(4 + 7 = 11)$. $\dfrac{4}{11}$ is rye @ \$2.07 and $\dfrac{7}{11}$ is bluegrass @ \$3.45.

Let x = the price of the mixture

Then $\dfrac{4}{11}$ of the price of rye plus $\dfrac{7}{11}$ of the price of bluegrass should be the price of the mixture:

$$x = \frac{4}{11}(207) + \frac{7}{11}(345)$$
$$x = 75.27 + 219.55$$
$$x = 294.82 \text{ cents}$$
$$x = \$2.95 \text{ (rounded to the nearest cent)}$$

9. 500 gallons of neutral grain spirits @ \$3.59 per gallon are worth a total of $500(3.59) = \$1795$.

1200 gallons of rye @ \$7.38 per gallon are worth a total of $1200(7.38) = \$8856$.

That means that the total value of the goods is \$1795 + \$8856 = \$10,651. 20% of that total is found by multiplying .20 (\$10,651) = \$2130.20.

Add that markup to the total value and you will get the amount for which the entire batch must be sold in order to make a 20% profit: \$2130.20 + \$10,651 = \$12,781.20.

To find what the distiller must charge the bottler, divide the total (\$12,781.20) by the total number of gallons (1700).

$$\frac{\$12,781.20}{1700} = \$7.52 \text{ (to the nearest penny)}$$

10. The ratio of the parts of the fruit cocktail is 5:3:1, which means that there are 9 parts altogether $(5 + 3 + 1)$; $\frac{5}{9}$ is pears, $\frac{3}{9}$ is peaches, and $\frac{1}{9}$ is pineapple. Accordingly, the price of the mix must be

$$\frac{5}{9}(.49) + \frac{3}{9}(.65) + \frac{1}{9}(1.12)$$

The mixture therefore cost the canner .2722 + .2167 + .1244 = .6133, or 61¢ per pound.

A 35% profit on \$.61 is calculated by finding 35% of \$.61 and then adding it onto the cost:

$.61(.35) = .214$, or 21¢. The canner should charge 82¢.

11. Since all units must be the same in order to compare them, we must first either convert the price of sesame seeds to a per pound value, or convert the other two to per ounce prices. The former is easier. Since there are 16 ounces in a pound, one pound of sesame seeds costs 16(.76) or \$12.16. If the mixture were prepared with 1 pound of sesame seeds, 3 pounds of cashews, and 4 pounds of raisins, it would cost 12.16(1) + 5.13(3) + 1.68(4), or \$34.27 to prepare 8 pounds

of the mixture. To find the cost of $\frac{1}{2}$ pound, divide the price of 8 pounds by 16. That gives a cost of \$2.14 per half-pound. The price, after 15% was added on, would be \$2.46.

PROBABILITY

While probability is not a major topic on the GED, you may have a small number of problems that ask you to compute some very basic probabilities. The section below tells you everything that you need to know about probability and the GED.

Probability is the branch of mathematics that gives you techniques for dealing with uncertainties. Intuitively, probability can be thought of as a numerical measure of the likelihood, or the chance, that an event will occur.

A probability value is always a number between 0 and 1. The nearer a probability value is to 0, the more unlikely the event is to occur; a probability value near 1 indicates that the event is almost certain to occur. Other probability values between 0 and 1 represent varying degrees of likelihood that an event will occur.

In the study of probability, an *experiment is* any process that yields one of a number of well-defined outcomes. By this, we mean that on any single performance of an experiment, one and only one of a number of possible outcomes will occur. Thus, tossing a coin is an experiment with two possible outcomes: heads or tails. Rolling a die is an experiment with six possible outcomes; playing a game of hockey is an experiment with three possible outcomes (win, lose, or tie).

In some experiments, all possible outcomes are equally likely. In such an experiment, with, say, n possible outcomes, we assign a probability of $\frac{1}{n}$ to each outcome. Thus, for example, in the experiment of tossing a fair coin, for which there are two equally likely outcomes, we would say that the probability of each outcome is $\frac{1}{2}$. In the experiment of tossing a fair die, for which there are six equally likely outcomes, we would say that the probability of each outcome is $\frac{1}{6}$.

How would you determine the probability of obtaining an even number when tossing a die? Clearly, there are three distinct ways that an even number can be obtained: tossing a 2, a 4, or a 6. The probability of each one of these three outcomes is $\frac{1}{6}$. The probability of obtaining an even number is simply the sum of the probabilities of these three favorable outcomes; that is to say, the probability of tossing an even number is equal to the probability of tossing a 2, plus the probability of tossing a 4, plus the probability of tossing a 6, which is $\frac{1}{6} + \frac{1}{6} + \frac{1}{6} = \frac{3}{6} = \frac{1}{2}$.

This result leads us to the fundamental formula for computing probabilities for events with equally likely outcomes:

The probability of an event occurring $= \dfrac{\text{The number of favorable outcomes}}{\text{The total number of possible outcomes}}$

In the case of tossing a die and obtaining an even number, there are six possible outcomes, three of which are favorable, leading to a probability of $\frac{3}{6} = \frac{1}{2}$.

Example 1

What is the probability of drawing one card from a standard deck of 52 cards and having it be a king? When you select a card from a deck, there are 52 possible outcomes, 4 of which are favorable. Thus, the probability of drawing a king is $\frac{4}{52} = \frac{1}{13}$.

Example 2

What is the probability of drawing one card from a standard deck of 52 cards and having it be a red face card? When you select a card from a deck, once again, there are 52 possible outcomes, and 6 of them are favorable (jack of hearts, queen of hearts, king of hearts, jack of diamonds, queen of diamonds, king of diamonds). Thus, the probability of drawing a red face card is $\frac{6}{52} = \frac{3}{26}$.

Two events are said to be *independent* if the occurrence of one does not affect the probability of the occurrence of the other. For example, if a coin is tossed and a die is thrown, obtaining heads on the coin and obtaining a 5 on the die are independent events. On the other hand, if a coin is tossed three times, the probability of obtaining heads on the first toss and the probability of obtaining tails on all three tosses are not independent. In particular, if heads is obtained on the first toss, the probability of obtaining three tails becomes 0.

When two events are independent, the probability that they both happen is the product of their individual probabilities. For example, the probability of obtaining heads when a coin is tossed is $\frac{1}{2}$, and the probability of obtaining 5 when a die is thrown is $\frac{1}{6}$; thus, the probability of both of these events happening is $\left(\frac{1}{2}\right)\left(\frac{1}{6}\right) = \frac{1}{12}$

In a situation where two events occur one after the other, be sure to correctly determine the number of favorable outcomes and the total number of possible outcomes.

Example 3

Consider a standard deck of 52 cards. What is the probability of drawing two kings in a row, if the first card drawn is replaced in the deck before the second card is drawn? What is the probability of drawing two kings in a row if the first card drawn is *not* replaced in the deck?

In the first case, the probability of drawing a king from the deck on the first attempt is $\frac{4}{52} = \frac{1}{13}$. If the selected card is replaced in the deck, the probability of drawing a king on the second draw is also $\frac{1}{13}$, and, thus, the probability of drawing two consecutive kings would be $\left(\frac{1}{13}\right)\left(\frac{1}{13}\right) = \frac{1}{169}$. On the other hand, if the first card drawn is a king and is not replaced, there are now only three kings in a deck of 51 cards, and the probability of drawing the second king becomes $\frac{3}{51} = \frac{1}{17}$. The overall probability, thus, would be $\left(\frac{1}{13}\right)\left(\frac{1}{17}\right) = \frac{1}{221}$.

Test Yourself

Directions: Solve the following.

1. A bag contains 7 blue marbles, three red marbles, and two white marbles. If one marble is chosen at random from the bag, what is the probability that it will be red? What is the probability that it will not be blue?

2. A woman's change purse contains a quarter, two dimes, and two pennies. What is the probability that a coin chosen at random will be worth at least ten cents?

3. A bag contains four white and three black marbles. One marble is selected, its color is noted, and then it is returned to the bag. Then a second marble is selected. What is the probability that both selected marbles were white?

4. Using the same set up as given in problem 3, what is the probability that both selected marbles will be white if the first marble is not returned to the bag?

5. A man applying for his driver's license estimates that his chances of passing the written test are $\frac{2}{3}$, and that his chances of passing the driving test are $\frac{1}{4}$. What is the probability that he passes both tests?

6. If two cards are selected at random from a standard deck of 52 cards, what is the probability that they will both be diamonds?

7. A bag contains 9 marbles, 3 of which are red, 3 of which are blue, and 3 of which are yellow. If three marbles are selected from the bag at random, what is the probability that they are all of different colors?

8. If you select three cards from a standard deck of 52 cards and they are all kings, what is the probability that the next card you select will also be a king?

Answers

1. There are 12 marbles in the bag. Since 3 of them are red, the probability of picking a red marble is $\frac{3}{12} = \frac{1}{4}$. There are 5 marbles in the bag that are not blue, so the probability of picking a marble that is not blue is $\frac{5}{12}$.

2. There are 5 coins in the purse, and 3 of them are worth at least ten cents. Thus, the probability that a coin chosen at random will be worth at least ten cents is $\frac{3}{5}$.

3. There are $7 \times 7 = 49$ ways in which two marbles can be selected. Since there are four ways to select a white marble on the first draw and four ways to select a white marble on the second draw, there are a total of $4 \times 4 = 16$ ways to select a white marble on two draws. Thus, the probability of selecting white on both draws is $\frac{16}{49}$.

4. The two selections can be made in $7 \times 6 = 42$ ways. Two white marbles can be selected in $4 \times 3 = 12$ ways. Thus, the desired probability is $\frac{12}{42} = \frac{2}{7}$.

5. Since these two events are independent, the probability of passing both is $\left(\frac{2}{3}\right) \times \frac{1}{4} = \frac{1}{6}$.

6. The probability of drawing a diamond from the full deck is $\frac{13}{52} = \frac{1}{4}$. After the first diamond has been removed, there are 51 cards in the deck, 12 of which are diamonds. The probability of selecting a diamond from this reduced deck is $\frac{12}{51}$. The probability, thus, of selecting two diamonds is $\frac{1}{4} \times \frac{12}{51} = \frac{1}{17}$.

7. After the first marble is selected, the bag has 8 marbles left, 6 of which are of a different color than that of the first marble selected. Thus, the probability that the second marble is of a different color is $\frac{6}{8}$. If the second marble is different, there are then 7 marbles in the bag, three of which are of the color not yet selected. The odds of drawing a marble of the third color is $\frac{3}{7}$. Overall, then, the probability of drawing three different colors is $\frac{6}{8} \times \frac{3}{7} = \frac{18}{56} = \frac{9}{28}$.

8. After three cards are selected, there are 49 cards left in the deck, of which only one is a king. Thus, the probability of drawing a king on the fourth draw is $\frac{1}{49}$.

STATISTICS

Statistics is the study of collecting, organizing, and analyzing data. The next chapter in this book is an entire unit on reading and interpreting graphs and charts, which is a crucial part of statistics. Prior to this, let's take a look at one of the most important statistical concepts: "computing measures of location," which is the fancy statistical way of saying "computing averages."

Measures of location describe the "centering" of a set of data; that is, they are used to represent the central value of the data. There are three common measures of central location. The one that is typically the most useful (and certainly the most common) is the *arithmetic mean,* which is computed by adding up all of the individual data values and dividing by the number of values.

Example 1

A researcher wishes to determine the average (arithmetic mean) amount of time a particular prescription drug remains in the bloodstream of users. She examines 5 people who have taken the drug and determines the amount of time the drug has remained in each of their bloodstreams. In hours, these times are: 24.3, 24.6, 23.8, 24.0, and 24.3. What is the mean number of hours that the drug remains in the bloodstream of these experimental participants?

To find the mean, we begin by adding up all of the measured values. In this case, $24.3 + 24.6 + 23.8 + 24.0 + 24.3 = 121$. We then divide by the number of participants (five) and obtain $\frac{121}{5}$ as the mean, or 24.2.

Example 2

Suppose the participant with the 23.8-hour measurement had actually been measured incorrectly, and a measurement of 11.8 hours obtained instead. What would the mean number of hours have been?

In this case, the sum of the data values is only 109, and the mean becomes 21.8.

This example exhibits the fact that the mean can be greatly thrown off by one incorrect measurement. Similarly, one measurement that is unusually large or unusually small can have great impact upon the mean. A measure of location that is not impacted as much by extreme values is called the *median.* The median of a group of numbers is simply the value in the middle when the data values are arranged in numerical order. This numerical measure is sometimes used in the place of the mean when we wish to minimize the impact of extreme values.

Example 3

What is the median value of the data from example 1? What is the median value of the modified data from example 2?

Note that in both cases, the median is 24.3. Clearly, the median was not impacted by the one unusually small observation in example 2.

In the event that there is an even number of data values, we find the median by computing the number halfway between the two values in the middle (that is, we find the mean of the two middle values).

Another measure of location is called the *mode.* The mode is simply the most frequently occurring value in a series of data. In the examples above, the mode is 24.3. The mode is determined in an experiment when we wish to know which outcome has happened the most often.

Test Yourself

> **Directions:** Solve the following.

1. During the twelve months of 1998, an executive charged 4, 1, 5, 6, 3, 5, 1, 0, 5, 6, 4, and 3 business luncheons at the Wardlaw Club. What was the mean monthly number of luncheons charged by the executive?

2. Brian got grades of 92, 89, and 86 on his first three math tests. What grade must he get on his final test to have an overall average of 90?

3. In order to determine the expected mileage for a particular car, an automobile manufacturer conducts a factory test on five of these cars. The results, in miles per gallon, are 25.3, 23.6, 24.8, 23.0, and 24.3. What is the mean mileage? What is the median mileage?

4. In problem 3 above, suppose the car with the 23.6 miles per gallon had a faulty fuel injection system and obtained a mileage of 12.8 miles per gallon instead. What would have been the mean mileage? What would have been the median mileage?

5. An elevator is designed to carry a maximum weight of 3000 pounds. Is it overloaded if it carries 17 passengers with a mean weight of 140 pounds?

Answers

1. The mean number of luncheons charged was

$$\frac{(4+1+5+6+3+5+1+0+5+6+4+3)}{12} = \frac{43}{12} = 3.58.$$

2. Let G = the grade on the final test. Then,

$$\frac{(92+89+86+G)}{4} = 90 \quad \text{Multiply by 4.}$$

$(92 + 89 + 86 + G) = 360$
$267 + G = 360$
$G = 93$
Brian must get a 93 on the final test.

3. The mean mileage is $\dfrac{(25.3 + 23.6 + 24.8 + 23.0 + 24.3)}{5} = \dfrac{121}{5} = 24.2$ miles per gallon. The median mileage is 24.3 miles per gallon.

4. The mean mileage would have been $\dfrac{(25.3 + 12.8 + 24.8 + 23.0 + 24.3)}{5} = \dfrac{110.2}{5} = 22.04$ miles per gallon. The median mileage would have been 24.3 miles per gallon, the same as it was in problem 3.

5. Since the mean is the total of the data divided by the number of pieces of data, that is mean = $\dfrac{\text{total}}{\text{number}}$, we have (mean)(number) = total. Thus, the weight of the people on the elevator totals $(17)(140) = 2380$. It is therefore not overloaded.

EXERCISES: ADDING AND SUBTRACTING WITH VARIABLES

1. $4x + 3x = $ _____
2. $9y - 3y = $ _____
3. $16r - 7r = $ _____
4. $11w + 13p = $ _____
5. $17g + 12g = $ _____
6. $19s - 4s = $ _____
7. $12c + c = $ _____
8. $13v - v = $ _____
9. $12l + 3b = $ _____
10. $f + f + f = $ _____
11. $5x + 5x = $ _____
12. $4q - 3q = $ _____
13. $13w - 13w = $ _____
14. $6n - n = $ _____
15. $5z + 7z = $ _____

ANSWERS

1. $7x$ 2. $6y$ 3. $9r$ 4. $11w + 13p$ 5. $29g$

6. $15s$ 7. $13c$ 8. $12v$ 9. $12l + 3b$ 10. $3f$

11. $10x$ 12. q 13. 0 14. $5n$ 15. $12z$

EXERCISES: ADDING AND SUBTRACTING SIGNED NUMBERS

1. $^+8 + {}^-9 = $ _____   2. $^-7 + {}^-11 = $ _____   3. $^-5 + {}^+12 = $ _____

4. $^+6 + {}^+9 = $ _____   5. $^-12 + {}^+7 = $ _____   6. $^+9 + {}^-15 = $ _____

7. $^-8 + {}^+8 = $ _____   8. $^+12 + {}^-15 = $ _____   9. $^-13 + {}^+20 = $ _____

10. $^+6 + {}^+14 = $ _____   11. $^+12 + {}^-9 = $ _____   12. $^-15 + {}^-13 = $ _____

13. $^-8 + {}^-15 = $ _____   14. $^-8 + {}^+15 = $ _____   15. $^+8 + {}^+15 = $ _____

16. $^+8 + {}^-15 = $ _____   17. $^-9 + {}^+7 + {}^+6 + {}^+4 = $ _____   18. $^+11 + {}^-5 + {}^+3 + {}^-8 = $ _____

19. $^-14 + {}^+5 + {}^-7 + {}^+12 = $ _____   20. $^+9 + {}^-12 + {}^-7 + {}^+5 + {}^-10 = $ _____

21. $^-9 - {}^-9 = $ _____   22. $^-6 - {}^+8 = $ _____   23. $^+7 - {}^-7 = $ _____

24. $^+12 - {}^+4 = $ _____   25. $^+18 - {}^-5 = $ _____   26. $^-14 - {}^+8 = $ _____

27. $^-17 - {}^-9 = $ _____   28. $^+13 - {}^-15 = $ _____   29. $^-17 - {}^+20 = $ _____

30. $^-11 - {}^-19 = $ _____   31. $^+10 - {}^+14 = $ _____   32. $^-6 - {}^-12 = $ _____

ANSWERS

1. ⁻1	2. ⁻18	3. ⁺7	4. ⁺15	5. ⁻5	6. ⁻6
7. 0	8. ⁻3	9. ⁺7	10. ⁺20	11. ⁺3	12. ⁻28
13. ⁻23	14. ⁺7	15. ⁺23	16. ⁻7	17. ⁺8	18. ⁺1
19. ⁻4	20. ⁻15	21. 0	22. ⁻14	23. ⁺14	24. ⁺8
25. +23	26. ⁻22	27. ⁻8	28. ⁺28	29. ⁻37	30. ⁺8
31. ⁻4	32. ⁺6				

EXERCISES: MULTIPLYING AND DIVIDING SIGNED NUMBERS

1. ⁺6 × ⁺8 = _____   2. ⁻5 × ⁻7 = _____   3. ⁺16 ÷ ⁻2 = _____

4. ⁻24 ÷ ⁺6 = _____   5. ⁻9 × ⁻9 = _____   6. ⁻60 ÷ ⁻10 = _____

7. ⁺7 × ⁻8 = _____   8. ⁻18 ÷ ⁻9 = _____   9. ⁻6 × ⁺8 = _____

10. ⁺36 ÷ ⁻6 = _____   11. ⁻4 × ⁻9 = _____   12. ⁻42 ÷ ⁺6 = _____

13. ⁻5 × ⁻9 = _____   14. ⁻28 ÷ ⁺4 = _____   15. ⁺6 × ⁻8 = _____

16. ⁺48 ÷ ⁻3 = _____   17. ⁻4 × ⁻6 = _____   18. ⁻12 ÷ ⁻3 = _____

ANSWERS

1. $^{+}48$	2. $^{+}35$	3. $^{-}8$	4. $^{-}4$	5. $^{+}81$
6. $^{+}6$	7. $^{-}56$	8. $^{+}2$	9. $^{-}48$	10. $^{-}6$
11. $^{+}36$	12. $^{-}7$	13. $^{+}45$	14. $^{-}7$	15. $^{-}48$
16. $^{-}16$	17. $^{+}24$	18. $^{+}4$		

EXERCISES: SOLVING PROPORTIONS

1. $\dfrac{x}{3} = \dfrac{6}{9}$ 2. $\dfrac{4}{x} = \dfrac{12}{18}$ 3. $\dfrac{5}{12} = \dfrac{x}{48}$

4. $\dfrac{6}{14} = \dfrac{9}{x}$ 5. $\dfrac{3}{8} = \dfrac{x}{24}$ 6. $\dfrac{4}{7} = \dfrac{x}{9}$

7. $\dfrac{4}{x} = \dfrac{11}{17}$ 8. $\dfrac{2x}{5} = \dfrac{9}{50}$ 9. $\dfrac{6}{3x} = \dfrac{5}{20}$

10. $\dfrac{5}{8} = \dfrac{12}{3x}$ 11. $\dfrac{x+3}{5} = \dfrac{8}{10}$ 12. $\dfrac{3}{x+2} = \dfrac{4}{20}$

13. $\dfrac{x-1}{4} = \dfrac{9}{2}$ 14. $\dfrac{6}{9} = \dfrac{6}{x+1}$ 15. $\dfrac{4}{7} = \dfrac{x-4}{56}$

ANSWERS

1. 2	**2.** 6	**3.** 20	**4.** 21	**5.** 9

6. $\dfrac{36}{7}$ or $5\dfrac{1}{7}$ **7.** $\dfrac{68}{11}$ or $6\dfrac{2}{11}$ **8.** $\dfrac{9}{20}$ **9.** 8   **10.** $\dfrac{96}{15}$ or $6\dfrac{2}{5}$

11. 1	**12.** 13	**13.** 19	**14.** 8	**15.** 36

SUMMING IT UP

- Algebra was developed specifically to help solve word problems, as well as to organize one's thinking in a number of mathematical situations.

- Two major elements of algebra are variables and constants.

- Motion problems are based on the following relationship:

  - Rate × Time = Distance

  - Rate is usually given in miles per hour, time is usually given in hours, and distance is given in miles.

- Percent problems may be solved by translating the relationship into an equation.

- A monomial is an expression that contains a single term. A polynomial is an expression that contains more than one term. Special polynomials are binomials (two terms) and trinomials (three terms).

Reading Graphs and Charts

OVERVIEW

- **Picture graphs**
- **Bar graphs**
- **Line graphs**
- **Circle graphs**
- **Summing it up**

A graph is a picture of mathematical data. The data usually refers to a single subject or group of subjects. There are many different types of graphs, the most common being pictographs, bar graphs, line graphs, and circle graphs (or pie charts). We shall look at each of these in this chapter.

Every graph contains a legend or key, which may be a small inset or may appear along the axes of the graph. The legend explains the meanings of the graph's markings. Always locate and read the legend or key before attempting to interpret the meaning of the graph itself. Two graphs may look identical yet have different keys, thereby making the interpretations of the graphs different from one another.

The axes of a graph are the horizontal and vertical lines that frame it.

PICTURE GRAPHS

Picture graphs are also known as histograms or pictographs. Each picture on a histogram represents a quantity of something. For example, a little picture of a person may stand for 500 or 1000 people; a little car may stand for 100 automobiles, and so on. There must be a key or legend to tell you what each picture means.

Directions: Questions 1 to 4 refer to the graph below.

Repairs Required During First Three Years for Selected TV Brands (per 100)

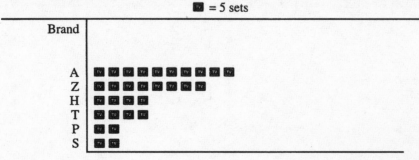

chapter 7

> **Q** How many more Brand A sets per hundred needed repair during the first three years than Brand T sets per hundred?
>
> **A** A careful look at the title of the graph and the key tells us that each TV set on the graph stands for 5 per 100 sold. The row for Brand A contains 6 more symbols than the row for Brand T. Multiply 6 (symbols) x 5 (number of sets each symbol stands for) to get 30 more sets per 100.

> **Q** Which brand(s) of TV shown on the graph appear to be the most reliable?
>
> **A** Those brands that required the fewest repairs would be the most reliable. Two brands, P and S, are tied for that honor, each with 10 repairs per hundred sets.

> **Q** What percent of Brand Z televisions required repair during the first three years?
>
> **A** Percent is a fraction of 100. Since the graph displays repairs per hundred sets sold, the graph actually shows percent figures, with each picture representing 5%. There are 8 symbols in the Brand Z row. That means that 8 x 5%, or 40%, is the percent of Brand Z sets requiring repair during their first three years.

> **Q** How many more Brand Z televisions were sold than Brand P during the three years covered by the graph?
>
> **A** The graph contains no information about sales records. Therefore, there is not enough information given to answer this question.

Occasionally, on GED multiple-choice questions, one of the five multiple-choice answers will be "Not enough information is given." Before you try to answer a question, it is a good idea to scan the multiple-choice answers to determine if "Not enough information" is an option.

Test Yourself

Directions: Answer the following questions. (Hint: Some questions cannot be answered on the basis of the information given.)

Questions 1 to 4 refer to the following graph.

Blood Samples Taken at Beth David Hospital

H = 10 samples

Day	
Sat	H H H H H H H H
Fri	H H H H H H H H H H H H
Thu	H H H H H H H H H H H H H H
Wed	H H H H H H H H H H H H H H H H
Tue	H H H H H H H H H H H H H H
Mon	H H H H H H H H H H H H H H H H H H

1. How many more samples were taken on Tuesday than on Thursday?

2. How many patients' blood samples revealed an illness?

3. On which two days were the same number of samples taken?

4. On which day were the samples taken half the number taken on Tuesday?

Questions 5 to 8 refer to the following graph.

Percent of Rejects per Week at ABCO Transmissions

= 1% of transmissions

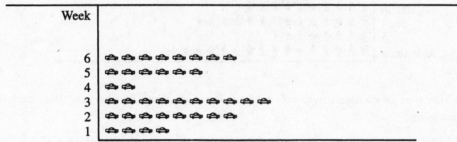

5. In which week was the percentage of rejects half that of week six?

6. What is the difference in the number of transmissions rejected in week three and those rejected in week one?

7. What percentage of transmissions was rejected in week three?

8. How many more transmissions were rejected in week five than in week one?

Questions 9 to 12 refer to the following graph.

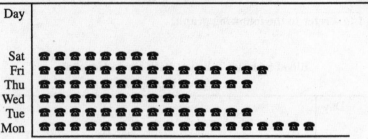

New Telephone Installations Last Week
☎ = 100 phones

9. How many more phones were installed on Thursday than on Saturday?

10. On which day of the week were the fewest phone installations done?

11. On which day of the week were 1000 installations done?

12. Are any phones installed on Sunday?

Questions 13 to 16 refer to the following graph.

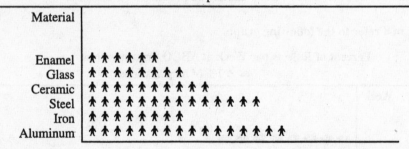

Survey of Preferred Cookware Materials
⚊ = 100 persons

13. Which type of cookware was preferred by fewer than half as many as those preferring aluminum?

14. Which types of cookware were preferred by exactly the same number of people?

15. How many more people preferred steel over glass cookware?

16. Which type of cookware is able to boil water quickest?

Answers

1.	20	**2.**	Not enough information	**3.**	Tuesday and Wednesday
4.	Saturday	**5.**	Week one	**6.**	Not enough information
7.	10	**8.**	Not enough information	**9.**	600
10.	Saturday	**11.**	Wednesday	**12.**	Not enough information
13.	Enamel	**14.**	Iron and glass	**15.**	600
16.	Not enough information				

BAR GRAPHS

The bar graph is the logical extension of the picture graph. It looks like a rectangle drawn around each row of pictures, with the pictures themselves then erased. Since there are no pictures to count, there are markings along one axis to tell you the quantity being represented. Bar graphs can be horizontal, like most picture graphs, or vertical. Unlike pictographs, bar graphs can be used to compare two or more different quantities at the same time, as in the graph below.

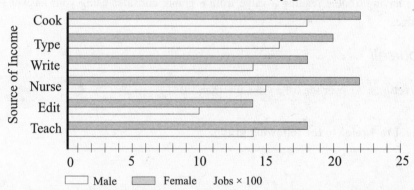

Part-Time Jobs in Lindaville (by gender)

Q How many more females than males teach part-time in Lindaville?

A To answer the question, we first need to examine the legend. There we find that the shaded bars are for females and the unshaded bars are for males. We also learn that the numbers written along the horizontal axis must be multiplied by 100. Put your finger (or better still, a straightedge) at the right end of the shaded "Teach" bar. Next, look down to the marking on the horizontal axis. It is 18. Multiplied by 100, that's 1800 female part-time teachers. Now do the same thing with the unshaded "Teach" bar. You should find 1400 male teachers. Subtract to find the difference: 1800 - 1400 = 400 more female teachers.

Q At which job are there as many females as there are males who write?

A Follow the unshaded "Write" bar to its right end. While holding your finger or straightedge there, find a shaded bar that ends at the same place. It's the shaded "Edit" bar.

> **Q** Do more females than males hold part-time jobs because more males than females in Lindaville hold full-time jobs?
>
> **A** The graph gives us figures on part-time jobs. It says nothing about full-time jobs, nor does it give us reasons or explanations for the information given. There is not enough information to answer this question.

> **Q** How many females in Lindaville work part-time as cooks or as typists?
>
> **A** Find the right end of the shaded "Cook" bar. Square a straightedge with the horizontal axis and move it to the right end of that bar. It ends at the 22 mark. Multiply 22 by 100, the number of jobs for which each mark stands: 22 x 100 = 2200. Now repeat the same thing for the shaded "Type" bar, and you'll get 2000. Add the two figures together: 2200 + 2000 = 4200 females who work as cooks or typists.

Of course, on the test itself, you will not be given a ruler to help you read values off of the graphs. If you are having trouble reading a value from a graph, consider using your answer sheet as a straightedge.

Test Yourself

Directions: Answer the following questions.

Questions 1 to 4 refer to the following graph.

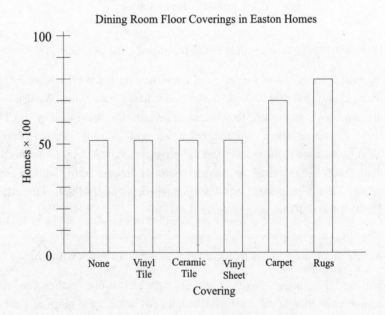

1. How many Easton homes have no floor covering in the dining room?

2. How many more Easton dining room floors have vinyl tile than ceramic tile?

3. How many Easton homes have carpeted living rooms?

4. How many Easton homes have either rugs or carpet in their dining rooms?

Questions 5 to 8 refer to the following graph.

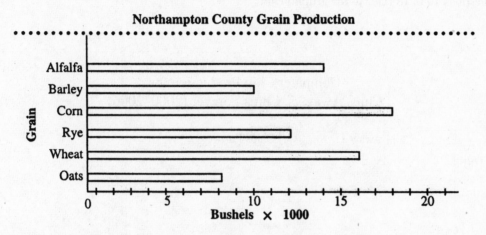

5. What is the difference between corn and oat production?

6. What is the total amount of grain produced in Northampton County?

7. How much grain is barley or rye?

8. What part of the grain production goes to cattle fodder?

Questions 9 to 13 refer to the graph below.

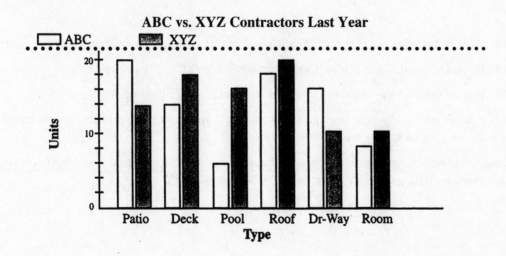

9. ABC contractors installed as many driveways last year as XYZ installed_____.

10. How many more patios did ABC contractors install last year compared to XYZ?

11. How many more roofs would you expect ABC to install this year compared to XYZ?

12. How many roofs were installed by both companies last year?

13. Which company did more jobs last year?

Questions 14 to 18 refer to the graph below.

Computer Chip Production for
Chips Away vs. Chips Forever (2000–2006)

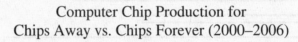

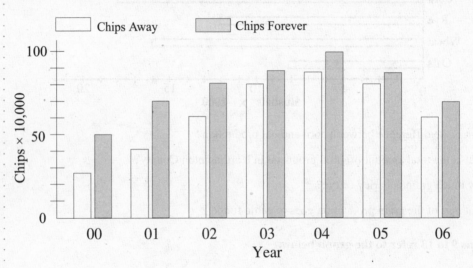

14. What is the difference between C.F.'s (Chips Forever's) production in 2006 and 2000?

15. In what year did C.A.'s (Chips Away's) production peak?

16. In what year was the output of each of the two companies furthest apart?

17. If the trend continues as the graph indicates, would you expect chip production to increase, decrease, or remain the same in 2008?

18. According to the graph, the gap between the two companies' production has narrowed over the years. By what year would you expect C.A. to overtake C.F.?

Answers

1.	5000	**2.**	0	**3.**	Not enough information
4.	15,000	**5.**	10,000 bushels	**6.**	78,000 bushels
7.	22,000 bushels	**8.**	Not enough information	**9.**	Pools
10.	6	**11.**	Not enough information	**12.**	38
13.	XYZ	**14.**	200,000	**15.**	2004
16.	2001	**17.**	Decrease	**18.**	Not enough information

LINE GRAPHS

You may think of a line graph as being formed by connecting the topmost points of vertical bars on a bar graph and then erasing the bars themselves. Or you may think of a line graph as being formed by plotting points at the intersection of the values along the horizontal and vertical axes and then connecting the points. **Line graphs can be used to display the same information as bar graphs**, but they are better suited than bar graphs for displaying continuous information, such as a range of temperatures over a period of time or stock-price variations. It is easier to visualize peaks and valleys on a line graph than on a bar graph, especially if more than one item is being considered.

Example

Questions 1 to 4 refer to the graph below.

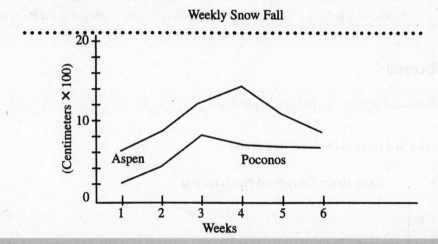

Q What is the difference in the amount of snowfall in Aspen and in the Poconos in Week #3?

A It helps to have two straightedges available. Move the first straightedge along the horizontal axis until it is on the Week 3 mark. Leave it there and bring the second straightedge down until it touches the "Aspen" line. Now read the snowfall on the vertical axis (each mark is 2 from the last). You should get 12. Multiply that 12 by 100 cm (from the legend):

$$12 \times 100 = 1200 \text{ cm}$$

Now move the second straightedge down until it intersects the "Pocono" line. Read 8 on the vertical axis. Multiplied by 100 cm, that's 800 cm. To find the difference, subtract:

$$1200 - 800 = 400 \text{ cm}$$

Q For which week is the difference between the snowfall in the Poconos and in Aspen the least?

A There is no need to calculate the answer to this question. The two lines are closest together at Week #6.

Q For which week is the difference between the snowfall in Aspen and in the Poconos greatest?

A Again, no need to measure. The lines are farthest apart at Week #4.

Q For which week(s) is the snowfall in the Poconos greater than the Week #1 snowfall in Aspen?

A Bring your straightedge down, keeping it square with the axes (against the vertical axis) until it reaches Aspen's snowfall for Week #1. The edge is now covering the "Poconos" line for 3,4, and 5. Hence, in those weeks, the snowfall in the Poconos is greater than the snowfall in Aspen.

Test Yourself

Directions: Answer the following questions.

Questions 1 to 4 refer to the following graph.

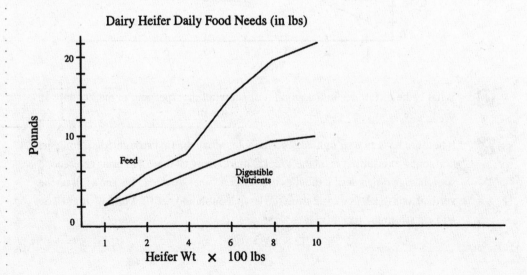

1. How many pounds of feed does a 600-lb heifer require daily?

2. What is the difference between feed and digestible nutrients needed by a 1000-lb heifer?

3. As a heifer increases in weight, does the proportion of its food that must be digestible nutrients increase, decrease, or remain the same?

4. At what weight must all of a heifer's daily feed be digestible nutrients?

Questions 5 to 8 refer to the following graph.

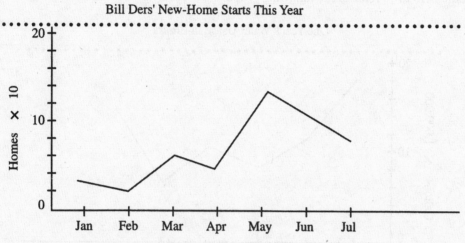

5. What was the difference in new-home starts between the best month and the poorest month?

6. Was the fear of not finishing before winter responsible for no new starts after July?

7. How many new homes were started in March?

8. During what months did the biggest downturn in home starts occur?

Questions 9 to 12 refer to the following graph.

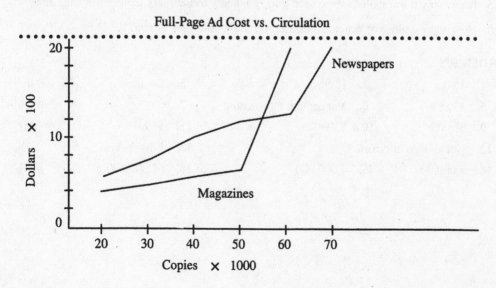

9. At what circulation does magazine advertising become more expensive than newspaper advertising?

10. What is the cost of a full-page ad in a newspaper with a circulation of 70,000?

11. What is the difference in cost between ads in newspapers and magazines with 40,000 circulation?

12. What is the cost of an ad in a newspaper with a circulation of 10,000?

Questions 13 to 16 refer to the following graph.

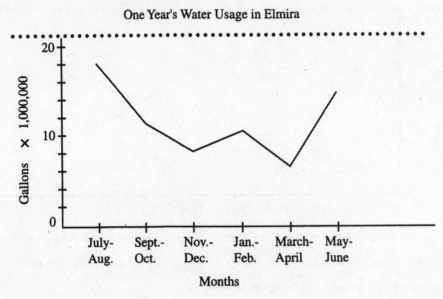

One Year's Water Usage in Elmira

13. For which period was usage three times higher than March and April?

14. How many gallons were used during November to December?

15. How many fewer gallons were used during January to February than in May and June?

16. How many gallons of water did Elmira use during the year shown?

Answers

1. 16 lbs	**2.** 12 lbs	**3.** Decrease	**4.** 100 lbs
5. 120	**6.** Not enough information	**7.** 60	**8.** June–July
9. 55,000	**10.** $2000	**11.** $400	or May–June
12. Not enough information		**13.** July–August	
14. 8,000,000	**15.** 4,000,000	**16.** 68,000,000	

CIRCLE GRAPHS

Circle graphs are also known as pie charts. (One look should tell you why.) They differ from other types of graphs in that they tell you how various parts of a whole are apportioned. Every pie chart in its entirety represents a particular whole. That whole could be a dollar, the total income of a company, the population of the world, or a person's income for a specific period of time. Each slice of the pie (segment of the circle) represents how a part of that whole is used or made up. When working with circle graphs, always keep in mind that the whole is 100%.

Example

Questions 1 to 3 refer to the graph below.

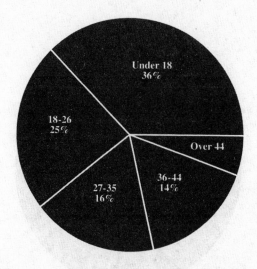

Rockin' Records' Sales by Age Group

Q   What percent of Rockin' Records' customers are younger than 27 years of age?

A   36 percent are shown as being under age 18, and 25 percent are shown as being aged 18 to 26. Add those two percentages together:

$36 + 25 = 61$

61 percent are younger than age 27.

Q   What percent of Rockin' Records' customers are over age 44?

A   To find this answer, we must first add up all the percentages that we have from the chart:

$36 + 25 + 16 + 14 = 91$

The group we are not interested in totals 91 percent of sales. Subtract that from 100 percent to get the over-44 group:

$100 - 91 = 9$

Nine percent are over age 44.

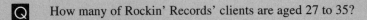

> **Q** How many of Rockin' Records' clients are aged 27 to 35?
>
> **A** The graph gives percentages, not numbers. If we knew how many clients there were altogether, then 16 percent of that number would be the answer to the question. With the information we have, however, we can't answer the question.

Test Yourself

Directions: Answer the following questions.

Questions 1 to 5 refer to the following graph.

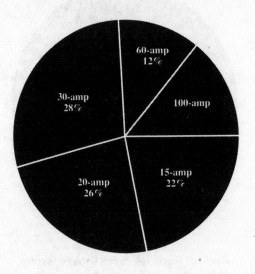

Circuit Breakers at Napa Industries

1. What is the difference in the percentages of 30- and 60-amp circuit breakers used at Napa Industries?

2. What percent of the breakers are 100-amp?

3. What percent of Napa breakers are lower than 30-amp?

4. What percent of Napa breakers can handle 220 volts?

5. If Napa used 300 breakers, how many would be 60-amp or higher?

Questions 6 to 10 refer to the following graph.

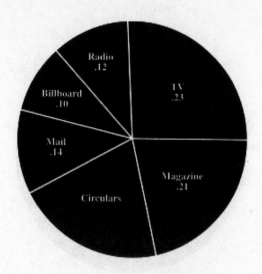

S.O. Badd's Advertising Dollar

6. What percent of Badd's advertising dollar is spent on TV and radio?

7. How many cents of each advertising dollar does Badd spend on circulars?

8. If Badd spends $10,000 on advertising, how many dollars are spent on magazine ads?

9. On which advertising medium does Badd spend the most?

10. How much money does Badd spend on billboard advertising, assuming they spend $10,000 total?

Questions 11 to 15 refer to the following graph.

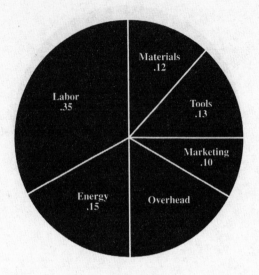

Where U.R. Manufacturing's Dollar Goes

11. How much does the average worker at U.R. earn per hour?

12. What percent of U.R.'s expenditures go to energy?

13. How many cents of each dollar does U.R. spend on marketing and overhead combined?

14. What does U.R. spend the least amount per dollar on?

15. If U.R. spends $200,000 per year altogether, how much does it spend on labor?

Answers

1. 16%	**2.** 12%	**3.** 48%	**4.** Not enough information	
5. 72	**6.** 35%	**7.** 20¢	**8.** $2100	**9.** TV
10. $1000	**11.** Not enough information	**12.** 15%	**13.** 25¢	
14. Marketing	**15.** $70,000			

EXERCISES: GRAPHS

Questions 1 to 5 refer to the graph below.

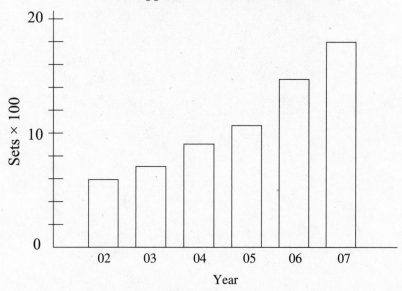

Irv's Appliances' TV Sales (2002–2007)

1. How many more TV sets were sold in 2007 than in 2002?

2. How many TV sets did Irv's sell in 2005?

3. Does Irv's sell other things besides TVs?

4. If the trend shown on the graph continues, is Irv's likely to sell more, fewer, or the same number of sets in 2008?

5. What was the total number of TV sets sold by Irv's from 2002 through 2007?

ANSWERS

1. 1200
2. 1100
3. Not enough information
4. More
5. 6600

SUMMING IT UP

- Graphs illustrate how things are compared and what trends exist.

- The four most common graphs are:

  ❶ Bar graphs—used to compare various quantities

  ❷ Line graphs—used to show trends over a period of time

  ❸ Circle graphs—used to show relationships

  ❹ Picture graphs—used to show comparison of quantities by using symbols

Geometry and Trigonometry

OVERVIEW

- Some very basic definitions
- Types of angles and angle measurement
- Triangles
- Quadrilaterals
- Other polygons
- Area and perimeter
- Circles
- Coordinate axes
- Graphing linear equations
- The Pythagorean theorem
- The distance between points on a graph
- Right triangle trigonometry
- Summing it up

Geometry is the study of the measurement of different types of figures, and their relationships. As a general rule, all geometric figures occupy space. The points, lines, and line segments that outline those figures, however, do not occupy any space. This distinction should become clearer in a short while.

Most of this section is concerned with plane geometry. A plane is a flat surface, and so plane geometry concerns those figures that may be drawn on a flat surface. A circle, for example, is a plane figure. A sphere (ball) is not.

Much of our knowledge of geometry we owe to the Greeks of about 2500 years ago. They studied the relationships of angles and sides of all types of plain and solid figures, with an eye toward their aesthetic as well as their practical applications. Much of what they discovered was applied in laying out and constructing buildings—including temples to the gods of Mount Olympus. The person responsible for organizing all of the discoveries into an organized volume of knowledge was a man known as Euclid. It is in honor of his achievement that for thousands of years people have studied "Euclidean Geometry."

chapter 8

SOME VERY BASIC DEFINITIONS

A **point** is a location in space. It occupies no space itself. Below, you see representations of two points, **A** and **B**. The dot in each case represents, **but is not actually**, a point. That is because the dots occupy space, and points (as just noted) do not. Note that a single uppercase letter is used to name a point.

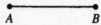

A **ray** is an infinite series of points. Infinite means "continuing forever without end." A ray has a single endpoint, and has direction. It is named by two points, the first of which is its endpoint, and the second of which is another point that lies on it. Note from the representation below that $\overrightarrow{CD}$ is not the same as $\overrightarrow{DC}$. Each has a different endpoint and goes in the opposite direction.

The arrowhead on the end of each ray indicates that it continues infinitely in the direction indicated.

A **line** is defined alternately as an infinite series of points (but unlike a ray) having no endpoint, or as two opposite rays with the same endpoint. It is represented as indicated below. The line represented below may be called either $\overleftrightarrow{EF}$ or $\overleftrightarrow{FE}$. Unlike the previous rays, both names indicate the same line.

Note that E and F can be any two points on the line.

Perhaps the geometric representation with which you have had the most experience is the **line segment**. A line segment, as in the two previous figures, is also an infinite series of points. It differs from the previous two, however, in that it has two endpoints. Often, a line is incorrectly defined as the shortest distance between two points. In fact, a line segment is the shortest distance between two points. A line segment is named by its endpoints, hence line segment $\overline{GH}$ is illustrated below:

When two rays have the same endpoint, they form an **angle**. That angle is named either by using the letter of the mutual endpoint of the rays, or by naming the three points in order with the mutual endpoint in the middle. Hence, below is pictured $\angle C$, $\angle BCD$, or $\angle DCB$.

The mutual endpoint of the two rays, at which the angle, BCD, is formed, is known as the **vertex** of the angle. Point C is the vertex of angle C. (The plural of vertex is vertices.)

Test Yourself

> **Directions:** Name each of the figures below. Also give the appropriate symbol and letter names.

1. $E \quad W \quad \longrightarrow$

2. $\longleftarrow \quad R \quad S$

3. $\longleftrightarrow P \quad Q$

4. $\bullet M$

5. $V \quad\quad\quad R$

6. $S \quad T \quad G$

7. $H \quad I \quad J$

8. $M \quad L$

9. $F \quad K$

10. $R \quad S$

Answers

1. ray, $\overrightarrow{EW}$

2. ray, $\overrightarrow{SR}$

3. line, $\overleftrightarrow{PQ}$ or $\overleftrightarrow{QP}$

4. point, M

5. line segment, $\overline{VR}$ or $\overline{RV}$

6. angle, $\angle S$, $\angle TSG$, or $\angle GST$

7. angle, $\angle I$, $\angle HIJ$, or $\angle JIH$

8. angle, $\angle M$

9. line segment, $\overline{KF}$ or $\overline{FK}$

10. line, $\overleftrightarrow{RS}$

TYPES OF ANGLES AND ANGLE MEASUREMENT

Imagine two rays, $\overrightarrow{OA}$ and $\overrightarrow{OB}$. At the start, point A is lying exactly on top of point B.

Figure 1
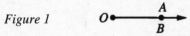

Pivoting at O, ray $\overrightarrow{OA}$ begins to rotate upward, forming an angle with $\overrightarrow{OB}$ at O. Before the rotation began (Figure 1) the angle at O had a measure of 0°. In Figure 2, that angle is greater than 0° but less than 90°.

Figure 2
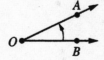

Figure 3 shows ray $\overrightarrow{OA}$ having rotated through a quarter of a circle. The distance through which it has rotated is measured as an angle of 90°.

Figure 3

In Figure 4, $\overrightarrow{OA}$ has rotated through an angle of more than 90° but less than 180°. Figure 5 shows a rotation of 180°, or a half circle. Therefore, degree measure is based on the amount of rotation around a pivotal point.

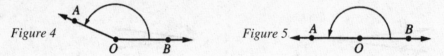

Figure 4   *Figure 5*

Based upon the amount of rotation, angles may be classified. An angle containing fewer than 90° is known as an **acute** angle. An acute angle is illustrated in Figure 2 (above). An angle containing exactly 90° is a **right** angle (see Figure 3, above). Figure 4 shows an **obtuse** angle—that is, one of greater than 90° but less than 180°. An angle of exactly 180° is often referred to as a **straight** angle.

Test Yourself

Directions: Classify each of the following as an acute, right, obtuse, or straight angle (It may be helpful in some cases to rotate your book in order to get a true picture.)

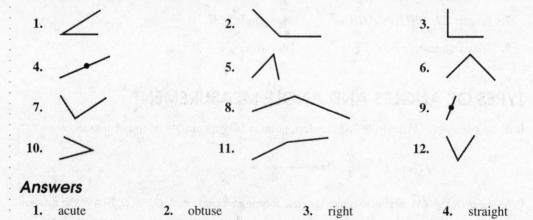

Answers

1.	acute	**2.**	obtuse	**3.**	right	**4.**	straight
5.	acute	**6.**	right	**7.**	right	**8.**	obtuse
9.	straight	**10.**	acute	**11.**	obtuse	**12.**	acute

TRIANGLES

Closed figures made up of line segments are known as **polygons**. The word, polygon, means many sides. The simplest of the polygons is the **triangle**—a closed Figure with three sides. It is impossible to have a closed Figure containing fewer than three sides. Try to make one, and prove the last statement to yourself.

Triangles may be classified in two ways—by angles or by sides. When classified according to sides, there are three types of triangles: **equilateral, isosceles,** and **scalene**.

An equilateral triangle is a triangle that has three sides of equal length.

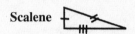
Equilateral

An isosceles triangle has two sides of equal length.

Isosceles

Those sides are known as the legs. The third side is known as the base.

A scalene triangle is a triangle that has no two sides equal in length.

Scalene

The markings on the sides of the triangles above are used to indicate equality of length (also known as **congruency**). Sides that are similarly marked are congruent.

When triangles are classified according to their angles, there are also three types: **acute, right,** and **obtuse**.

An acute triangle is one that contains acute angles, and acute angles only.

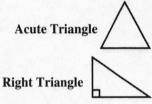
Acute Triangle

A right triangle is a triangle that contains exactly one right angle. Notice the way in which the right angle is marked. This is a standard marking used to indicate that an angle is a right angle. The corners of this page are right angles.

Right Triangle

An obtuse triangle is one that contains exactly one obtuse angle. As with the right triangle, those angles other than the one from which the Figure gets its name must be acute angles.

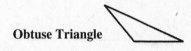
Obtuse Triangle

You may wish to note at this time that an equilateral triangle is also equiangular. That is to say, all three angles are congruent (contain the same size degree measure). Also, an isosceles triangle contains two congruent angles—those opposite the congruent sides.

The sum of the measures of all the angles of any triangle is equal to 180°. If you wish to prove this to yourself, you may cut out a triangle from a piece of paper, and mark each of the angles in some way (1). Then cut or tear the triangle apart, making sure not to damage any of the original angles

(2). Finally piece the three angles together with the vertices touching, and you will see that the angles total up to a straight angle (3). This will be true of any triangle, no matter whether it is acute, right, or obtuse. Try it with as many different types of triangles as you like until you are convinced.

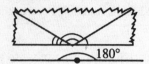

Test Yourself

Directions: Classify each triangle according to its sides.

1.    2.   3.

Directions: Classify each triangle according to its angles.

4.   5.   6.

7.

Directions: Find the number of degrees in angle x.

8.   9.   10.

11. 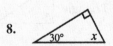   12. 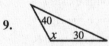   13.

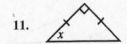  

Answers

1.	scalene	2.	isosceles	3.	equilateral	4.	acute
5.	right	6.	obtuse	7.	equiangular	8.	60°
9.	110°	10.	70°	11.	45°	12.	65°
13.	60°						

It might be easier to understand the answers to numbers 10 through 12 by noting that all three of those triangles are marked as isosceles. The angles opposite the sides marked as equal must also be equal. Number 13 is an equilateral triangle, which by definition means that all angles must be equal.

QUADRILATERALS

After the triangle, the next polygon in complexity is the quadrilateral. A quadrilateral is any four-sided polygon. Quadrilaterals, like triangles, may be classified according to certain features. One with no special features is simply known as a **quadrilateral** (1).

Incorporating one pair of parallel sides into a quadrilateral makes a Figure known as a **trapezoid** (2).

If the other pair of legs of a trapezoid are also made parallel, a new quadrilateral is formed which is known as a **parallelogram** (3).

From the basic parallelogram, it is possible to go in either of two directions. A right angle may be added, in which case the parallelogram becomes a **rectangle** (4). Otherwise, all sides of the parallelogram may be made equal, in which case the parallelogram is known as a **rhombus** (4a).

If a rectangle is made equilateral, the resulting Figure is a **square**. If a right angle is added to a rhombus, the resulting Figure is also a **square** (5). Examine the diagram closely, to see how the various quadrilaterals relate.

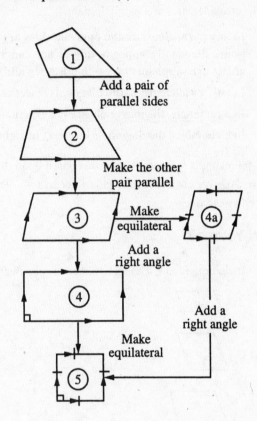

There are certain properties that are shared by all quadrilaterals. All have four sides and four vertices. Each quadrilateral is also capable of being cut by a diagonal into two triangles:

 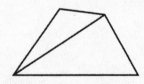 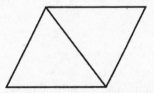

Since we have seen that the total degree measure of a triangle's angles is 180°, and since we have seen that any quadrilateral can be cut into two triangles, it is reasonable to conclude that the total number of degrees in the angles of a quadrilateral is 360°.

Special Features of Parallelograms

Bear in mind that as you progress down the chart of quadrilaterals (above), each Figure incorporates the features of the one before and then adds a special characteristic of its own. All the figures on the chart are quadrilaterals. The square, rhombus, and rectangle are all special parallelograms, retaining all the characteristics of a parallelogram.

> **In any parallelogram, the opposite sides are parallel and congruent. In any parallelogram, diagonally opposite angles are congruent. In any parallelogram, consecutive angles are supplementary—that is, they add up to 180°.**
>
> **In any parallelogram, the diagonals bisect each other (cut each other in half).**
>
> **In a rectangle, the diagonals are congruent. (Remember a square is a rectangle.)**
>
> **In a rhombus, the diagonals intersect at right angles.**

The following is a group of exercises based upon the characteristics of quadrilaterals discussed above. Analyze each problem, and, if necessary, refer to the rules stated above and immediately preceding them.

Test Yourself

Directions: State the name that most specifically identifies each type of figure.

1. a)
    b)
    c)
    d)
    e)

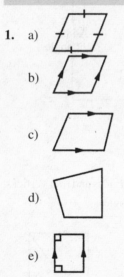

Directions: Find the value of x in each figure.

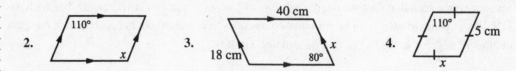

2.   3.   4.

5.

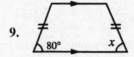

6. $AC = 12$
    $BD = x$
    $x = ?$

7.

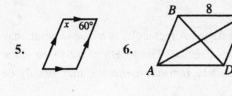

8.

9.

Answers

1. a) rhombus   b) parallelogram   c) trapezoid   d) quadrilateral   e) rectangle
2. 110°   3. 18cm   4. 5cm   5. 120°
6. not enough information   7. 90°   8. 85°   9. 80°

OTHER POLYGONS

A five-sided polygon is known as a pentagon, a six-sided polygon is known as a hexagon, and an eight-sided polygon is known as an octagon. It is at least theoretically possible to have a polygon of any number of sides. Technically, a polygon of an infinite number of sides is a circle. In a circle, each of the infinite number of points making up its circumference (the distance around it) may be thought of as a separate side. Circles will be discussed separately in a later section.

Any polygon has the same number of sides as it has vertices. That is to say, a triangle has three sides and three vertices, a quadrilateral four sides and four vertices, and a pentagon five sides and five vertices. Therefore, an *n*-gon has *n* sides and *n* vertices. A **regular polygon** is defined as **a polygon whose sides, and thus whose angles, are all congruent**.

To calculate the sum of the degrees in all the interior angles of a regular polygon, divide the polygon up into triangles by drawing as many diagonals as can be drawn from a single vertex. In the pentagon pictured below, all diagonals drawn originated at vertex *V*.

The number of triangles is then multiplied by 180° to find the number of degrees of angle measure in the figure. In this case, $3 \cdot 180° = 540°$. To find the number of degrees in each angle, divide the total by the number of vertices (in this case 5). $\frac{540}{5}$ yields a Figure of 108° per interior angle of the pentagon.

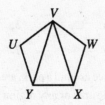

AREA AND PERIMETER

The **perimeter** of a Figure is the distance around that figure. A rectangle, as a case in point, may be considered as the fence around a rectangular-shaped region. The length of the fence is the perimeter of the rectangle. For a rectangle or a parallelogram, the perimeter may readily be computed as twice the length plus twice the width:

$$P = 2L + 2W$$

The perimeter of a square or rhombus is even easier to compute, since its length and width are the same. Simply multiply the length of any side by 4:

$$P = 4s$$

For other figures, however, some figuring may have to be done, based upon the information provided. Try the exercises below, and you should get a pretty good idea of how the process works.

Test Yourself

Directions: Find the perimeter of each figure.

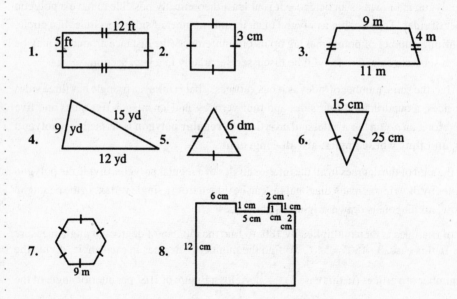

Answers

1.	34 ft	2.	12 cm	3.	28 m	4.	36 yd
5.	18 dm	6.	65 cm	7.	54 m	8.	57 cm

Just as we described perimeter as the fence around a park, area is a way of measuring the park itself—i.e. the region enclosed by the fence. The area of any region is found by dividing that region up into small squares and then finding how many of those squares fit into the region. For that reason, area is always expressed in square units, such as square inches, square feet, square meters, etc.

Area of a Rectangle

To find the area of a rectangular region that is 3 meters wide by 5 meters long, we follow the three steps below:

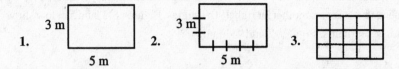

Counting up the little squares, we find that the area is 15 square meters (or 15 m²). Of course, if you look back at step 1, you might see another way in which we could have found the same result. We could have multiplied the base of the rectangle by the height (we formerly referred to these dimensions as length and width but will now change to base and height since, as you will see, these names have wider applicability).

$$A = bh$$
$$A = 5 \cdot 3$$
$$A = 15 \mathrm{m}^2$$

Area of a Square

A square is, as previously noted, a special case of the rectangle (that is, a rectangle that happens to be equilateral). Since that is the case, the same formula for finding its area must apply. That is to say, a square's base times its height will give you its area.

Since a square's base is the same length as its height, a simpler formula is possible. Examine the square in the diagram, and you will notice that its side has been marked as being s units long. Since the base and the height must each be s units long, then we may substitute s for both b and h in the formula for area. Thus we derive a special formula for the area of the square. While this is a handy formula of which to be aware, memorizing it is hardly necessary, since it can easily be derived from the rectangle area formula. You will find that the same is true for special formulas for most of the other areas that we will look at. Most are derived from the basic area formula. If you can clearly see how that derivation takes place and can understand why it works, then you should be readily able to derive the appropriate formula when the need arises.

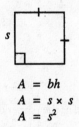

$$A = bh$$
$$A = s \times s$$
$$A = s^2$$

Height and Perpendicular Defined

We have previously discussed in passing the term "height," without ever bothering to formally define it. From here on, it will be necessary to have a formal definition of the term. Before defining height, it is necessary to examine another term: perpendicular. **Perpendicular** lines and perpendicular line segments intersect, or cross one another, at right (90°) angles. Figures 1, 3, and 5 below show perpendicular lines and/or segments. Figures 2 and 4 do not.

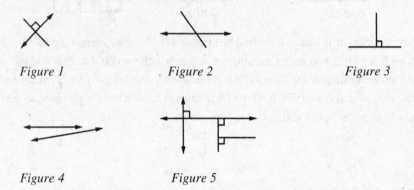

Figure 1   *Figure 2*   *Figure 3*

Figure 4   *Figure 5*

The symbol $\perp$ is used to indicate perpendicularity and is read "is perpendicular to." Now consider the case of point P and segment AB (Figure 6). How far is point P from $\overline{AB}$? Any of an infinite number of segments may be drawn from P to $\overline{AB}$ (Figure 7). Only one of them, however, would represent the shortest distance. The shortest distance from P to AB is the perpendicular distance from P to $\overline{AB}$ (Figure 8).

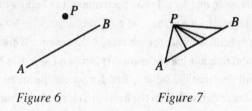

Figure 6   *Figure 7*

This distance PQ is also defined as "the distance" from the point to the line.

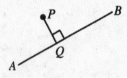

Figure 8

The height of a geometric figure is defined as the perpendicular distance from the base of the figure to the opposite vertex. In the case of the rectangle, which we have already discussed, any side qualifies as the height with respect to an adjacent side, since any pair of adjacent sides of a rectangle are perpendicular to one another (Figure 9). In a parallelogram that is not a rectangle, however, no side is the height to another. Rather, the height must be drawn in, as in Figure 10b.

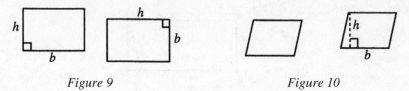

Figure 9   *Figure 10*

A triangle has three different heights, each of which corresponds to a different base. The base of any figure, as you should be able to infer from the diagram (Figure 11), does not necessarily refer to the side upon which the figure is resting. Rather, it refers to the side that is perpendicular to the height, and may change according to convenience. (This is not true of an isosceles triangle, in which the base is defined as the third side (the other two being the legs) (Figure 12), or of a trapezoid, in which the bases are defined as the two parallel sides (Figure 13)—the other two sides being the legs.)

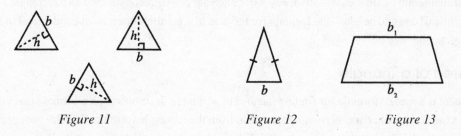

Figure 11   *Figure 12*   *Figure 13*

Area of a Parallelogram

The formula for the area of a parallelogram is rather easily developed by considering a series of diagrams. Start out by considering the parallelogram in Figure 1. Next, let us cut that parallelogram into two figures: The first is a trapezoid (I) and the second is a right triangle (II). You can see this in Figure 2.

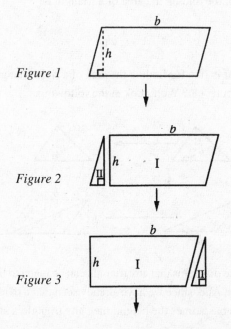

Figure 1

Figure 2

Figure 3

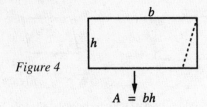

Figure 4

$$A = bh$$

Figure 3 shows the triangle and the trapezoid juxtaposed. Since the hypotenuse (longest side) of the right triangle was originally parallel to the non-perpendicular leg of the trapezoid, it will align perfectly with that leg in its new position. We can therefore put the two figures together (Figure 4) to form a new Figure that is identical in area to the original parallelogram. That new Figure is, of course, a rectangle. Can you see why the base of the rectangle is identical to the base of the original parallelogram? Can you see why the height of the rectangle is identical to the height of the parallelogram? In that case—since any parallelogram can be reconstructed as a rectangle with the identical base and height—the formula for the area of a parallelogram must be identical to that for a rectangle: $A = bh$.

Area of a Triangle

There is a special formula for finding the area of a triangle. It is, however, a formula that you need not memorize, since it is derived, once again, from the rectangle formula. Consider any rectangle. Draw the diagonal of that rectangle, and you will have two triangles:

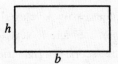 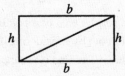

Then, it should be rather obvious that the area of either of the two triangles is half that of the rectangle. Hence, the formula for finding the area of a triangle is:

$$A = \frac{1}{2}bh$$

Now, you well might say, that is fine for finding the area of a right triangle, but what about oddly shaped triangles—scalene, acute, etc? Well, look at the following:

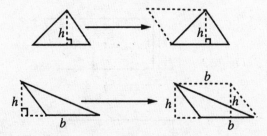

You can readily see from these diagrams that any triangle can be made to be half of a parallelogram with the same base and height. And, since we have already seen that a parallelogram's area may be computed by multiplying the base times the height, then any triangle's area may be computed by finding half the parallelogram's area, or half its base times its height. You may also notice from the diagram—and if you have not, make sure that you do—that the height of an obtuse triangle falls outside the triangle itself. Since the shortest distance between a point and a line is the perpendicular distance, by extending the triangle's base (the line that the base is a line segment of) we can legitimately find that distance.

The height of a triangle is also known as its **altitude**.

Area of a Trapezoid

The two parallel sides of a trapezoid are known as its **bases**. A trapezoid's area can be rather complex to determine. The information required in order to determine a trapezoid's area is the height of the Figure as well as the length of both bases. From this information, coupled with the rectangle area formula, it is possible to derive a formula that may be applied to find the area of any trapezoid. If you follow the series of diagrams, you should be able to see how the trapezoid area formula is derived. For openers, we have a trapezoid with an altitude of h units, and the bases respectively designed as b_1 and b_2.

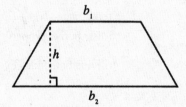

Figure 1

In Figure 2, you will notice that the second altitude has been constructed to the base. It is the same length as the first altitude (h) since the two bases are parallel, and therefore are always the same distance apart.

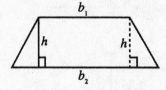

Figure 2

In Figure 3, the Figure is cut along the altitudes to form a rectangle and two right triangles. Notice that the rectangle has dimensions h and b_1. The triangles both have an altitude h, but the lengths of the bases of the two right triangles are somewhat less obvious.

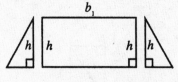

Figure 3

In Figure 4, the two right triangles are combined into a single triangle of altitude h. It can now be seen that the length of the base of the new triangle is the difference of the trapezoid's two bases ($b_2 - b_1$). We can now proceed to find the areas of the two figures:

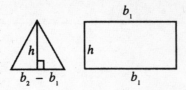

The triangle's area is $\frac{1}{2}h(b_2 - b_1)$.   The rectangle's area is $b_1 h$.

Figure 4

Therefore, the area of the trapezoid is found by combining the area of the rectangle with the area of the triangle:

$A = \frac{1}{2}h(b_2 - b_1) + b_1 h$

$2A = h(b_2 - b_1) + 2b_1 h$ (by doubling both sides)

$2A = b_2 h - b_1 h + 2b_1 h$ (distribute)

$2A = b_2 h + b_1 h$ (combine— $b_1 h + 2b_1 h$)

$2A = h(b_2 + b_1)$ (distributive property)

$A = \frac{1}{2}h(b_2 + b_1)$ (divide both sides by 2)

And there's the formula!

Of course, you do not need to memorize the steps that were used to develop the area formulas. However, understanding the steps will help you remember how to use the formulas when you need to.

Test Yourself

Directions: Find the area enclosed by each figure.

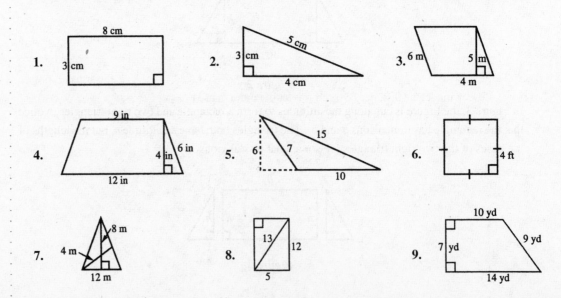

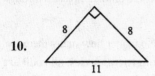
10.

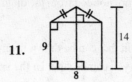
11.

12.

Answers

1.	24 cm²	**2.**	6 cm²	**3.**	20 m²	**4.**	42 in²
5.	30	**6.**	16 ft²	**7.**	48 m²	**8.**	60
9.	84 yd²	**10.**	32	**11.**	92	**12.**	40

CIRCLES

The Circle is the one Figure that GED geometry considers as not composed of line segments or rays. Indeed, it is sufficiently unique to merit at least one section all to itself. It was the Greeks who first discovered a relationship between the longest distance across a Circle and the distance around the circle. The longest distance across a Circle is known as its **diameter**. It is labelled d in the Circle below. The distance around a Circle (labelled C) is known as the **circumference**.

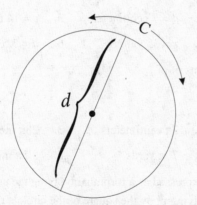

The Greeks discovered that if they rolled a circular wheel along a straight line until one complete turn had been made, the distance covered was a little more than 3 times the diameter. Today, the best minds of modern science and the most sophisticated computers have tried to find an exact number for the ratio of a circle's circumference to its diameter. They have found the ratio to equal 3.141592 ... and then some. In other words, they have found it to be three and a little bit more. This relationship has been named after the Greek letter π, known as pi, and is often approximated by the decimal 3.14 or by the fraction $\frac{22}{7}$. Thus, the formula for the circumference of a Circle is:

$$C = \pi d$$

Any line segment that connects the center of a Circle to the circle itself is known as a **radius** of that Circle (plural is **radii**). A radius is half the length of a diameter, or, conversely, a diameter is the length of two radii. (In case you had not noticed from the diagram above, the diameter passes through the center of the circle.)

There is no limit to the number of radii that can be drawn in any circle. No matter how many there are, however, there is one thing of which you can be sure: **In the same or congruent circles, all radii are congruent.**

Since a diameter is equal in length to two radii, there is an alternate formula for finding the circumference of a circle:

$$C = 2\pi r$$

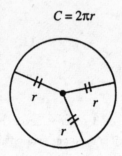

Test Yourself

> **Directions:** Find the circumference of a circle when

1. $r = 2$ inches
2. $d = 7$ centimeters
3. $r = 14$ meters
4. $d = 21$ feet
5. $r = 35$ yards
6. $d = 3.5$ millimeters

(Express your answers in terms of pi.)

Answers

1. 4π inches
2. 7π centimeters
3. 28π meters
4. 21π feet
5. 70π yards
6. 3.5π millimeters

A circle's area may also be expressed as a formula involving the use of the quantity, pi. The area of a circle is found by multiplying pi by the square of the circle's radius, hence the formula:

$$A = \pi r^2$$

Using this formula, if we know the radius of a circle to be 5, its area is 25π. Similarly, if we know the area of a circle to be 49π, then we can determine the radius of that circle to be 7.

For the following exercises, leave the answer in terms of pi (where appropriate).

Test Yourself

> **Directions:** Find the area of a Circle of

1. radius 3
2. radius 9
3. radius 11
4. diameter 12
5. diameter 14

Directions: Find the radius of a circle when its

6.   area = 64π **7.**   area = 100π

8.   circumference = 32π **9.**   circumference = 25π

Answers

1. 9π	**2.** 81π	**3.** 121π	**4.** 36π	**5.** 49π
6. 8	**7.** 10	**8.** 16	**9.** $12\frac{1}{2}$	

The diameter of a circle has already been mentioned as a measure of the longest distance across a circle. A diameter is also the longest chord in a circle. A chord is a line segment that connects two points on the circle. In the circle in Figure 1 below, chords *AB* and *CD* are drawn. The endpoints of $\overline{AB}$ are also the endpoints of arc *AB* (written $\overparen{AB}$). Similarly, *C* and *D* are the endpoints of $\overparen{CD}$. What other arcs can you name that are part of Circle *O*?

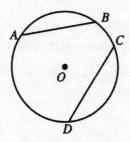

Figure 1

In the same or congruent circles, congruent chords cut off congruent arcs. In the same or congruent circles, congruent arcs cut off congruent chords.

A rather interesting and useful fact about chords concerns chords that intersect within a circle. You will notice that in Circle *P*, the lengths of segments *EI*, *FI*, *GI*, and *HI* are marked. If you multiply the lengths of the segments of chord *EF* together, you will get 24. What do you get by multiplying the lengths of the segments of $\overline{GH}$ together?

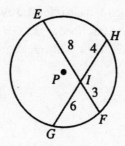

Circle P

When two chords intersect in a circle, the product of the lengths of segments of one equals the product of the lengths of segments of the other.

Arcs of a circle are measured in degrees, just as angles are. In fact, there are 360 arc degrees in a circle, just as there are 360 angle degrees. An angle formed by two radii of a circle is known as a **central angle**. By definition, it is equal in degree measure to the measure of its intercepted arc. Look at angle *POQ* in Circle *O*. It is a central angle and has a degree measure of 75°. Its intercepted arc, $\overarc{PQ}$, has a measure of 75°.

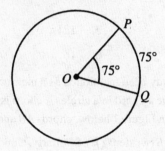

Circle *O*

A second type of angle in a circle is formed by two chords that share a common endpoint. Witness angle *MVW* in Circle *S*. Angle *MVW* is known as an **inscribed angle**. It can be shown that its measure is equal to half the measure of its intercepted arc. Since arc $\overarc{MW}$ is shown as containing 140°, angle *MVW* must contain half of that amount, or 70°.

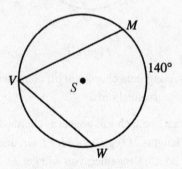

Circle *S*

In Circle *Q*, you see several angles. Each of them is formed by two chords with a common endpoint. Therefore, each of the angles drawn is an inscribed angle. $\overline{GH}$ is a diameter of Circle *Q*. That means that $\overline{GH}$ cuts Circle *Q* into two halves. How many arc degrees are there in each arc $\overarc{GH}$?

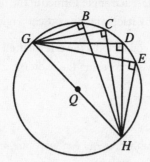

Circle *Q*

Now look at all the angles inscribed in Circle *Q*. Since each of them is cutting off an arc of 180°, and since each of them is an inscribed angle, then each of them must be a right angle.

Any angle inscribed in a semiCircle is a right angle.

When two chords cross within a circle, angles are formed at the point of intersection. Angles formed by two intersecting lines or segments that are opposite each other are called **vertical angles**. In or out of a circle, vertical angles are congruent. Angle *MRP* ≅ angle *QRN*. Similarly, angle *MRQ* ≅ angle *PRN*.

Let us now examine the two pairs of angles formed when chords *YZ* and *VW* intersect in Circle *T*.

Angle *VXY* and its vertical companion, *WXZ*, intercept arcs $\overset{\frown}{YV}$ and $\overset{\frown}{WZ}$. Those arcs contain 70° and 50° respectively. By adding their degree measures together and dividing by 2, we find that each of the angles in question has a measure of 60°.

The other pair of angles, *YXW* and *VXZ* intercept arcs of 180° and 60° respectively, for a total of 240°. Divide by 2 and find that each of the angles in question contains 120°. The rule may be stated as follows:

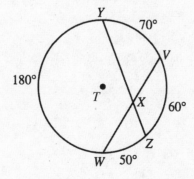

Circle *T*

When two chords intersect in a circle, each angle formed at the point of intersection is measured by $\frac{1}{2}$ the sum of the arcs intercepted by itself and its vertical angle.

The following questions will allow you to apply the information discussed in this section. Feel free to refer to the rules and diagrams while working the exercises.

Test Yourself

Directions: Given Circle O, with diameter FM, measure of $\overarc{FC} = 60°$, measure of $\overarc{CR} = 40°$. Find:

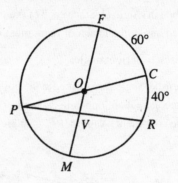

Circle O

1. The measure of $\angle FOC$.

2. The measure of $\angle CPR$.

3. The measure of $\angle POM$.

4. The measure of $\overarc{MR}$, $\overarc{MP}$ and $\overarc{FP}$.

5. The measure of $\angle MVR$ and $\angle FVR$.

6. In the same circle, draw $\angle FRM$. What is its degree measure?

Directions: Given Circle Q with the angles and arcs as marked, find the measures of:

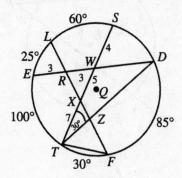

Circle *Q*

7. $\overset{\frown}{SD}$

8. $\angle EDT$

9. $\angle LXS$

10. $\angle SWX$

11. $\angle TFX$

12. $\overline{WD}$

13. $\angle DZF$

14. $\angle ESF$

Answers*

1. 60°

2. 20°

3. 60°

4. $m\overset{\frown}{MR} = 80°$, $m\overset{\frown}{MP} = 60°$, $m\overset{\frown}{FP} = 120°$

5. $m\angle MVR = 100°$, $m\angle FVR = 80°$

6. 90°

7. 60°

8. 50°

9. 45°

10. 180°

11. $62\frac{1}{2}°$

12. 8

13. 105°

14. 65° (when drawn)

A line outside the Circle that touches the Circle at a single point only is known as a **tangent**. $\overline{RS}$ is tangent to Circle *O* at point *T*. $\overline{PM}$ and $\overline{PV}$ are not tangents. They touch Circle *O* at two points, and, hence are known as **secants**. Secants may be thought of as chords that extend beyond the circle. Notice that secant *PM* terminates at *M* on the circle, while $\overline{PV}$ continues on past the Circle to *U*. Both segments are, nonetheless, secants.

*m stands for "measure of"

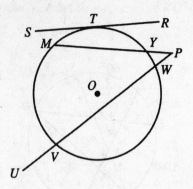

Circle *O*

In Circle *Q* you can see two tangents drawn to the Circle from an outside point. Two tangents drawn to a Circle from an outside point are always congruent. A radius drawn to the point of tangency (as is *QC*) is perpendicular to the tangent. That is to say, a right angle is formed at *C*.

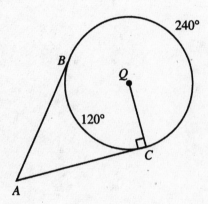

Circle *Q*

When two secants intersect outside a circle, the angle formed may be measured by taking half of the difference of their intercepted arcs. The angle formed by two tangents meeting outside the Circle may be found in the same way. The angle formed by the two secants to Circle *P* is $\frac{1}{2}(110 - 20)$ or 45°. Find the measure of angle *BAC* in the Figure that contains Circle *Q*.

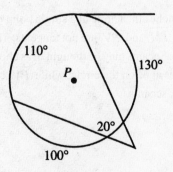

Circle *P*

Test Yourself

Directions: Complete the following.

Questions 1 and 2 refer to Circle *O* below.

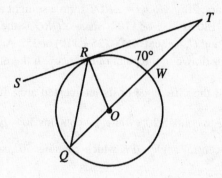

Circle *O*

1. The rule that applies to finding the angle formed by two secants or two tangents intersecting outside a Circle applies to an angle formed by an intersecting secant and tangent as well. Find m∠*STQ* in Circle *O*.

2. In Circle *O*, find the measure of ∠*QRO*. Also find m∠*SRO*.

Questions 3–7 all refer to Circle *P*.

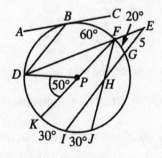

Circle *P*

3. Find the measure of ∠*E*.

4. Find the measure of ∠*GHJ*.

5. Find the degree measure of $\overparen{BD}$.

6. Find the measure of ∠*ABD*.

7. Find the measure of ∠*FDP*.

Answers

1. Since $\overline{QW}$ is a diameter, arc $\overparen{QW}$ contains $180°$. That means that $\overparen{QR} + 70° = 180°$, so $\overparen{QR} = 110°$. $\angle STQ$ then is found by taking half the difference of its two intercepted arcs: $\frac{1}{2}(110 - 70) = \frac{1}{2}(40) = 20°$.

2. $\angle Q$ is an inscribed angle measured by half $\overparen{RW}$, or $35°$. $\angle ROT$ is a central angle, and therefore equal to its intercepted arc, $70°$. $\angle ROQ + \angle ROT$ form a straight angle containing $180°$. That makes $\angle ROQ$ equal to $180° - 70°$, or $110°$. Since $\angle QRO$ is the third angle of the triangle containing angles ROQ and Q, it equals $180 - (35 + 110)$ or $35°$. Angle SRO is an angle formed by a tangent and a radius drawn to the point of tangency. It therefore contains $90°$.

3. $\angle E$ is measured by half the difference of its intercepted arcs. Its intercepted arcs are $\overparen{FG}$ and $\overparen{DKI}$. We know $\overparen{FG}$ to contain $20°$, while $\overparen{KI}$ contains $30°$. $\overparen{DKI}$ is formed by $\overparen{DK} + \overparen{KI}$. $\overparen{DK}$ is intercepted by central angle DPK which contains $50°$, so $\overparen{DK}$ contains $50°$. $\overparen{DKI}$, therefore, contains $50 + 30$, or $80°$. $m\angle E = \frac{1}{2}(80 - 20) = \frac{1}{2}(60) = 30°$.

4. To find $m\angle GHJ$ we first need the measure of $\overparen{GJ}$. $\overparen{FGJIK}$ is a semicircle bounded by diameter $\overline{FK}$. Since $80°$ of $\overparen{FGJIK}$ are already accounted for, $\overparen{GJ}$ must contain $100°$. We have already determined that $\overparen{DK}$ contains $50°$, and so can readily find $m\overparen{BD}$ which is part of the arc cut off by $\angle GHJ$'s vertical angle, $\angle FHI$. $m\overparen{BD} = 70°$. Since $\angle GHJ$ cuts off a $100°$ arc, and its vertical angle cuts off a $210°$ arc (add $m\overparen{FB} + m\overparen{BD} + m\overparen{DK} + m\overparen{KI}$), then $m\angle GHJ = \frac{1}{2}(100 + 210) = \frac{1}{2}(310) = 155°$.

5. We have already found $\overparen{BD}$ to contain $70°$.

6. $\angle ABD$ is measured by half its intercepted arc, $\overparen{BD}$, hence is $35°$.

7. $\angle FDP$ is an angle of triangle FDP. $\angle FPD$ is a central angle measured by $\overparen{FB} + \overparen{BD}$, and so is $60 + 70$, or $130°$. $\angle DFK$ is an inscribed angle measured by $\frac{1}{2}\overparen{KD}$ and so is $25°$. Adding the measures of the two angles just found together, we find that $155°$ of the $180°$ allotted for all the angles of the triangle have been used up. $180 - 155 = 25°$. $m\angle FDP = 25°$.

COORDINATE AXES

On a plane surface, any point can be located by giving two coordinates, a horizontal and a vertical one. This system assumes that all flat surfaces can be covered by a grid of intersecting lines that form little square boxes. Each point where two lines intersect on that grid is assigned a pair of numbered **coordinates**. The lines from which all numbering begins are called **axes** (axis is the singular of axes).

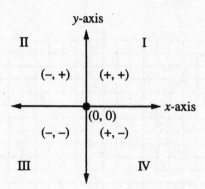

The vertical axis is known as the "y-axis." The y-axis intersects the horizontal "x-axis" at the point with coordinates (0,0). Beginning at that point, known as the **origin,** all other points on the grid are assigned a pair of coordinates in the form (x,y). By convention, the coordinate written first is the x coordinate. It tells the horizontal distance of a point from the origin, as well as its direction from the origin. A point with a positive x coordinate is to the right of the origin, while one with a negative x coordinate is to the left of the origin. The y coordinate tells the vertical distance of a point above (positive) or below (negative) the origin. The signs written as **ordered pairs** on the axes above indicate which coordinate is positive and which is negative on each portion of the grid. (The four sections of the grid are known as **quadrants**.) In the first quadrant, both x and y are positive. In the second quadrant (upper left) x is negative and y is positive. Both x and y are negative in the third quadrant. The fourth quadrant is positive for x and negative for y.

Consider the point with coordinates (3,5). This point is located by counting three spaces to the right of the y-axis and then 5 spaces up from the x-axis. You can see it marked on the grid below. Figure out how (⁻2,3) and (4, ⁻4) were arrived at. What are the coordinates of point A? Have you actually figured out A or are you peeking ahead? Well, in either case, its coordinates are (⁻5, ⁻3).

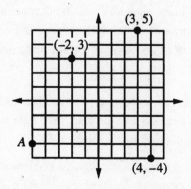

Test Yourself

> **Directions:** Following is a grid with a number of points marked off and indicated by letters. Give the coordinates of each lettered point. It may help you to remember that the origin has coordinates (0,0). Name each lettered point by an ordered pair of coordinates.

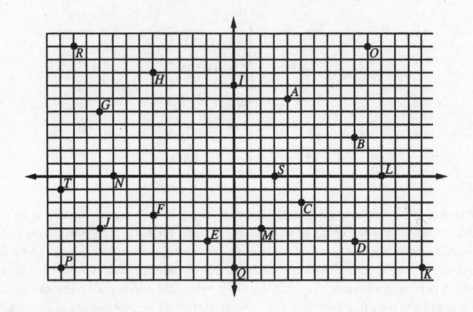

Answers

A. (4,6)	*B.* (9,3)	*C.* (5,⁻2)	*D.* (9,⁻5)
E. (⁻2,⁻5)	*F.* (⁻6,⁻3)	*G.* (⁻10,5)	*H.* (⁻6,8)
I. (0,7)	*J.* (⁻10,⁻4)	*K.* (14,⁻7)	*L.* (11,0)
M. (2,⁻4)	*N.* (⁻9, 0)	*O.* (10, 10)	*P.* (⁻13,⁻7)
Q. (0,⁻7)	*R.* (⁻12,10)	*S.* (3,0)	*T.* (⁻13,⁻1)

GRAPHING LINEAR EQUATIONS

Any equation may be graphed. Many equations that do not contain any whole number exponents higher than 1 have graphs that are straight lines. Hence, they are known as **linear** equations. Any linear equation may be written in what is called the **slope and y-intercept form**:

$$y = mx + b$$

In the equation above, x and y stand for, respectively, the x and y coordinates of any point on the graph of a given equation, m stands for the slope of the graph, and b for its y-intercept. The meanings of each of the last two terms we will now discuss separately.

The **y-intercept** is the name given to the point at which a graph crosses the y (vertical) axis. At this point, the value of the x coordinate must be 0. Think about that for a moment. If the graph (the line) is touching the y-axis at a particular point, then it is neither to the right nor to the left of the y-axis. Its distance from the y-axis, and hence its x-coordinate, must be 0. See Figure 1. The y-intercepts of the three graphs shown on the grid are, from top to bottom, 3, 0, and ⁻2. Note the coordinates of each y-intercept (as shown in the figure).

Now, assume that a linear equation has been written in slope and y-intercept form ($y = mx + b$). It has already been pointed out that b indicates the y-intercept. Here is why. Consider the equation, $y = 3x - 4$.

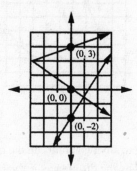

Figure 1

Notice that ⁻4 is in the b position, and, hence should be the y-intercept—if you are inclined to take what has been said so far on faith. You may recall that the x and y in the standard form of the linear equation stand for the coordinates of any point on the graph of the equation. But, since we know that at the y-intercept the x coordinate must be zero, let us substitute 0 for the x in the equation and see what happens:

$$y = 3x - 4$$
$$y = 3 \cdot 0 - 4$$
$$y = 0 - 4$$
$$y = {}^{-}4$$

Well, what do you know? The y-intercept is ⁻4 after all!

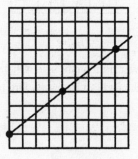

Figure 2

Now that we have discussed y-intercept, it is time to take a look at the meaning of m (in $y = mx + b$): **slope**. A look back at Figure 1 will reveal to you that the graphs of the equations that are pictured are lines that move from left to right at different angles. There are several different ways to calculate the steepness of the incline. One of those ways is known as slope. Slope is a measure of how much a line rises for every space it moves to the right on the grid. The graph in Figure 2 is rising 3 spaces for every 4 it moves to the right. It therefore has a slope of $\frac{3}{4}$.

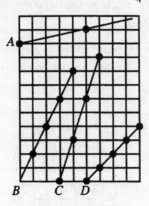

Figure 3

Slope may be computed by referring to any two points on the graph of an equation. It is calculated as the difference in y values divided by the difference in x values. While slope may be calculated between any two points on a graph, it is most accurately computed by using two points that clearly are on the intersection of two grid lines. Otherwise, the slope would be no more than an approximation. The slope of A is $\frac{1}{5}$, B's is 2, C's is 3, and D's is 1.

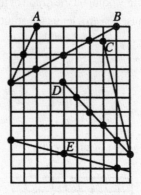

Figure 4

A line that runs from upper left to lower right on the grid will have a negative slope. That is because for every square the line moves to the right, it is descending rather than rising. The slope of A is 2, B is $\frac{1}{2}$, C is ¯4, D is ¯1, and E is $\frac{^-1}{4}$.

Writing an Equation from the Slope and *Y*-Intercept

If one can tell the slope and *y*-intercept of an equation by just examining it, then it stands to reason that one should be able to determine the equation of a line if the slope and *y*-intercept are known. Consider the model example below, and you will see how easy it is.

Example

Find the equation of the line with a slope of ⁻3 and a *y*-intercept of 6.

Remember that the standard form for a linear equation is:

$$y = mx + b$$

But we know that m = the slope = ⁻3 and b = the *y*-intercept = 6. By substituting the known values for m and b, we get:

$$y = {}^-3x + 6$$

That is the equation of the line with the given slope and *y*-intercept.

Use the example above to help you to do the exercises below. Once you get the hang of it, you will discover that it becomes almost a mechanical operation.

Test Yourself

Directions: Find the equation of the line that has a slope and *y*-intercept as indicated.

1. $m = 5, b = 7$
2. $m = 9, b = \dfrac{1}{2}$
3. $m = \dfrac{1}{4}, b = {}^-8$
4. $m = {}^-3, b = 2$
5. slope is ⁻2, *y*-intercept is ⁻6
6. slope is $\dfrac{2}{3}$, *y*-intercept is 5
7. *y*-intercept is ⁻18, slope is $\dfrac{4}{3}$
8. slope is ⁻15, $b = 9$
9. slope is $\dfrac{4}{7}$, the line passes through (0,9)
10. The line passes through (0,6) and has a slope of $\dfrac{{}^-1}{4}$.

Answers

1. $y = 5x + 7$
2. $y = 9x + \dfrac{1}{2}$
3. $y = \dfrac{1}{4}x - 8$
4. $y = {}^-3x + 2$
5. $y = {}^-2x - 6$
6. $y = \dfrac{2}{3}x + 5$
7. $y = \dfrac{4}{3}x - 18$
8. $y = {}^-15x + 9$
9. $y = \dfrac{4}{7}x + 9$
10. $y = \dfrac{{}^-1}{4}x + 6$

You should have noted that in exercises 9 and 10, the points whose coordinates were given were the *y*-intercepts, since the *x*-coordinate of each was 0. (0,*y*) for any value of *y* represents a *y*-intercept.

Finding an Equation from Two Points

It is possible to determine the equation of a line if you know the coordinates of two points on that line. In order to understand how this works, you must consider once more the meaning of slope and how it is derived. Remember that slope is the difference in y values over the difference in x values. Applying that to any two points, subtracting their y-coordinates and then putting the result over the difference in their x-coordinates yields the slope of that line. Consider a graph that passes through points with coordinates $(5,7)$ and $(3,9)$. Since you know two points on the graph, you find the slope of the graph. Beginning with the second point, we can find the difference in y values to be $9 - 7 = 2$. The difference in x values is $3 - 5 = {}^-2$. The slope, therefore, is $\frac{2}{{}^-2} = {}^-1$. If the slope had been calculated using the points in different order, we would have found $\frac{7 - 9}{5 - 3} = \frac{{}^-2}{2} = {}^-1$. You see then that either way we get a slope of ${}^-1$.

We may now substitute the value that we have found for the slope into the equation, $y = mx + b$:

$$y = {}^-1x + b, \text{ or}$$
$$y = {}^-x + b$$

Now, for any point on the graph, the x and y coordinates of that point may be substituted into the equation, since any point on the graph must satisfy the equation. We may choose either $(5,7)$ or $(3,9)$. If we choose the point $(5,7)$, then the value of x is 5 and the value of y is 7. Remember, the coordinates of any point are in the order (x,y). After substituting, the equation now looks like this:

$$7 = {}^-1(5) + b$$

It is now possible to solve for b and get the result $b = 12$. Suppose, for a moment, that we had chosen to select x and y using point $(3,9)$. The equation then would have to read:

$$9 = -1(3) + b$$

Solving that equation for b would have yielded a value $b = 12$. You'll notice that b comes out the same regardless of which point's coordinates are used.

Now that we know the value of m to be -1 and the value of b to be 12, all that is necessary is to substitute those values into the slope and y-intercept form of the equation, and we will have the equation that we are looking for:

$$y = mx + b$$
$$y = -1x + 12$$
$$y = -x + 12 \ \ldots \text{ and there you have it.}$$

Follow the example below, referring to the previous page if necessary.

Example

Find the equation of the line that passes through the points (3,5) and (-2,7).

First find the slope:

$$m = \frac{5 - 7}{3 - (-2)} = \frac{-2}{5} = -\frac{2}{5}$$

Then rewrite the slope and *y*-intercept form
including the slope that was found:

$$y = -\frac{2}{5}x + b$$

Now substitute the *x* and *y* values from

either point. Here, we use (3,5):

$$5 = -\frac{2}{5}(3) + b$$

Then solve the equation for *b*:

$$5 + \frac{6}{5} = b$$

$$b = 6\frac{1}{5}$$

Finally, substitute the values found for *b* and for
m into the standard form:

$$y = mx + b$$
$$y = -\frac{2}{5}x + 6\frac{1}{5}$$

Now you should be ready for some exercises. Use the example above as your guide. After you have done a few you should have the hang of it.

Test Yourself

Directions: Find the equation of the line that passes through the points named.

1. (3,6) and (5,8)   2. (2,4) and (5,10)   3. (0,7) and (2,11)

4. (2,8) and (6,10)   5. (7,10) and (9,2)   6. (5,6) and (7,4)

7. (4,2) and (8,1)   8. (4,4) and (8,7)   9. (0,1) and (5,2)

10. (2,4) and (4,8)   11. (3,0) and (4,3)   12. (7,2) and (9,4)

13. (4, -2) and (-1,3)   14. (6, -3) and (9, -3)   15. (-2, -3) and (-6, -8)

Answers

1. $y = x + 3$ 2. $y = 2x$ 3. $y = 2x + 7$

4. $y = \frac{1}{2}x + 7$ 5. $y = -4x + 38$ 6. $y = -x + 11$

7. $y = -\frac{1}{4}x + 3$ 8. $y = \frac{3}{4}x + 1$ 9. $y = \frac{1}{5}x + 1$

10. $y = 2x$ 11. $y = 3x - 9$ 12. $y = x - 5$

13. $y = -x + 2$ 14. $y = -3$ 15. $y = \frac{5}{4}x - \frac{1}{2}$

THE PYTHAGOREAN THEOREM

The ancient Greek mathematician, Pythagoras, came up with a fascinating discovery about 2300 years ago (give or take a century). He discovered that the square formed on the hypotenuse of a right triangle is equal in area to the sum of the squares formed on the other two sides of that triangle. His discovery is illustrated by the Figure you see here. Notice that sides of the triangle are labelled as having lengths of a, b, and c units. That means that the areas of the squares may be related as follows

$$a^2 + b^2 = c^2$$

Given any right triangle, if the lengths of two sides are known, the length of the third side may be found by means of this formula. Examine the following two examples and you will see how it works.

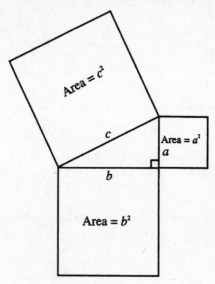

Example 1

The two legs of a right triangle are 3 inches and 4 inches long respectively. Find the length of the hypotenuse.

$a^2 + b^2 = c^2$	First write the formula ...
$(3)^2 + (4)^2 = c^2$	... then substitute into it
$9 + 16 = c^2$	Square the 3 and the 4 (raise them to the 2nd power) ...
$c^2 = 25$	... then add them together.
$c = 5$	Take the square root of each side.

The hypotenuse is 5 inches long.

Example 2

Find the length of side *AB* in the triangle pictured at the right. You may leave the answer in radical form.

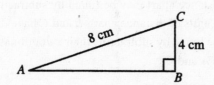

$$a^2 + b^2 \quad = c^2$$

$$(4)^2 + b^2 \quad = (8)^2$$
$$16 + b^2 \quad = 64$$
$$b^2 \quad = 48$$
$$b = \sqrt{48} \quad = 4\sqrt{3} \text{ cm}$$

First write the formula. Note, however, that this time we know the length of the hypotenuse, which means that we know the value of *c*! That fact must be taken into account when substituting. Now we square the 4 and the 8 and ... ... solve the equation for *b*.

Test Yourself

Directions: Each of the following questions refer to the figure below. The lengths of two sides of the triangle are given. Find the length of the third side. You may leave your answer in simplest radical form (where applicable).

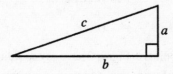

1. $a = 6, b = 8, c = ?$ 2. $a = 5, b = 12, c = ?$

3. $a = 15, c = 25, b = ?$ 4. $a = 20, c = 52, b = ?$

5. $a = 7, b = 9, c = ?$ 6. $a = 4, b = 11, c = ?$

7. $a = 9, c = 14, b = ?$ 8. $a = 4, c = 10, b = ?$

9. $b = 9, c = 12, a = ?$ 10. $c = 15, a = 9, b = ?$

11. $c = 16, b = 10, a = ?$ 12. $a = 7, b = 10, c = ?$

Answers

1. 10   2. 13   3. 20   4. 48

5. $\sqrt{130}$ 6. $\sqrt{137}$ 7. $\sqrt{115}$ 8. $2\sqrt{21}$

9. $3\sqrt{7}$ 10. 12   11. $2\sqrt{39}$ 12. $\sqrt{149}$

THE DISTANCE BETWEEN POINTS ON A GRAPH

Figure 1 shows a portion of the first quadrant of a grid with three lettered points in proper relationship to one another. Since points *A* and *B* are both on the same vertical line (have the same *x*-coordinate), their distance apart may be found by subtracting their *y*-coordinates (8 - 3). *A* and *B* are, therefore, 5 units of distance apart. *B* and *C* have the same *y*-coordinates, and so their distance apart may be found by subtracting their *x*-coordinates (10 - 4). They are 6 units apart. But how far apart are *A* and *C*?

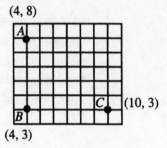

Figure 1

If segment *AC* were drawn, it would form the hypotenuse of right triangle *ABC*. The distance point *A* is from point *C* can then be found using the Pythagorean theorem:

$$AC^2 = (5)^2 + (6)^2$$
$$AC = \sqrt{61} \text{ units}$$

Now look at Figure 2. No pair of points in Figure 2 have any coordinate in common. Yet the distance between any pair of points in the Figure may be found in a manner similar to that used in finding the length of $\overline{AC}$.

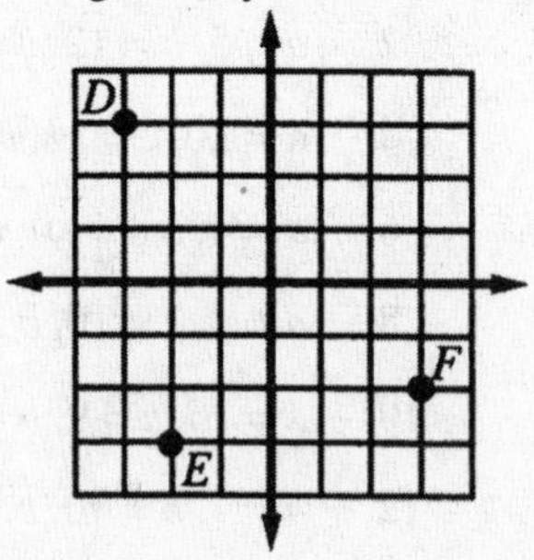

Figure 2

Example

Find the lengths of the sides of the triangle formed by connecting points *D*, *E*, and *F* in Figure 2.

First note the coordinates of the points:

$$D: (^-3, 3) \ E: (^-2, ^-3) \ F: (3, ^-2)$$

Let us first find the length of segment *DE*. The horizontal distance between *D* and *E* is found by subtracting the horizontal coordinates of the two points: $^-3 - (^-2) = ^-1$. Since the minus sign in $^-1$ simply shows direction, we can safely ignore it, and say that the horizontal distance is 1 unit. The vertical distance is found by subtracting their vertical coordinates $3 - (^-3)$ which = 6.

For all intents and purposes, we are now seeking to find the hypotenuse of a right triangle with sides 1 and 6. Look at Figure 3, and you just might see that imaginary triangle. In fact, you might even see imaginary right triangles around segments *DF* and *EF*. As long as you are using your imagination, you might as well see them as being shaded. ✦

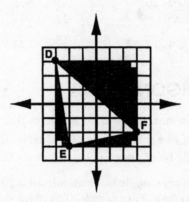

DE^2, then, is equal to $(1)^2 + (6)^2$, if you recall your Pythagorean Theorem. That means that DE is $\sqrt{37}$ long.

To find *DF*, find the difference between the vertical and horizontal coordinates of the two points. They are 5 and 6 respectively.

$$DF^2, \text{ then, is equal to } (5)^2 + (6)^2$$
$$= 25 + 36$$
$$= 61$$
$$DF = \sqrt{61} \text{ units}$$

To find *EF*, find the differences between the vertical coordinates and the horizontal coordinates. Then plug those differences into the Pythagorean formula and find that $EF = \sqrt{26}$.

Test Yourself

Directions: Find the distance between the points with coordinates as given.

1. (3,7) and (3,⁻9)   2. (5,12) and (5,18)   3. (2,15) and (⁻11,15)

4. (⁻8,⁻5) and (⁻13,⁻5)   5. (4,11) and (7,11)   6. (8,6) and (4,9)

7. (⁻2,10) and (10,5)   8. (3,7) and (5,⁻3)   9. (⁻1,⁻4) and (5,8)

10. (9,6) and (8,1)   11. (0,0) and (10,24)   12. (5,7) and (8,0)

13. (12,3) and (6,5)   14. (⁻3,4) and (5,⁻8)   15. (6,⁻8) and (7,⁻2)

Answers

1. 16   2. 6   3. 13

4. 5   5. 3   6. 5

7. 13   8. $2\sqrt{26}$ 9. $6\sqrt{5}$

10. $\sqrt{26}$ 11. 26   12. $\sqrt{58}$

13. $2\sqrt{10}$ 14. $4\sqrt{13}$ 15. $\sqrt{37}$

RIGHT TRIANGLE TRIGONOMETRY

On the GED, you may have a small number of problems that ask you to use right triangle trigonometry to find the length of a missing side of a triangle. Even if you have never learned trigonometry before, with a little bit of practice, you will easily be able to do these problems.

As you will read below, since every triangle has three sides, it is possible to form 6 different ratios between pairs of sides of any triangle. For example, if the three sides of a triangle are called a, b, and c, the three ratios would be $\frac{a}{b}, \frac{b}{a}, \frac{a}{c}, \frac{c}{a}, \frac{c}{b}$, and $\frac{b}{c}$. As you will see, the values of the ratio of one side of a triangle to another can be related to the measures of the angles within the triangle.

The Trigonometric Ratios

Every right triangle contains two acute angles. With respect to each of these angles, it is possible to define six ratios, called trigonometric ratios, each involving the lengths of two of the sides of the triangle. For example, consider the following triangle *ABC*.

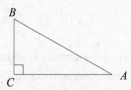

In this triangle, side *AC* is called the side adjacent to angle *A*, and side *BC* is called the side opposite angle *A*. Similarly, side *AC* is called the side opposite angle *B*, and side *BC* is called the side adjacent to angle *B*. Of course, side *AB* is referred to as the hypotenuse with respect to both angles *A* and *B*.

The six trigonometric ratios with respect to angle A, along with their standard abbreviations, are given below:

Sine of angle A = sin A = $\dfrac{\text{opposite}}{\text{hypotenuse}} = \dfrac{BC}{AB}$

Cosine of angle A = cos A = $\dfrac{\text{adjacent}}{\text{hypotenuse}} = \dfrac{AC}{AB}$

Tangent of angle A = tan A = $\dfrac{\text{opposite}}{\text{adjacent}} = \dfrac{BC}{AC}$

Cotangent of angle A = cot A = $\dfrac{\text{adjacent}}{\text{opposite}} = \dfrac{AC}{BC}$

Secant of angle A = sec A = $\dfrac{\text{hypotenuse}}{\text{adjacent}} = \dfrac{AB}{AC}$

Cosecant of angle A = csc A = $\dfrac{\text{hypotenuse}}{\text{opposite}} = \dfrac{AB}{BC}$

The last three ratios are actually the reciprocals of the first three, in particular:

cot A = $\dfrac{1}{\tan A}$

sec A = $\dfrac{1}{\cos A}$

csc A = $\dfrac{1}{\sin A}$

Also note that:

$$\frac{\sin A}{\cos A} = \tan A, \text{ and } \frac{\cos A}{\sin A} = \cot A.$$

Example 1

Consider right triangle DEF below, whose sides have the lengths indicated. Find sin D, cos D, tan D, sin E, cos E, and tan E.

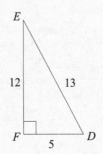

NOTE

The sine of D is equal to the cosine of E, and the cosine of D is equal to the sine of E.

$$\sin D = \frac{EF}{ED} = \frac{12}{13} \qquad \sin E = \frac{DF}{ED} = \frac{5}{13}$$

$$\cos D = \frac{DF}{ED} = \frac{5}{13} \qquad \cos E = \frac{EF}{ED} = \frac{12}{13}$$

$$\tan D = \frac{EF}{DF} = \frac{12}{5} \qquad \tan E = \frac{DF}{EF} = \frac{5}{12}$$

Example 2

In right triangle ABC, $\sin A = \frac{4}{5}$. Find the values of the other 5 trigonometric ratios.

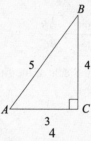

Since the sine of A = opposite over hypotenuse = $\frac{4}{5}$, we know that $BC = 4$ and $AB = 5$. We can use the Pythagorean theorem to determine that $AC = 3$. Then:

$$\cos A = \frac{3}{5}, \tan A = \frac{4}{3}, \cot A = \frac{3}{4}, \sec A = \frac{5}{3}, \csc A = \frac{5}{4}.$$

The actual values for the trigonometric ratios for most angles are irrational numbers, whose values can most easily be found by looking in a trigonometry table or using a calculator. There are, however, a few angles who ratios can be obtained exactly. The ratios for 30°, 45°, and 60° can be determined from the properties of the 30-60-90 right triangle and the 45-45-90 right triangle. First of all, note that the Pythagorean theorem can be used to determine the following side and angle relationships in 30-60-90 and 45-45-90 triangles.

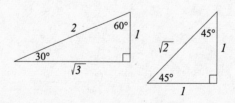

From these diagrams, it is easy to see that

$$\sin 30° = \frac{1}{2}, \cos 30° = \frac{\sqrt{3}}{2}, \tan 30° = \frac{1}{\sqrt{3}} = \frac{\sqrt{3}}{3}$$

$$\sin 60° = \frac{\sqrt{3}}{2}, \cos 60° = \frac{1}{2}, \tan 60° = \sqrt{3} \quad \sin 45° = \cos 45° = \frac{1}{\sqrt{2}} = \frac{\sqrt{2}}{2}, \tan 45° = 1$$

Example 3

From point *A*, which is directly across from point *B* on the opposite sides of the banks of a straight river, the measure of angle *CAB* to point *C*, 35 meters upstream from *B*, is 30°. How wide is the river?

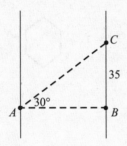

To solve this problem, note that

$\tan A = \dfrac{\text{opposite}}{\text{adjacent}} = \dfrac{BC}{AB} = \dfrac{35}{AB}$. Since the measure of angle A is 30°, we have

$\tan 30° = \dfrac{35}{AB}$. Then:

$$AB = \frac{35}{\tan 30°} = \frac{35}{\dfrac{\sqrt{3}}{3}} =$$

Therefore, the width of the river is $\dfrac{105}{\sqrt{3}}$ meters, or approximately 61 meters wide.

EXERCISES: GEOMETRY

What type of triangle is each of the triangles below?

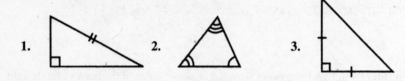

1.   2.   3.

Find the total number of degrees of angle measure in each figure.

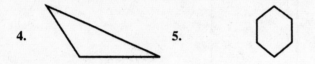

4.   5.

Find the area of each figure.

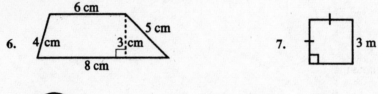

6.   7.

8.

9.   Write the equation of a line that passes through the points (-4,2) and (6,5).

10.   Find the distance between the points (3,6) and (-9,11).

Answers

1. right
2. scalene
3. right-isosceles
4. 180°
5. 720°
6. 21 cm^2
7. 9 m^2
8. 9π cm^2
9. $y = \dfrac{3}{10}x + 3\dfrac{1}{5}$
10. 13

SUMMING IT UP

- Geometry is the science of measurement.

- A line in geometry is always assumed to be a straight line.

- When two rays share a common endpoint, they form an angle. The rays are called sides of the angle, and the point is called the vertex. The symbol for angle is ∠.

- When two lines intersect, four angles are formed.

- A triangle is a polygon of three sides.

- Circles are closed curves with all points on the curve equally distant from the center. The circle is named by its center point.

- A line segment from the center of a circle to any point on the circle is called the radius.

- The perimeter of a circle is called the circumference.

PART IV

TWO PRACTICE TESTS

✂

ANSWER SHEET PRACTICE TEST 2

Part 1

1. ① ② ③ ④ ⑤   7. ① ② ③ ④ ⑤   14. ① ② ③ ④ ⑤   20. ① ② ③ ④ ⑤
2. ① ② ③ ④ ⑤   8. ① ② ③ ④ ⑤   15. ① ② ③ ④ ⑤   24. ① ② ③ ④ ⑤
3. ① ② ③ ④ ⑤   11. ① ② ③ ④ ⑤   17. ① ② ③ ④ ⑤   25. ① ② ③ ④ ⑤
5. ① ② ③ ④ ⑤   12. ① ② ③ ④ ⑤   18. ① ② ③ ④ ⑤
6. ① ② ③ ④ ⑤   13. ① ② ③ ④ ⑤   19. ① ② ③ ④ ⑤

4.   9.   10.
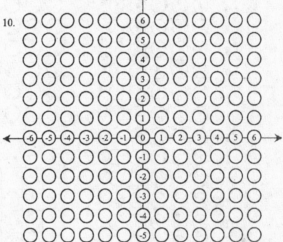

16.   21.   22.   23.

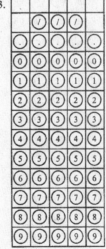

answer sheet

Part 2

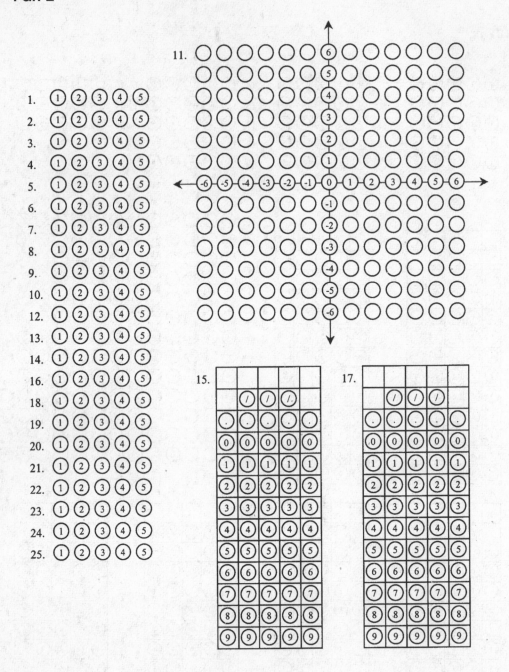

1. ① ② ③ ④ ⑤
2. ① ② ③ ④ ⑤
3. ① ② ③ ④ ⑤
4. ① ② ③ ④ ⑤
5. ① ② ③ ④ ⑤
6. ① ② ③ ④ ⑤
7. ① ② ③ ④ ⑤
8. ① ② ③ ④ ⑤
9. ① ② ③ ④ ⑤
10. ① ② ③ ④ ⑤
12. ① ② ③ ④ ⑤
13. ① ② ③ ④ ⑤
14. ① ② ③ ④ ⑤
16. ① ② ③ ④ ⑤
18. ① ② ③ ④ ⑤
19. ① ② ③ ④ ⑤
20. ① ② ③ ④ ⑤
21. ① ② ③ ④ ⑤
22. ① ② ③ ④ ⑤
23. ① ② ③ ④ ⑤
24. ① ② ③ ④ ⑤
25. ① ② ③ ④ ⑤

PRACTICE TEST 2

90 Minutes • 50 Questions Total

PART 1—25 QUESTIONS (A CALCULATOR IS PERMITTED): 45 MINUTES

PART 2—25 QUESTIONS (A CALCULATOR IS NOT PERMITTED): 45 MINUTES

Directions: The Mathematics Test consists of questions intended to measure general mathematics skills and problem-solving ability. The questions are based on short readings that often include a graph, chart, or figure. Work carefully, but do not spend too much time on any one question. Be sure you answer every question. You will not be penalized for incorrect answers.

Formulas you may need are given on the following pages. Only some of the questions will require you to use a formula. Record your answers on the separate answer sheet. Be sure that all information is properly recorded.

There are three types of answers found on the answer sheet:

❶ Type 1 is a regular format answer that is the solution to a multiple-choice question. It requires shading in 1 of 5 bubble choices.

❷ Type 2 is an alternate format answer that is the solution to the standard grid "fill-in" type question. It requires shading in bubbles representing the actual numbers, including a decimal or division sign where applicable.

❸ Type 3 is an alternate format answer that is the solution to a coordinate plane grid problem. It requires shading in the bubble representing the correct coordinate of a graph.

Type 1: Regular Format, Multiple-Choice Question

To record your answers for multiple-choice questions, fill in the numbered circle on the answer sheet that corresponds to the answer you select for each question in the test booklet.

Q Jill's drug store bill totals $8.68. How much change should she get if she pays with a $10.00 bill?

  (1) $2.32

  (2) $1.42

  (3) $1.32

  (4) $1.28

  (5) $1.22   ① ② ● ④ ⑤

A **The correct answer is (3).** Therefore, you should mark answer space (3) on your answer sheet.

Type 2: Alternate Format, Standard Grid Question

To record the answer to the previous example, "1.32," using the Alternate Format, Standard Grid, see below:

Standard Grid

Type 3: Alternate Format, Coordinate Plane Grid Question

To record your answer, fill in the numbered circle on the answer sheet that corresponds to the correct coordinate in the graph. For example:

Q A system of two linear equations is given below.

$$x = -3y$$
$$x + y = 4$$

What point represents the common solution for the system of equations?

A The correct answer is **(6, –2).** The answer should be gridded as shown below.

Coordinate Plane Grid

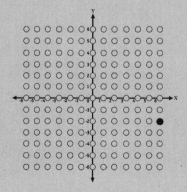

FORMULAS

Description	Formula
AREA (A) of a:	
square	$A = s^2$; where s = side
rectangle	$A = lw$; where l = length, w = width
parallelogram	$A = bh$; where b = base, h = height
triangle	$A = \frac{1}{2} bh$; where b = base, h = height
circle	$A = \pi r^2$; where π = 3.14, r = radius
PERIMETER (P) of a:	
square	$P = 4s$; where s = side
rectangle	$P = 2l + 2w$; where l = length, w = width
triangle	$P = a + b + c$; where a, b, and c are the sides
Circumference (C) of a circle	$C = \pi d$; where π = 3.14, d = diameter
VOLUME (V) of a:	
cube	$V = s^3$; where s = side
rectangular container	$V = lwh$; where l = length, w = width, h = height
cylinder	$V = \pi r^2 h$; where π = 3.14, r = radius, h = height
square pyramid	Volume = $\frac{1}{3} \times$ (base edge)$^2 \times$ height
cone	Volume = $\frac{1}{3} \times \pi \times$ radius$^2 \times$ height; π is approximately equal to 3.14.
Pythagorean theorem	$c^2 = a^2 + b^2$; where c = hypotenuse, a and b are legs of a right triangle
distance (d) between two points in a plane	$d = \sqrt{(x_2 - x_1)^2 + (y_2 - y_1)^2}$; where (x_1, y_1) and (x_2, y_2) are two points in a plane
slope of a line (m)	$m = \dfrac{y_2 - y_1}{x_2 - x_1}$ where (x_1, y_1) and (x_2, y_2) are two points in a plane
trigonometric ratios	given an acute angle with measure x of a right triangle, $\sin x = \dfrac{\text{opposite}}{\text{hypotenuse}}$, $\cos x = \dfrac{\text{adjacent}}{\text{hypotenuse}}$, $\tan x = \dfrac{\text{opposite}}{\text{adjacent}}$
mean	mean $= \dfrac{x_1 + x_2 + \cdots + x_n}{n}$; where the x's are the values for which a mean is desired, and n = number of values in the series
median	median = the point in an ordered set of numbers at which half of the numbers are above and half of the numbers are below this value
simple interest (i)	$i = prt$; where p = principal, r = rate, t = time
distance (d) as function of rate and time	$d = rt$; where r = rate, t = time
total cost (c)	$c = nr$; where n = number of units, r = cost per unit

You may use a scientific calculator for Part 1. (A Casio FX-260 Scientific Calculator will be provided at your Official GED Testing Center.)

CALCULATOR DIRECTIONS

You may practice with your calculator, using the following directions.

CALCULATOR DIRECTIONS

To prepare the calculator for use the *first* time, press the [ON] (upper-rightmost) key. "DEG" will appear at the top-center of the screen and "0" at the right. This indicates the calculator is in the proper format for all your calculations.

To prepare the calculator for *another* question, press the [ON] or the red [AC] key. This clears any entries made previously.

To do any arithmetic, enter the expression as it is written. Press [ON] (equals sign) when finished.
EXAMPLE A: 8 – 3 + 9

First press [ON] or [AC].
Enter the following:
[8] [–] [3] [+] [9] [=]
The correct answer is 14.

If an expression in parentheses is to be multiplied by a number, press [x] (multiplication sign) between the number and the parenthesis sign.
EXAMPLE B: 6(8 + 5)

First press [ON] or [AC].
Enter the following:
[6] [x] [(] [8] [+] [5] [)] [=]
The correct answer is 78.

To find the square root of a number
• enter the number;
• press the [SHIFT] (upper-leftmost) key ("SHIFT" appears at top-left of the screen);
• press [x^2] (third from the left on top row) to access its second function: square root.
DO NOT press [SHIFT] and [x^2] at the same time.
EXAMPLE C: $\sqrt{64}$

First press [ON] or [AC].
Enter the following:
[6] [4] [SHIFT] [x^2] [=]
The correct answer is 8.

To enter a negative number such as -8,
• enter the number without the negative sign (enter 8);
• press the "change sign" ([+/-]) key which is directly above the [7] key.
All arithmetic can be done with positive and/or negative numbers.
EXAMPLE D: -8 – -5

First press [ON] or [AC].
Enter the following:
[8] [+/-] [–] [5] [+/-] [=]
The correct answer is -3.

Part 1

Directions: You may now begin Part 1 of the Mathematics Test. Bubble in the correct response to each question on Part 1 of your answer sheet.

1. Danny bought 3 sodas for 95 cents each, and a newspaper for 65 cents. How much change would he receive from a ten dollar bill?

   (1) $3.50
   (2) $4.80
   (3) $5.20
   (4) $6.50
   (5) $8.40

2. A sporting goods store normally discounts all merchandise 16%. At a special sale, it is taking an additional $\frac{1}{5}$ off its discount price. During the special sale, how much would you expect to pay for a baseball glove with a list price of $56?

   (1) $47.04
   (2) $44.80
   (3) $37.63
   (4) $50.20
   (5) $35.84

3. 2.54 centimeters = 1 inch

   From the fact stated above, 1 centimeter is about equal to

   (1) .4 inches
   (2) 1.54 inches
   (3) 2.54 inches
   (4) .6 inches
   (5) .7 inches

4. Nancy wishes to make a macramé wall hanging for her living room. The directions call for 12 pieces of jute 12.5 meters long, 18 pieces 8.25 meters long, and 24 pieces 7 meters long. How much jute will she need for her wall hanging?

   Mark your answer in the circles in the grid on your answer sheet.

5. Peter is reading the book *Ivanhoe* that is 436 pages long. He has read 60 pages and it has taken him $1\frac{1}{2}$ hours. If he continues to read at the same pace, how much longer will it take him to finish the book?

   (1) 9.4 hours
   (2) 9.9 hours
   (3) 10.9 hours
   (4) 11.4 hours
   (5) 11.9 hours

6. A cassette box has dimensions as shown below.

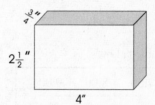

   Rusty wants to ship his collection of cassettes to his brother in California. What is the maximum number of cassettes Rusty can put into a carton with a capacity of 180 cubic inches?

   (1) 12
   (2) 18
   (3) 24
   (4) 30
   (5) 36

7. Mrs. Gabaway wants to telephone her friend in Boston. The day rate is $.48 for the first minute and $.34 for each additional minute. The evening rate discounts the day rate by 35%. If Mrs. Gabaway is planning a 45-minute chat, to the nearest penny, how much would she save if she took advantage of the evening rate by calling after 5 PM?

Mark your answer in the circles in the grid on your answer sheet.

8. Janet withdrew amounts of $2,356 and $1,131 from her savings account. After the withdrawals, she was left with a balance of $11,516. What was her balance before the withdrawals?

(1) $8,029

(2) $15,003

(3) $6,974

(4) $10,461

(5) $18,490

9. Look at the table below.

Time	7:00	8:00	9:00
Distance	24 km	48 km	72 km

If a ship traveled away from port at a steady speed as shown in the table above, how far from port was it at 8:35?

(1) 56 kilometers

(2) 58 kilometers

(3) 60 kilometers

(4) 62 kilometers

(5) Not enough information

10. Quadrilateral ABCD is a square. The coordinates of point A are (3, 2), the coordinates of point B are (-3, 2), and the coordinates of points C are (-3, -4).

On the coordinate plane on your answer sheet, mark the location of point D.

Questions 11 to 15 refer to the following graph.

Karen's Take-Home Dollar

11. How much of each dollar does Karen spend on food and clothing?

(1) $.35

(2) $.25

(3) $.15

(4) $.20

(5) Not enough information is given

12. Suppose Karen earns $800 per week after taxes. How much money does she save each week?

(1) $28

(2) $280

(3) $56

(4) $112

(5) Not enough information is given

13. Suppose Karen brings home $800 per week after taxes. How much does she spend per week on entertainment?

(1) $56

(2) $120

(3) $12

(4) $24

(5) Not enough information is given

14. If Karen spends $90 per week on clothing, how much money does she take home each week?

    (1) $400

    (2) $550

    (3) $600

    (4) $700

    (5) $72

15. If Karen brings home $450 per week, which of the following might she pay weekly for electricity?

    (1) $24

    (2) $38

    (3) $48

    (4) $60

    (5) $72

16. Suzanne bought 4 record albums from Marvin's Music Emporium when it had a "Going Out of Business" sale. Two albums originally sold for $6.95 each; the other two for $8.95 each. Every album in the store was discounted by 30% How much did Suzanne spend?

    Mark your answer in the circle in the grid on your answer sheet.

17. What is the average height of a player on the 10th Street Basketball Team if the heights of the individual players are 5´8", 6´1", 5´10", 6´3", and 5´9"?

    (1) 5´8"

    (2) 5´9"

    (3) 5´10"

    (4) 5´11"

    (5) 6´

18. Hazel received a chain letter in the mail that instructed her to send $5 to the first name on the list. Hazel sent the $5, crossed off the receiver's name, and added her name to the bottom of the list that at any time consists of 6 names. She then sent copies to 10 of her friends, who were instructed to do likewise. Assuming everyone followed directions, by the time Hazel's name rose to the first position on the list, how much money would she receive?

    (1) $5,000

    (2) $50,000

    (3) $500,000

    (4) $5,000,000

    (5) $50,000,000

Questions 19 and 20 refer to the figure below.

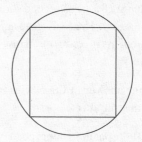

19. As shown in the figure, a square is inscribed in a circle of radius 5. What is the length of a side of the square?

    (1) $5\sqrt{2}$

    (2) $10\sqrt{2}$

    (3) $5\sqrt{3}$

    (4) 10

    (5) 12

20. What is the area of the square in the figure?

    (1) 25

    (2) $25\sqrt{2}$

    (3) 50

    (4) $50\sqrt{2}$

    (5) 100

21. In right triangle PQR, cot P = $\dfrac{5}{12}$. Find the value of sec P.

Mark your answer in the circles in the grid on your answer sheet.

22. The shrew has a heartier appetite than any other animal. A shrew, weighing only $\dfrac{1}{8}$ of an ounce, is capable of eating $3\dfrac{1}{2}$ ounces of food in 8 days. How much can the shrew consume in one day?

Mark your answer in the circles in the grid on your answer sheet.

23. Bill invested a sum of money at 9% and a second sum at 18%. The second sum was $450 less than the first. He received an annual return of $162 from his investments. How much was placed at 18%?

Mark your answer in the circles in the grid on your answer sheet.

24. Work O'Holic put in 12 hours per day, Monday through Friday, 8 hours on Saturday and 4 hours on Sunday to produce and package the liniment his factory needed. He is paid $8.40 an hour for a 40 hour week and time and a half for anything over 40 hours. How much did Mr. O'Holic gross for the week?

(1) $436.80

(2) $650.40

(3) $736.80

(4) $739.20

(5) $840.00

25. How many cubic inches of liquid can the cylindrical can below hold?

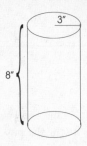

(1) 48π

(2) 64π

(3) 72π

(4) 96π

(5) 108π

Part 2

Directions: You may not return to Part 1 or use your calculator for this part. Bubble in the correct response to each question on Part 2 of your answer sheet.

1. Brian wishes to fence in a rectangular garden plot shown below. If the width of the garden is two feet less than half its length, the number of feet of fencing he will need can be represented as

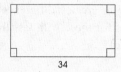

34

(1) $34 + 15$

(2) 34×15

(3) $2(34 \times 15)$

(4) $2(34) + 2(17) - 2$

(5) $2(34) + 2(15)$

2. If there are 10 millimeters in a centimeter and 10 centimeters in a decimeter, how many millimeters are there in a decimeter?

(1) 1

(2) 10

(3) 100

(4) 1000

(5) 10,000

3. Penelope saved half of her birthday cake for her grandparents and Aunt Lucy who were coming for a visit. She divided what was left of the cake into 3 equal pieces. What portion of the original cake was each piece?

(1) $\frac{1}{3}$

(2) $\frac{1}{4}$

(3) $\frac{1}{5}$

(4) $\frac{1}{6}$

(5) $\frac{1}{7}$

4. As the diagram below shows, a 12-foot tall lamp-post casts an 8-foot long shadow at the same time that a tree nearby casts a 28-foot shadow. If T represents the height of the tree, which of the following equations could be solved to determine the value of T?

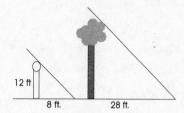

12 ft

8 ft.   28 ft.

(1) $\frac{8}{12} = \frac{28}{T}$

(2) $\frac{8}{12} = \frac{T}{28}$

(3) $\frac{12}{8} = \frac{28}{T}$

(4) $\frac{12}{28} = \frac{8}{T}$

(5) $\frac{8}{28} = \frac{T}{12}$

5. $527(316 + 274)$ has the same value as which of the following?

(1) $316(527) + 274$

(2) $316 + 527 + 274$

(3) $527(316) + 527(274)$

(4) $316(527) + 316(274)$

(5) $527(274) + 316(274)$

6. If Elaine spends an average of 13 minutes on each interview at her temporary employment agency, approximately how many prospective employees can she interview in an 8-hour workday?

(1) $\dfrac{8 \times 13}{60}$

(2) $8 \times 60 \times 13$

(3) $\dfrac{8 \times 60}{13}$

(4) $\dfrac{13}{8}(60)$

(5) $\dfrac{60 \times 13}{8}$

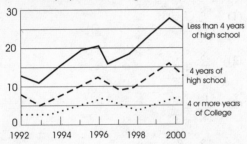

Unemployed Civilians, ages 18–24

The graph illustrates the unemployment rate by age and amount of education. The next 3 questions refer to the graph above.

7. According to the graph, in which year was unemployment the highest?

(1) 1996

(2) 1997

(3) 1998

(4) 1999

(5) 2000

8. What was the difference in percent of unemployment between high school graduates and those with less than 4 years of high school in 1997?

(1) 8%

(2) 10%

(3) 12%

(4) 15%

(5) 18%

9. What was the greatest difference in unemployment rates for those with less than 4 years of high school?

(1) 55%

(2) 10%

(3) 15%

(4) 20%

(5) 25%

10. Twice the sum of 3 and a number is 1 less than 3 times the number. If the letter N is used to represent the number, which of the following equations could be solved in order to determine the number?

(1) $2(3) + N = 3N - 1$

(2) $2(3 \times N) = 3N - 1$

(3) $2(3 + N) = 3N - 1$

(4) $2(3 + N) - 1 = 3N$

(5) $2(3 + N) = 1 - 3N$

11. A line segment has endpoints (–1, –2) and (3, 4).

On the coordinate plane on your answer sheet, mark the midpoint of the line segment.

12. If a car averages 60 mph on a cross country trip, how long would it take to go 1500 miles?

(1) 20 hours

(2) 25 hours

(3) 30 hours

(4) 35 hours

(5) 40 hours

13. If $x^2 + 5x + 6 = 0$, then $x =$

    (1) $^-2$ and $^-3$

    (2) $^-3$ only

    (3) $^+3$ and $^+2$

    (4) $^+5$ only

    (5) $^+2$ only

14. Which of the following is the radius of the largest ball that will fit into the box shown below?

    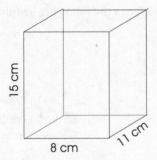

    (1) 15 centimeters

    (2) 11 centimeters

    (3) 8 centimeters

    (4) 5 centimeters

    (5) 4 centimeters

15. Stock in North American Electric fluctuated in price with a high of $67\frac{3}{4}$ and a low of $63\frac{5}{8}$. Find the difference between the high and the low price.

    Mark your answer in the circles in the grid on your answer sheet.

16. In order to determine the expected mileage for a particular car, an automobile manufacturer conducts a factory test on five of these cars. The results, in miles per gallon, are 25.3, 23.6, 24.8, 23.0, and 24.3. What is the median mileage?

    Mark your answer in the circles in the grid on your answer sheet.

17. James plans to cut 58 meters of fencing into 8 pieces of equal length. How long will each piece be?

    (1) 7.25 meters

    (2) 7.5 meters

    (3) 8 meters

    (4) 64 meters

    (5) 464 meters

18. Bill and Bert drove 530 miles to a secluded lake for a week of fishing. They drove for 10 hours, alternating between 60 mph on the highways and 40 mph through the towns. For how much of their trip were the men traveling 40 mph?

    (1) $3\frac{1}{2}$ hours

    (2) 4 hours

    (3) $4\frac{1}{2}$

    (4) $5\frac{1}{2}$

    (5) Not enough information is given

19. Mr. Scalici's office floor has measurements as shown below. The vinyl tiles that he chose for his flooring came in 1-foot squares, packed 12 to a carton. How many cartons would he need to buy in order to tile the entire floor?

    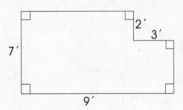

    (1) 2 cartons

    (2) 3 cartons

    (3) 4 cartons

    (4) 5 cartons

    (5) 6 cartons

20. Roast beef is selling at the local supermarket for \$4.80 per pound. Alessandra buys $\frac{2}{3}$ of a pound. How much does she pay for it?

(1) \$4.13

(2) \$5.47

(3) \$4.80

(4) \$2.40

(5) \$3.20

21. There are 155 children signed up for a class field trip. The number of girls exceeds the number of boys by 17. If B represents the number of boys, which of the following equations could be solved to determine the number of boys signed up for the class trip?

(1) B + (B − 17) = 155

(2) 155 + B = B + 17

(3) B + (B + 17) = 155

(4) 155 + B = B - 17

(5) B + (B − 17) = 138

22. A triangular shaped building lot has a front on the road that is 20 meters long, as shown in the diagram below. What is the area of the lot?

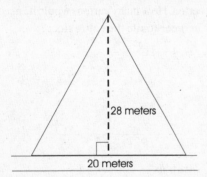

(1) 140 m²

(2) 280 m²

(3) 500 m²

(4) 750 m²

(5) 1,000 m²

23. A vending machine contains \$21 in dimes and nickels. Altogether there are 305 coins. If N represents the number of nickels, which of the following equations could be solved in order to determine the value of N?

(1) .10(N − 305) + .5N = 21

(2) 10(305 − N) + 5N = 21

(3) .10N + .05(305 − N) = 21

(4) .10 (N −21) + 5N = 305

(5) .10(305 − N) + .05N = 21

Questions 24-25 refer to the following graph:

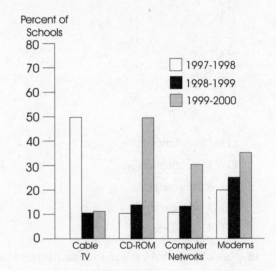

24. In the year that 15% of the public schools had CD-ROMs, what percent had modems?

(1) 5%

(2) 10%

(3) 20%

(4) 25%

(5) 40%

25. In 1997-1998, there were 1,500 schools. How many more schools used cable TV than used modems?

(1) 25

(2) 350

(3) 375

(4) 425

(5) 450

ANSWER KEY AND EXPLANATIONS

Part 1

1. 4	**6.** 3	**11.** 1	**16.** 22.26	**21.** 2.6
2. 3	**7.** 5.40	**12.** 3	**17.** 4	**22.** .4375
3. 1	**8.** 2	**13.** 5	**18.** 4	**23.** 450
4. 466.5	**9.** 4	**14.** 3	**19.** 1	**24.** 4
5. 1	**10.** 3,-4	**15.** 1	**20.** 3	**25.** 3

1. **The correct answer is (4).** 3 sodas at $0.95 each cost a total of $2.85. With the $0.65 newspaper, the total is $3.50. The change from a $10.00 bill would be $10.00 - $3.50 = $6.50.

2. **The correct answer is (3).** A 16% discount means that you would normally pay 84% of $56 for the glove, or $.84 \times 56 = \$47.04$.

   Now take another $\frac{1}{5}$ off $47.04. That means divide $47.04 by 5, multiply it by $\frac{1}{5}$, or multiply it by .20 to find out how much more to take off, or, if you are taking off another 20% ($\frac{1}{5}$) you are paying 80% of $47.04 = $.80 \times \$47.04 =$ $37.63.

3. **The correct answer is (1).** Make a proportion:

$$\frac{2.54 \, cm}{1 \, inch} = \frac{1 \, cm}{x \, inch}$$

$$2.54x = 1$$

$$x = .3937, \text{ or } .4 \text{ inch}$$

4. $12(12.5) + 18(8.25) + 24(7) = 150 + 148.5 + 168$ $= 466.5$ meters. Therefore, the number 466.5 should be coded on your answer sheet, as shown below.

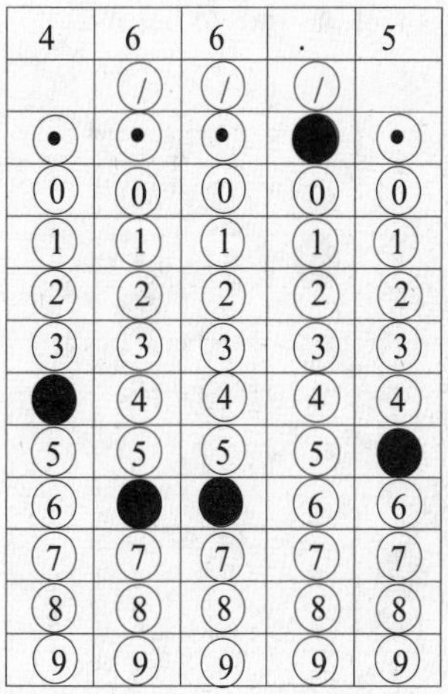

4	6	6	.	5
	/	/	/	
⊙	⊙	⊙	●	⊙
0	0	0	0	0
1	1	1	1	1
2	2	2	2	2
3	3	3	3	3
●	4	4	4	4
5	5	5	5	●
6	●	●	6	6
7	7	7	7	7
8	8	8	8	8
9	9	9	9	9

5. **The correct answer is (1).** Form a proportion:

$$\frac{60 \, pages}{90 \, min} = \frac{436 \, pages}{x \, min} \text{ (time needed to read the entire book)}$$

$$\frac{60}{90} = \frac{2}{3}, \text{ therefore } \frac{2}{3} = \frac{436}{x}$$

$$2x = 1308$$

$$x = 654 \text{ minutes}$$

Dividing by 60, we get 10 hours, 54 minutes, or 10.9 hours.

10.9 − 1.5 = 9.4 hours (time needed to finish the book)

6. **The correct answer is (3).** The volume of each cassette box is $2.5 \times 4 \times \frac{3}{4} = 2.5 \times 4 \times 0.75 = 7.5$ cubic inches. The number of cassette boxes that will fit into a crate that is 180 cubic inches is $180 \div 7.5 = 24$.

7. A day call would cost $0.34 × 44 minutes + $0.48, 0r $15.44. If the evening rate discounts the day rates by 35%, an evening call would cost 65% of $15.44, or 0.65 × $15.44 = $10.04. The saving is $15.44 - $10.04, or $5.40. Therefore, the number $5.40 should be coded on your answer sheet, as shown below.

5	.	4	0	
	/	/	/	
●	●	●	●	●
⓪	⓪	⓪	●	⓪
①	①	①	①	①
②	②	②	②	②
③	③	③	③	③
④	④	●	④	④
●	⑤	⑤	⑤	⑤
⑥	⑥	⑥	⑥	⑥
⑦	⑦	⑦	⑦	⑦
⑧	⑧	⑧	⑧	⑧
⑨	⑨	⑨	⑨	⑨

8. **The correct answer is (2).** The balance before the withdrawals would equal the remaining balance plus the amounts of the two withdrawals. Thus, the solution is $11,516 + $2,356 + $1,131 = $15,003.

9. **The correct answer is (4).** The ship is traveling at 24 km/hr. In 35 minutes it will go $\frac{35}{60}$ of 24 kilometers. $\frac{35}{60} \times \frac{24}{1} = 14$ kilometers.

48 kilometers that the ship was out at 8:00 + 14 = 62 kilometers.

10. When point A(3, 2), and point B(-3, 2), when connected, form a horizontal line segment of length 6, each side of the square must be of length 6. The missing corner is 6 units below (3, 2), which puts it at (3, -4). Therefore, the point (3, -4) must be entered on your answer sheet as shown below.

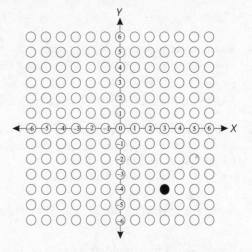

11. **The correct answer is (1).** Add together the amount spent on food (.20) and the amount spent on clothes (.15):

.20 + .15 = $.35 spent on food and clothing

12. **The correct answer is (3).** First find out how much of each dollar she saves. To do that, add up all of the known amounts on the pie and subtract from $1.00:

$1.00 − (.35 + .08 + .15 + .20 + .15) = $.07

That means that she saves $.07 of each dollar she takes home. Now, multiply the amount she saves per dollar by the number of dollars she takes home:

$800 × .07 = $56.00

Thus, she saves $56 per week.

13. **The correct answer is (5).** While the amount Karen spends for entertainment is in the "Other" segment of the pie, we don't know

that it's the only thing covered in that segment, hence there is not enough information to answer the question.

14. **The correct answer is (3).** Karen spends $0.15 out of each dollar on clothing. If she's spending $90 on clothing a week, set up a proportion comparing her weekly clothing expense to her clothing expense per dollar:

Weekly clothing: weekly income as clothing expense per dollar:

Mathematically, letting x = weekly income,

that's $\dfrac{90}{x} = \dfrac{.15}{1}$

$.15x = 90$

$\quad x = 600$

She takes home $600 per week.

15. **The correct answer is (1).** $.08 of every dollar goes for utilities. Electricity is a utility: 0.08 x 450 = $36 means she spends $36 per month on all her utilities. The only possible answer is $24, since she can't spend more on one utility than she spends on all utilities.

16. $2 \times \$6.95 = \$13.90$. $2 \times \$8.95 = \$17.90$. Adding them together makes a total list price of $31.80. With a 30% discount, Suzanne pays 70% of list. $.70 \times \$31.80 = \$22.26$. Therefore, the number $22.26 should be coded on your answer sheet, as shown below.

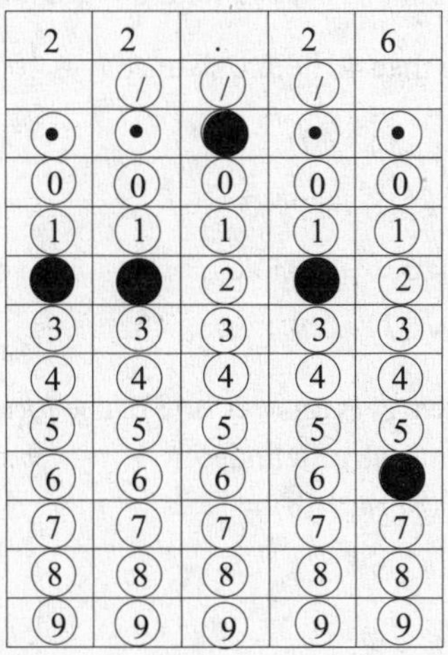

17. **The correct answer is (4).** To find the average, add all the values given, then divide by the number of values. To add the values in this problem, first change all the heights to inches:

$5'8'' = 68''$

$6'1'' = 73''$

$5'10'' = 70$

$6'3'' = 75''$

$5'9'' = \underline{69''}$

$355'' \div 5'' = 71''$

$71'' \div 12'' = 5'11''$

18. **The correct answer is (4).** Hazel writes 10 letters with her name at the bottom of the list of 6. Each of those 10 writes 10 (=100) letters with Hazel's name in 5th place. Each of those 100 writes 10 (=1,000) letters with Hazel's name in 4th place. Each of those 1,000 writes 10 (=10,000) letters with Hazel's name in 3rd place. Each of those 10,000 writes 10 (=100,000) letters with Hazel's name in 2nd place. Each of those 10 writes 10 (=1,000,000) letters with Hazel's name in 1st place. Now Hazel gets 1,000,000 $5 bills, or $5,000,000. P.S. This scheme is illegal, but it's fun to think about, isn't it?

19. **The correct answer is (1).** The diameter of the circle, which is 10, is equal to the length of the diagonal of the square. If the length of the side of the square is equal to S, then, by the Pythagorean theorem,

$S^2 + S^2 = 10^2$

$2S^2 = 100$

$S^2 = 50$

$S = \sqrt{50} = 5\sqrt{2}$

20. **The correct answer is (3).** Since the square is $5\sqrt{2}$ by $5\sqrt{2}$, its area is $(5\sqrt{2})^2 = 50$.

21. The cotangent of an angle is defined as the ratio $\dfrac{\text{adjacent}}{\text{opposite}}$. Thus, triangle PQR can be treated as if it is a right triangle with legs of 5 and 12. The side adjacent to angle P would be 5, and the side opposite would be 12. The Pythagorean theorem can be used to compute

the hypotenuse of 13. The secant of an angle is defined as the ratio $\dfrac{\text{hypotenuse}}{\text{adjacent}}$. Thus, $\sec P = \dfrac{13}{5}$. In order to code this on the answer sheet, it can be written as a decimal, which would be 2.6. Thus, 2.6 can be coded on the answer sheet, as shown below:

2	.	6		
	/	/	/	
•	●	•	•	•
0	0	0	0	0
1	1	1	1	1
●	2	2	2	2
3	3	3	3	3
4	4	4	4	4
5	5	5	5	5
6	6	●	6	6
7	7	7	7	7
8	8	8	8	8
9	9	9	9	9

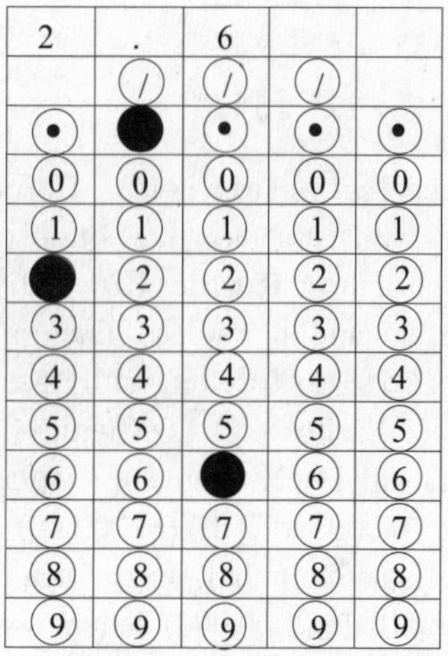

22. To solve this problem, write $3\dfrac{1}{2}$ as 3.5, and divide by 8. $3.5 \div 8 = .4375$. Thus, .4375 must be coded on the answer sheet as shown below:

.	4	3	7	5
	/	/	/	
●	•	•	•	•
0	0	0	0	0
1	1	1	1	1
2	2	2	2	2
3	3	●	3	3
4	●	4	4	4
5	5	5	5	●
6	6	6	6	6
7	7	7	●	7
8	8	8	8	8
9	9	9	9	9

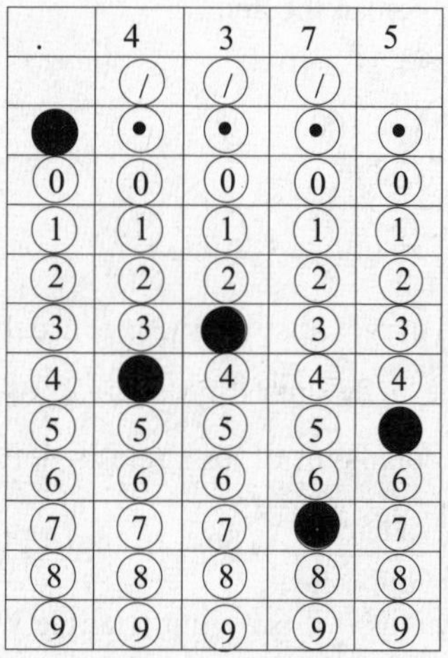

23. Bill invested x dollars at 9% and $x - 450$ at 18%. His annual return from both investments is represented by the equation:

$$.09x + .18(x - 450) = 162$$
$$9x + 18(x - 450) = 16{,}200$$
$$9x + 18x - 8{,}100 = 16{,}200$$
$$27x = 24{,}300$$
$$x = 900$$

That means that $450 was invested at 18%. Thus, $450 must be coded on the answer sheet as shown below:

4	5	0		
	/	/	/	
•	•	•	•	•
0	0	●	0	0
1	1	1	1	1
2	2	2	2	2
3	3	3	3	3
●	4	4	4	4
5	●	5	5	5
6	6	6	6	6
7	7	7	7	7
8	8	8	8	8
9	9	9	9	9

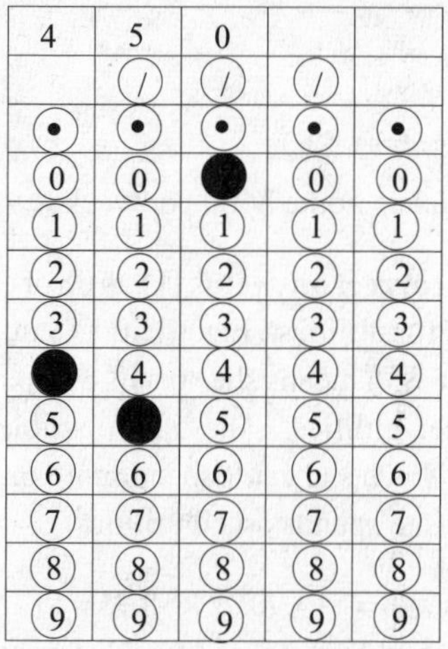

24. The correct answer is (4). $5 \times 12 = 60$ hours Monday through Friday, + 12 hours for the weekend. That is a total of 72 hours. 72 is 32 more than 40 hours, so the pay is 40($8.40) and $32(1\dfrac{1}{2} \times \$8.40)$.

$$= 336 + 32(12.60)$$
$$= 336 + 403.20$$
$$= \$739.20$$

25. The correct answer is (3). The formula for the volume of a cylinder is $V = \pi r^2 h$. Thus, $V = \pi(3)^2(8) = \pi(9)(8) = 72\pi$.

Part 2

1. 5	**6.** 3	**11.** 1,1	**16.** 24.3	**21.** 3
2. 3	**7.** 5	**12.** 2	**17.** 1	**22.** 2
3. 4	**8.** 1	**13.** 1	**18.** 1	**23.** 5
4. 1	**9.** 3	**14.** 5	**19.** 4	**24.** 4
5. 3	**10.** 3	**15.** 4.125	**20.** 5	**25.** 5

1. **The correct answer is (5).** The length of the rectangular garden is 34 feet. If the width is two feet less than half its length, then the width is 17 − 2 = 15 feet. In order to fence in the garden, Brian would need 34 + 34 + 15 + 15 feet, which is the same as 2(34) + 2(15).

2. **The correct answer is (3).** 10 x10 = 100 millimeters

3. **The correct answer is (4).**

$$\frac{\frac{1}{2}}{3} = \frac{1}{2} \times \frac{1}{3} = \frac{1}{6}$$

4. **The correct answer is (1).** The only proportion that is set up correctly is the first one, in which the comparison is "length of lamp-post shadow : height of lamp-post as length of tree shadow : height of tree."

5. **The correct answer is (3).** This is an illustration of the distributive property.

527(316 + 274) = 527(316) + 527(274)

6. **The correct answer is (3).** Multiply 8 × 60 to find the number of minutes in each workday. Then divide by 13 to find number of interviews per day.

$$\frac{8 \times 60}{13}$$

7. **The correct answer is (5).** The highest peaks on all three graphs occur in 2000.

8. **The correct answer is (1).** In 1997, the high school graduate line is at about 9%, while the line for those with less than 4 years of high school is at 17%. The difference, then, is about 8%. The answer of 6% is the closest choice given.

9. **The correct answer is (3).** The greatest difference occurs between the low in 1993, and the high in 2000. The low is about 12%, while the high is about 27%. The difference is found by subtracting 12 from 27 and getting 15%.

10. **The correct answer is (3).** The word "sum" indicates addition, so twice the sum of 3 and a number is represented by 2(3 + N). Then, "less than" indicates subtraction, so 1 less than 3 times the number is represented by 3N − 1.

11. The midpoint of a line segment has an x-coordinate which can be found by taking the average (mean) of the x-coordinates of the endpoints, and a y-coordinate which can be found by taking the average (mean) of the y-coordinates of the endpoints. Thus, the x-coordinate of the midpoint is the average of −1 and 3, which is 1. The y-coordinate of the midpoint is the average of −2 and 4, which is 1. Therefore,

the midpoint is (1, 1), which must be coded on the answer sheet as shown below.

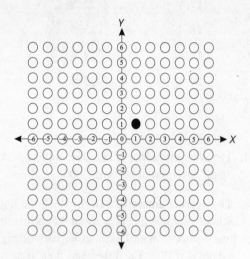

12. The correct answer is (2).

$d = rt$

$1500 = 60t$

$t = 25$ hours

13. The correct answer is (1).

$x^2 + 5x + 6 = 0$

$(x + 3)(x + 2) = 0$

$x + 3 = 0$ or $x + 2 = 0$

$x = -3$ or $x = -2$

14. The correct answer is (5). The largest ball that can fit in the box can have a diameter of no more than the smallest dimension of the box, which is 8 cm. If the diameter is 8 centimeters, then the radius is 4 cm.

15. $67\dfrac{3}{4} - 63\dfrac{5}{8} = 67\dfrac{6}{8} - 63\dfrac{5}{8} = 4\dfrac{1}{8} = 4.125.$

Therefore, the number 4.125 must be coded on the answer sheet as shown below:

4	.	1	2	5
	○/	○/	○/	
●	●	●	●	●
○0	○0	○0	○0	○0
○1	○1	●1	○1	○1
○2	○2	○2	●2	○2
○3	○3	○3	○3	○3
●	○4	○4	○4	○4
○5	○5	○5	○5	●
○6	○6	○6	○6	○6
○7	○7	○7	○7	○7
○8	○8	○8	○8	○8
○9	○9	○9	○9	○9

16. The median mileage is simply the mileage in the middle when the 5 mileages are written in numerical order. This number is 24.3. Therefore, 24.3 must be coded on the answer sheet as shown below:

2	4	.	3
		○/	○/
●	●	●	●
○0	○0	○0	○0
○1	○1	○1	○1
●2	○2	○2	○2
○3	○3	○3	●3
○4	●4	○4	○4
○5	○5	○5	○5
○6	○6	○6	○6
○7	○7	○7	○7
○8	○8	○8	○8
○9	○9	○9	○9

17. The correct answer is (1).

$$\frac{58}{8} = 7.25 \text{ meters}$$

18. The correct answer is (1).

Let t = time at 40 mph

Then $10 - t$ = time at 60 mph

$$d = rt$$
$$530 = 40t + 60(10 - t)$$
$$530 = 40t + 600 - 60t$$
$$20t = 70$$

$$t = 3\frac{1}{2} \text{ hours}$$

19. The correct answer is (4). If Mr. Scalici's office were a 9' × 7' rectangle, its area would be 63 square feet. However, a 2' × 3' = 6 square foot rectangle has been "cut out of" his floor space, leaving him with an office space of 57 square feet. Since each carton contains 12 square feet of tiles, he would need 5 cartons (which would contain 60 square feet) to tile his office.

20. The correct answer is (5).

21. The correct answer is (3). If B represents the number of boys, then the number of girls would be represented by B + 17. Since the number of girls plus the number of boys added together totals up to 155, it must be true that B + (B + 17) = 155.

22. The correct answer is (2). The formula for the area of a triangle is $\frac{1}{2}$ bh. In the triangle pictured, the base is 20, and the height is 28, so the area would be $\frac{(20)(28)}{2} = 280$ square meters.

23. The correct answer is (5). If N represents the number of nickels, then the number of dimes must be 305 – N. Each nickel is worth $0.05, so the total value of the nickels in the machine is .05N. Similarly, the total value of the dimes in the machine is .10(305 – N). The value of the nickels plus the value of the dimes added together is $21, so .10(305 – N) + .05N = 21.

24. The correct answer is (4). 15% of the public schools had CD-ROMs in 1998-1999. At the same time, 25% had modems.

25. The correct answer is (5). 50% of the schools used cable TV in 1997–1998. Since there are 1,500 schools, this represents 750 schools. At the same time, 20% of the schools, or 300 schools used modems. Therefore, 750 – 300 = 450 more schools used cable TV than used modems.

ANSWER SHEET PRACTICE TEST 3

Part 1

1. ① ② ③ ④ ⑤
2. ① ② ③ ④ ⑤
3. ① ② ③ ④ ⑤
5. ① ② ③ ④ ⑤
6. ① ② ③ ④ ⑤

7. ① ② ③ ④ ⑤
8. ① ② ③ ④ ⑤
11. ① ② ③ ④ ⑤
12. ① ② ③ ④ ⑤
13. ① ② ③ ④ ⑤

14. ① ② ③ ④ ⑤
15. ① ② ③ ④ ⑤
17. ① ② ③ ④ ⑤
18. ① ② ③ ④ ⑤
19. ① ② ③ ④ ⑤

20. ① ② ③ ④ ⑤
24. ① ② ③ ④ ⑤
25. ① ② ③ ④ ⑤

4.

9.

10.

16.

21.

22.

23.

Part 2

1. ① ② ③ ④ ⑤
2. ① ② ③ ④ ⑤
3. ① ② ③ ④ ⑤
4. ① ② ③ ④ ⑤
5. ① ② ③ ④ ⑤
6. ① ② ③ ④ ⑤
7. ① ② ③ ④ ⑤
8. ① ② ③ ④ ⑤
9. ① ② ③ ④ ⑤
10. ① ② ③ ④ ⑤
12. ① ② ③ ④ ⑤
13. ① ② ③ ④ ⑤
14. ① ② ③ ④ ⑤
16. ① ② ③ ④ ⑤
18. ① ② ③ ④ ⑤
19. ① ② ③ ④ ⑤
20. ① ② ③ ④ ⑤
21. ① ② ③ ④ ⑤
22. ① ② ③ ④ ⑤
23. ① ② ③ ④ ⑤
24. ① ② ③ ④ ⑤
25. ① ② ③ ④ ⑤

11.

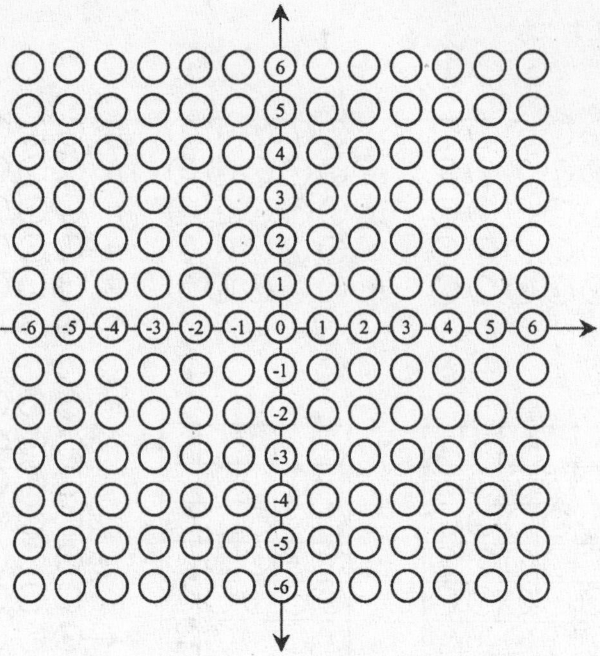

15.

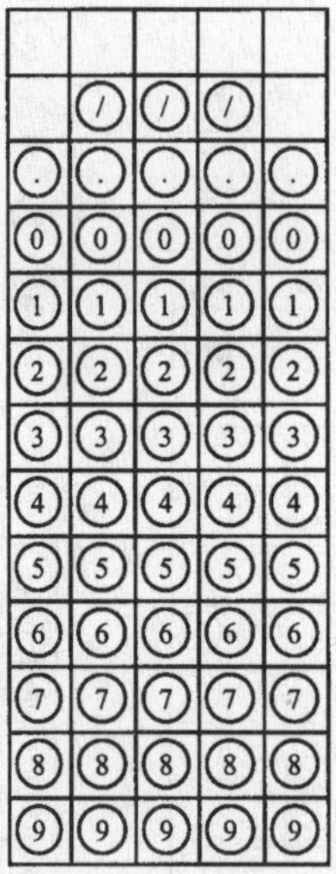

17.

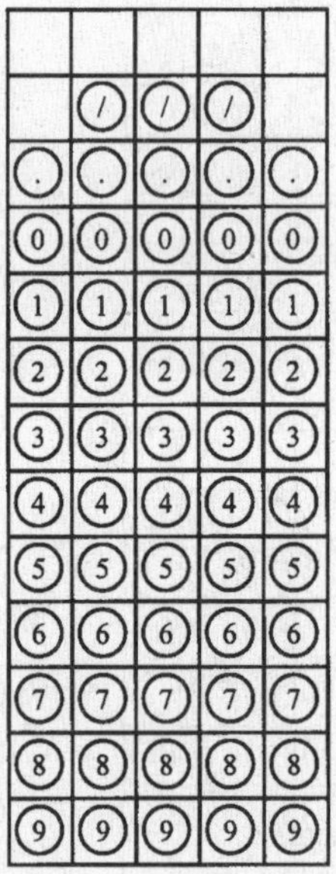

90 Minutes • 50 Questions Total

PART 1—25 QUESTIONS (A CALCULATOR IS PERMITTED): 45 MINUTES

PART 2—25 QUESTIONS (A CALCULATOR IS NOT PERMITTED): 45 MINUTES

Directions: The Mathematics Test consists of questions intended to measure general mathematics skills and problem-solving ability. The questions are based on short readings that often include a graph, chart, or figure. Work carefully, but do not spend too much time on any one question. Be sure you answer every question. You will not be penalized for incorrect answers.

Formulas you may need are given on the following pages. Only some of the questions will require you to use a formula. Record your answers on the separate answer sheet. Be sure that all information is properly recorded.

There are three types of answers found on the answer sheet:

❶ Type 1 is a regular format answer that is the solution to a multiple-choice question. It requires shading in 1 of 5 bubble choices.

❷ Type 2 is an alternate format answer that is the solution to the standard grid "fill-in" type question. It requires shading in bubbles representing the actual numbers, including a decimal or division sign where applicable.

❸ Type 3 is an alternate format answer that is the solution to a coordinate plane grid problem. It requires shading in the bubble representing the correct coordinate of a graph.

Type 1: Regular Format, Multiple-Choice Question

To record your answers for multiple-choice questions, fill in the numbered circle on the answer sheet that corresponds to the answer you select for each question in the test booklet.

Q Jill's drug store bill totals $8.68. How much change should she get if she pays with a $10.00 bill?

(1) $2.32

(2) $1.42

(3) $1.32

(4) $1.28

(5) $1.22

 ① ② ● ④ ⑤

A The correct answer is (3). Therefore, you should mark answer space (3) on your answer sheet.

Type 2: Alternate Format, Standard Grid Question

To record the answer to the previous example, "1.32," using the Alternate Format, Standard Grid,
see below:

Standard Grid

1	.	3	2	

Type 3: Alternate Format, Coordinate Plane Grid Question

To record your answer, fill in the numbered circle on the answer sheet that corresponds to the correct
coordinate in the graph. For example:

Q A system of two linear equations is given below.

$$x = -3y$$
$$x + y = 4$$

What point represents the common solution for the system of equations?

A **The correct answer is (6, –2).** The answer should be gridded as shown below.

Coordinate Plane Grid

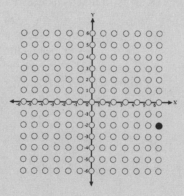

FORMULAS

Description	Formula
AREA (A) of a:	
square	$A = s^2$; where s = side
rectangle	$A = lw$; where l = length, w = width
parallelogram	$A = bh$; where b = base, h = height
triangle	$A = \frac{1}{2} bh$; where b = base, h = height
circle	$A = \pi r^2$; where π = 3.14, r = radius
PERIMETER (P) of a:	
square	$P = 4s$; where s = side
rectangle	$P = 2l + 2w$; where l = length, w = width
triangle	$P = a + b + c$; where a, b, and c are the sides
Circumference (C) of a circle	$C = \pi d$; where π = 3.14, d = diameter
VOLUME (V) of a:	
cube	$V = s^3$; where s = side
rectangular container	$V = lwh$; where l = length, w = width, h = height
cylinder	$V = \pi r^2 h$; where π = 3.14, r = radius, h = height
square pyramid	Volume = $\frac{1}{3} \times$ (base edge)$^2 \times$ height
cone	Volume = $\frac{1}{3} \times \pi \times$ radius$^2 \times$ height; π is approximately equal to 3.14.
Pythagorean theorem	$c^2 = a^2 + b^2$; where c = hypotenuse, a and b are legs of a right triangle
distance (d) between two points in a plane	$d = \sqrt{(x_2 - x_1)^2 + (y_2 - y_1)^2}$; where (x_1, y_1) and (x_2, y_2) are two points in a plane
slope of a line (m)	$m = \dfrac{y_2 - y_1}{x_2 - x_1}$ where $(x_1,\ y_1)$ and $(x_2,\ y_2)$ are two points in a plane
trigonometric ratios	given an acute angle with measure x of a right triangle, $\sin x = \dfrac{\text{opposite}}{\text{hypotenuse}}$, $\cos x = \dfrac{\text{adjacent}}{\text{hypotenuse}}$, $\tan x = \dfrac{\text{opposite}}{\text{adjacent}}$
mean	mean $= \dfrac{x_1 + x_2 + \cdots + x_n}{n}$; where the x's are the values for which a mean is desired, and n = number of values in the series
median	median = the point in an ordered set of numbers at which half of the numbers are above and half of the numbers are below this value
simple interest (i)	$i = prt$; where p = principal, r = rate, t = time
distance (d) as function of rate and time	$d = rt$; where r = rate, t = time
total cost (c)	$c = nr$; where n = number of units, r = cost per unit

You may use a scientific calculator for Part 1. (A Casio FX-260 Scientific Calculator will be provided at your Official GED Testing Center.)

CALCULATOR DIRECTIONS

You may practice with your calculator, using the following directions.

CALCULATOR DIRECTIONS

To prepare the calculator for use the *first* time, press the ON (upper-rightmost) key. "DEG" will appear at the top-center of the screen and "0" at the right. This indicates the calculator is in the proper format for all your calculations.

To prepare the calculator for *another* question, press the ON or the red AC key. This clears any entries made previously.

To do any arithmetic, enter the expression as it is written. Press ON (equals sign) when finished.

EXAMPLE A: 8 – 3 + 9

First press ON or AC .
Enter the following:

8 – 3 + 9 =

The correct answer is 14.

If an expression in parentheses is to be multiplied by a number, press x (multiplication sign) between the number and the parenthesis sign.

EXAMPLE B: 6(8 + 5)

First press ON or AC .
Enter the following:

6 x (8 + 5) =

The correct answer is 78.

To find the square root of a number

- enter the number;
- press the SHIFT (upper-leftmost) key ("SHIFT" appears at top-left of the screen);
- press x² (third from the left on top row) to access its second function: square root.

DO NOT press SHIFT and x² at the same time.

EXAMPLE C: √64

First press ON or AC .
Enter the following:

6 4 SHIFT x² =

The correct answer is 8.

To enter a negative number such as -8,

- enter the number without the negative sign (enter 8);
- press the "change sign" (+/-) key which is directly above the 7 key.

All arithmetic can be done with positive and/or negative numbers.

EXAMPLE D: -8 – -5

First press ON or AC .
Enter the following:

8 +/- – 5 +/- =

The correct answer is -3.

Part 1

Directions: You may now begin Part 1 of the Mathematics Test. Bubble in the correct response to each question on Part 1 of your answer sheet.

1. A plumber completed five jobs yesterday. On the first job she earned $36.45, the second $52.80, the third $42.81, the fourth $49.54, and the fifth $48.90. What was the average amount she earned for each job?

   (1) $38.50

   (2) $39.75

   (3) $40.80

   (4) $42.50

   (5) $46.10

2. Ed bought his gasoline at a station which recently converted its pumps to measuring gasoline in liters. If his tank took 34.0 liters, approximately how many gallons did it take? One gallon is equal to 3.785 liters.

   (1) 8

   (2) 9

   (3) 9.25

   (4) 10

   (5) 11.5

3. Using $\frac{22}{7}$ as an approximation for π, find the circumference of the circle shown below.

   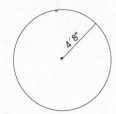

   (1) 14'10"

   (2) 340"

   (3) 176"

   (4) 29'4"

   (5) 21'2"

4. Jonathan drove his 66-year-old grandmother and 9-year old little brother to the movie theater on Saturday afternoon. He treated all three of them to the matinee. The prices read as follows: Adults $3.50, Children Under 12 $1.50, Senior Citizens 20% discount. How much change did Jonathan receive from his $10.00 bill?

   Mark your answer in the circles in the grid on your answer sheet.

5. Brian got grades of 92, 89, and 86 on his first three math tests. What grade must he get on his final test to have an overall average of 90?

   (1) 89

   (2) 90

   (3) 92

   (4) 93

   (5) 94

6.

   Between which 10 years does the chart show the greatest percent of increase of women in the labor force?

   (1) 1950 - 1960

   (2) 1960 - 1970

   (3) 1970 - 1980

   (4) 1980 - 1990

   (5) 1990 - 2000

7. Amanda buys a package of bacon for $1.98, a dozen eggs for $1.29, paper plates for $1.25, and napkins for $.75. There is a 5% sales tax on non-food items. How much change did Amanda receive from a $10.00 bill?

Mark your answer in the circles in the grid on your answer sheet.

8. In 1990, the average salary of a New Jersey teacher was $30,588. Ten years later, teachers in New Jersey were averaging $61,008. Find the nearest whole percentage of increase in salary from 1990 to 2000.
   (1) 50%
   (2) 100%
   (3) 99%
   (4) 86%
   (5) 75%

9. At the ballpark, 23,000 customers consumed 630 pounds of hot dogs. At that rate of consumption, how many pounds of hot dogs would be needed for a crowd of 57,500.
   (1) 1175
   (2) 1225
   (3) 1575
   (4) 1625
   (5) 1875

10. On the coordinate plane on your answer sheet, mark the center of the circle $(x - 2)^2 + (y + 3)^2 = 9$.

11. Henry's VW Rabbit Diesel gets 50 mpg. Diesel fuel costs an average of $1.10 per gallon. This summer, Henry and his family drove 1200 miles to Niagara Falls for vacation. How much did Henry pay for fuel?
    (1) $26.40
    (2) $52.80
    (3) $105.60
    (4) $132.20
    (5) $264.00

12. What is the area of the circular ring formed by two concentric circles of radii 6 and 8 inches, as shown in the diagram below?

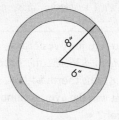

   (1) 2π
   (2) 4π
   (3) 14π
   (4) 28π
   (5) 56π

Questions 13-15 refer to the following graphs:

Top Purchasers of U.S. Exports

1999
Total U.S. Exports – $510 Billion

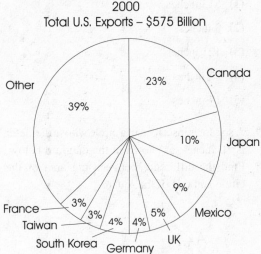

2000
Total U.S. Exports – $575 Billion

13. Approximately how much money did the United Kingdom spend on the purchase of U.S. exports in 2000?

    (1) $2.9 billion
    (2) $3.1 billion
    (3) $29 billion
    (4) $31 billion
    (5) $33 billion

14. How much more did Germany spend on U.S. exports than Taiwan in 1999?

    (1) $23 billion
    (2) $17.25 billion
    (3) $5.1 billion
    (4) $2.3 billion
    (5) $575,000,000

15. How many countries spent less than $25 billion for U.S. exports in 2000?

    (1) 2
    (2) 3
    (3) 4
    (4) 6
    (5) 8

16. Tom and Johnny leave the house at 6 AM to go camping. Colin decides to go with them, but to his dismay, when he reaches Tom's house, he learns that he and Johnny had left an hour ago. If Colin drives 65 mph, how many hours will it take him to overtake his friends who are traveling at 45 mph?

Mark your answer in the circles in the grid on your answer sheet.

17. The sailfish is built for speed and can swim through the water at speeds of 68 mph. Approximately how many kilometers can it travel in an hour? 1 kilometer = .62 miles.

    (1) 42 km/hr
    (2) 68 km/hr
    (3) 96 km/hr
    (4) 109 km/hr
    (5) 110 km/hr

18. On a map, 1 inch represents three miles. How many inches are needed to represent a road that is actually 171 miles long?

    (1) 513 inches

    (2) 3 inches

    (3) 121 inches

    (4) 57 inches

    (5) 17 inches

19. A square is inscribed in a circle whose diameter is 10 inches, as shown in the diagram below. Find the difference between the area of the circle and that of the square.

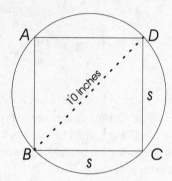

    (1) 50 - 25π sq. in.

    (2) 25π - 50 sq. in.

    (3) 100 - 25π sq. in.

    (4) 25π - 100 sq. in

    (5) 50 - 100π sq. in.

20. A discount toy store general takes $\frac{1}{5}$ off the list price of their merchandise. During the holiday season, an additional 15% is taken off the list price. Mrs. Johnson bought a sled listed at $62.00 for Joey and a doll house listed at $54.00 for Jill. To the nearest penny, how much did Mrs. Johnson pay?

    (1) $75.40

    (2) $78.88

    (3) $82.64

    (4) $92.80

    (5) $112.52

21. A woman bought $4\frac{1}{2}$ yards of ribbon to decorate curtains. If she cut the ribbon into 8 equal pieces, how long was each piece?

    Mark your answer in the circles in the grid on your answer sheet.

22. In right triangle DEF below, what is the value of tan D?

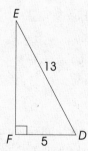

    Mark your answer in the circles in the grid on your answer sheet.

23. Find the slope of the line containing the points (4, 5) and (6, 12).

    Mark your answer in the circles in the grid on your answer sheet.

24. A lawn mower is on sale for $111.20. That represents a discount of 20% from its normal price. What is the lawn mower's normal price?

    (1) $22.24
    (2) $133.44
    (3) $139.00
    (4) $200.16
    (5) $224.00

25. Two ships leave the same harbor at the same time and travel in opposite directions, one at 30 km/hr and the other at 50 km/hr. After how many hours will they be 360 kilometers apart?

    (1) $2\frac{1}{2}$

    (2) $3\frac{1}{2}$

    (3) $4\frac{1}{2}$

    (4) $5\frac{1}{2}$

    (5) $6\frac{1}{2}$

Part 2

Directions: You may not return to Part 1 or use your calculator for this part. Bubble in the correct response to each question on Part 2 of your answer sheet.

1. Antoinette told Alice that her street was exactly 6.4 miles from the railroad trestle. At the trestle, Alice noticed that her odometer read 5488.9 miles. What will her odometer read once she has reached Antoinette's street?

    (1) 5494.13
    (2) 5494.3
    (3) 5495.3
    (4) 5495.4
    (5) 5495.14

2. A coat that lists for $240 is on sale for $180. By what percent had the coat been discounted?

    (1) 40%
    (2) 35%
    (3) 30%
    (4) 25%
    (5) 20%

3. Which of the following expressions is equivalent to 82(9) + 82(12)?

    (1) $(82 + 9) + (82 + 12)$
    (2) $82(9 + 12)$
    (3) $(82 + 9)(82 + 12)$
    (4) $82(108)$
    (5) $82(9) + 9(12)$

4. Mr. Tretola owns the rectangular table whose top is pictured below. He would like to purchase a table cloth which overlaps each of the four sides by 12 inches. What will the perimeter of his new table cloth be?

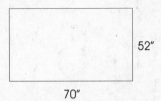

      52"

    70"

  (1) 244 inches

  (2) 170 inches

  (3) 340 inches

  (4) 268 inches

  (5) 292 inches

Questions 5 and 6 are based on the rectangular solid shown below:

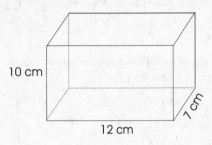

10 cm

7 cm

12 cm

5. What is the volume of the rectangular solid?

  (1) 840 cm^2

  (2) 84 cm^3

  (3) 840 cm^3

  (4) 8.4 m^3

  (5) 8.4 m^2

6. What is the surface area of the rectangular solid?

  (1) 274 cm^2

  (2) 274 cm^3

  (3) 548 cm^2

  (4) 548 cm^3

  (5) 840 cm^2

7. $x^2 - 9x - 22 = 0$. The value of x is

  (1) $^-$2 and 11

  (2) $^-$2 and $^-$11

  (3) 2 and $^-$11

  (4) 11 only

  (5) 2 only.

8. Hans Dishpan can wash approximately 240 dishes in one hour's time. How many dishes would Hans average in 5 minutes?

  (1) 15

  (2) 20

  (3) 24

  (4) 40

  (5) 42

Questions 9 and 10 refer to the following graph.

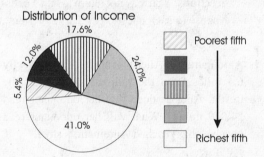

Distribution of Income

On the circle graph, the distribution of available income in the United States for a particular year is represented. The population has been divided into 5 equal parts, starting from the poorest fifth and ending with the richest.

9. What is the difference between the income available to the richest $\frac{1}{5}$ of the population and that available to the poorest $\frac{3}{5}$?

   (1) 11%
   (2) 6%
   (3) 18%
   (4) 12.9%
   (5) 5.6%

10. Which of the following ratios best illustrates the relationship in income distribution between the poorer $\frac{4}{5}$ of the population and the richest $\frac{1}{5}$?

    (1) 34:66
    (2) 41:59
    (3) 76:42
    (4) 6:4
    (5) 59:41

11. Consider the equation $y = 7x - 3$. On the grid on your answer page mark the y-intercept of this equation.

12. Mrs. Cogswell went to the butcher store to purchase link sausages for herself and her neighbor. There were 57 sausages on the chain purchased. When Mrs. Cogswell split the sausage chain into two pieces, one piece had 3 fewer sausages than the other. The chain with the most sausages went to the neighbor. If S represents the number of sausages in Mrs. Cogswell's chain, which of the following equations could be solved to find the value of S?

    (1) $S + (S - 3) = 57$
    (2) $S - 57 = (S - 3)$
    (3) $S(S + 3) = 57$
    (4) $S + (S + 3) = 57$
    (5) $S(S - 3) = 57$

13. Christine earned $84 by working over the summer. The money was deposited in a bank at the beginning of an interest period at a simple yearly rate of 6%. If Christine left her money in the bank for 4 entire years, which of the following computations could be performed to determine how much money she would have in the account after 4 years?

    (1) $84(.06)^4$
    (2) $84(1.06)^4$
    (3) $84 (1.06)^5$
    (4) $84(.06)^5$
    (5) $84 + (1.06)^4$

14. One cubic foot of water weighs 62.4 pounds. A swimming pool holds $18\frac{1}{4}$ cubic feet of water. Which of the following computations could be performed to compute the number of pounds of water the pool can hold?

    (1) $\dfrac{18.25}{62.4}$
    (2) $62.4 \times (18\frac{1}{4})^3$
    (3) 18.25×62.4
    (4) $\dfrac{62.4}{3} \times \left(18\frac{1}{4}\right)$
    (5) $\dfrac{18.25 \times 62.4}{3}$

15. The snail can creep at speeds up to 0.03 mph, but the snail has also been observed to travel as slowly as 0.0036 mph. Find the difference between the snail's fastest and slowest speeds.

    Mark your answer in the circles in the grid on your answer sheet.

16. Mitch needs cord to section off his bean bushes from the rest of his garden. The bean section is 3 feet wide and half again as long. How much cord will he need?

    Mark your answer in the circles in the grid on your answer sheet.

Questions 17 to 21 refer to the following graph.

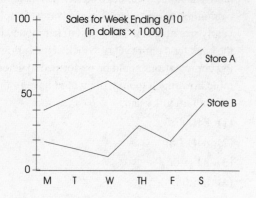

17. On which day was the gap between the two stores' sales greatest?

(1) Tuesday

(2) Wednesday

(3) Thursday

(4) Friday

(5) Saturday

18. What was store B's largest sales volume for a single day?

(1) $4,900

(2) $8,000

(3) $48,000

(4) $45,000

(5) Not enough information is given

19. Sales dipped for Store A between

(1) Monday and Wednesday

(2) Tuesday and Thursday

(3) Wednesday and Thursday

(4) Wednesday and Friday

(5) Thursday and Saturday

20. What is the difference between Store A's best and worst days?

(1) $30,000

(2) $35,000

(3) $40,000

(4) $45,000

(5) $50,000

21. On which day did Store B draw the greatest number of shoppers?

(1) Monday

(2) Tuesday

(3) Thursday

(4) Saturday

(5) Not enough information is given

22. There are 25 coins in your piggy bank totaling $3.85. If you only save dimes and quarters, which of the following equations could be solved to find the number of quarters, Q, in the piggy bank?

(1) $10(25 - Q) + 25Q = 3.85$

(2) $.10(25 - Q) + .25Q = 385$

(3) $.10(Q + 25) + .25Q = 3.85$

(4) $.10(25 - Q) + .25Q = 3.85$

(5) $10(Q - 25) + .25Q = 3.85$

Questions 23 and 24 are based on the graph shown below.

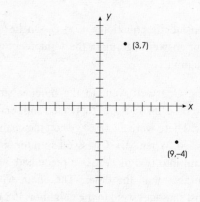

23. How long is the line segment that connects the two points shown in the graph?

(1) $4\sqrt{3}$

(2) $\sqrt{119}$

(3) $\sqrt{157}$

(4) $3\sqrt{71}$

(5) $\sqrt{179}$

24. What is the midpoint of the line segment that connects the two points shown in the graph?

(1) (12, 3)

(2) (6, 11)

(3) (6, 1)

(4) $(6, 1\frac{1}{2})$

(5) $(6\frac{1}{2}, 1\frac{1}{2})$

25. When 9 is added to a certain number, the result is the same as when twice the number is diminished by 6. If the number is represented by N, which of the following equations can be solved to determine the value of N?

(1) N + 9 = 2N - 6

(2) N + 9 = 2(6 - N)

(3) N + 9 = 6 - 2N

(4) N + 9 = 2(N - 6)

(5) N - 9 = 2(N + 6)

ANSWER KEY AND EXPLANATIONS

Part 1

1. 5	**6.** 2	**11.** 1	**16.** 2.25	**21.** .5625
2. 2	**7.** 4.63	**12.** 4	**17.** 5	**22.** 2.4
3. 4	**8.** 3	**13.** 3	**18.** 4	**23.** 3.5
4. 2.20	**9.** 3	**14.** 3	**19.** 2	**24.** 3
5. 4	**10.** 2, -3	**15.** 3	**20.** 1	**25.** 3

1. **The correct answer is (5).** To find the average, add all the values given, then divide by the number of values.

    $36.45
    $52.80
    $42.81
    $49.54
    <u>$48.90</u>
    $230.50 ÷ 5 = $46.10

2. **The correct answer is (2).** Divide 34.0 by 3.785 to get 8.98, which is closest to 9 liters.

3. **The correct answer is (4).** To begin, 4' 8" is equal to $4(12) + 8 = 56$". The formula for the circumference of a circle is $C = 2\pi r$. Substituting r = 56" yields $C = 2\left(\dfrac{22}{7}\right)(56") = 352"$. This is equal to 29 feet, 4 inches.

4. If Jonathan drove, he must be an adult, so he paid $3.50 for his ticket, and $1.50 for his brother's. His grandmother's ticket cost 80% of $3.50 = $.8 \times 3.50 = 2.80. Add the three ticket prices together and get $7.80. Jonathan got

$2.20 in change. Therefore, the number $2.20 must be coded on your answer sheet, as shown below.

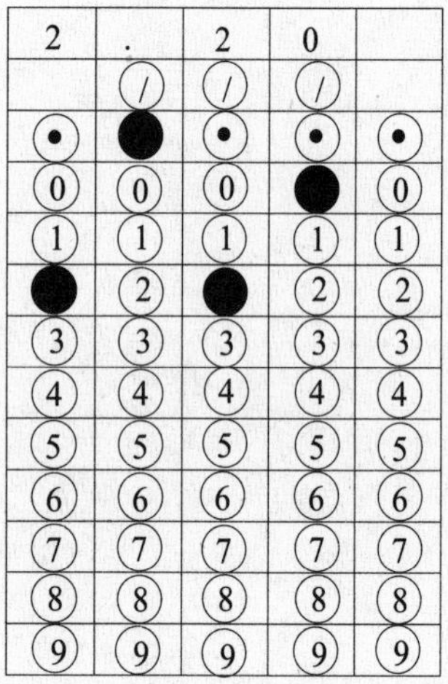

5. **The correct answer is (4).** Let x represent the score that Brian must get on his last test. Then, $\dfrac{92 + 89 + 86 + x}{4} = 90$. Multiplying both sides by 4,

    $92 + 89 + 86 + x = 360$, or

    $267 + x = 360$, so that

    $x = 93$.

6. The correct answer is (2). In order to answer this question, you must use your calculator to compute 5 percents of increase:

$$1950-1960: \quad \frac{32.1-28.8}{28.8} = 0.115 = 11.5\%$$

$$1960-1970: \quad \frac{36.7-32.1}{32.1} = 0.143 = 14.3\%$$

$$1970-1980: \quad \frac{40.3-36.7}{36.7} = 0.098 = 9.8\%$$

$$1980-1990: \quad \frac{42.0-40.3}{40.3} = .042 = 4.2\%$$

$$1990-2000: \quad \frac{47.5-42.0}{42.0} = 0.131 = 13.1\%$$

Therefore, the greatest percent of increase is from 1960 – 1970.

7. The napkins and paper plates add up to $2.00. 5% tax on $2.00 is $.05 \times 2 = \$0.10$. Now, add $2.00 + .10 + 1.98 + 1.29 = \$5.37$.

$10.00 - 5.37 = \$4.63$ change.

Therefore, the number $4.63 must be coded on your answer sheet, as shown below.

4	.	6	3	
	①	①	①	
●	⬤	●	●	●
⓪	⓪	⓪	⓪	⓪
①	①	①	①	①
②	②	②	②	②
③	③	③	⬤	③
⬤	④	④	④	④
⑤	⑤	⑤	⑤	⑤
⑥	⑥	⬤	⑥	⑥
⑦	⑦	⑦	⑦	⑦
⑧	⑧	⑧	⑧	⑧
⑨	⑨	⑨	⑨	⑨

8. The correct answer is (3). The percent of increase in salary is computed by

$$\frac{61,008-30,588}{30,588} = \frac{30,420}{30,588} = 0.9945 =$$

$99.45\% \approx 99\%$.

9. The correct answer is (3). This is a problem for which proportion will yield a quick solu-

tion: $\dfrac{630}{23,000} = \dfrac{x}{57,500}$

$$23,000x = 57,500(630)$$

$$23x = 57.5(630)$$

$$23x = 36225$$

$$x = 1575 \text{ pounds}$$

10. The center of the circle $(x-2)^2 + (y+3)^2 = 9$ is at (2, -3). Therefore, you should make a dot on the grid on your answer sheet at (2, -3), as shown below.

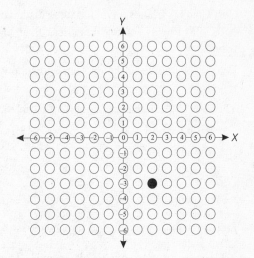

11. The correct answer is (1).

$$\frac{1200 \text{ miles}}{50 \text{ miles per gallon}} = 24 \text{ gallons used. Then}$$

multiply $1.10 by 24 to get a total cost of $26.40.

12. The correct answer is (4). The area of the larger circle is $A = \pi r^2 = \pi(8)^2 = 64\pi$

The area of the smaller circle is $A = \pi r^2 = \pi(6)^2 = 36\pi$.

The area between the two circles is $64\pi - 36\pi = 28\pi$.

13. The correct answer is (3). The United Kingdom purchased 5% of $575 billion = $575 billion $\times .05 = \$28.75$ billion, which is closest to $29 billion.

14. The correct answer is (3). Germany spent $510 billion × .04 = $20.4 billion. Taiwan spent $510 × .03 = $15.3 billion. The difference is 20.4 − 15.3 = $5.1 billion.

15. The correct answer is (3). The quickest way to answer this problem is to find what percent $25 billion is of $575 billion. Since 25 ÷ 575 = 0.043 = 4.3%. So, any country that purchases less than 4.3% of the US exports, would have spent less than $25 billion. There are 4 such countries, France, Taiwan, South Korea, and Germany.

16. Recall that $D = rt$.

Colin and his friends travel the same distance, but Colin travels for one hour less than his friends.

Let t = the number of hours Tom and Johnny travel.

Then, $t − 1$ = the number of hours Colin travels.

$$45t = 65(t − 1)$$
$$45t = 65t − 65$$
$$−20t = -65$$

$$t = 3\frac{1}{4} \text{ hours}$$

$t − 1 = 2\frac{1}{4}$ hours. Thus, it will take Colin $2\frac{1}{4}$ = 2.25 hours to catch up to his friends. This answer must be coded on the answer sheet as shown below.

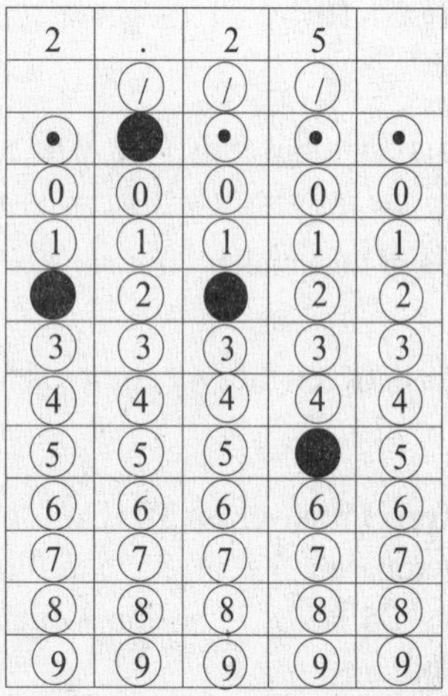

17. The correct answer is (5). Since a kilometer is less than a mile, it will travel more than 68 kilometers. Multiplying by a fraction (.62) will give a smaller number, so we must divide

$$\frac{68}{.62} = 109.67 = 110 \text{ kilometers.}$$

Alternate solution: $\dfrac{68}{x} = \dfrac{.62}{1}$

$$.62x = 68$$
$$x = 109.67 = 110 \text{ kilometers}$$

18. The correct answer is (4). A proportion can be established between the scale of the map in inches and the actual distance in miles:

$$\frac{1}{3} = \frac{x}{171}$$
$$3x = 171$$
$$x = 57 \text{ inches}$$

19. The correct answer is (2). We are given that the diagonal of the square is 10. The Pythagorean Theorem tells us that the length of a side of the square is $\dfrac{10}{\sqrt{2}}$. The area of the square is then $\dfrac{10}{\sqrt{2}} \times \dfrac{10}{\sqrt{2}} = \dfrac{100}{2} = 50$. The diameter of the circle is 10, so the radius is 5, and the area is $\pi r^2 = \pi(5)^2 = 25\pi$. The difference between the area of the circle and that of the square is, then, $25\pi - 50$.

20. The correct answer is (1). $\dfrac{1}{5}$ is 20%. If both are taken off the list price, then the discount is actually 35%. The total purchases are $116. If 35% is coming off that, then 65% is being paid. .65(116) = $75.40.

21. Divide $4\frac{1}{2}$ by 8: $4.5 \div 8 = 0.5625$. This answer must be coded onto your answer sheet as shown below:

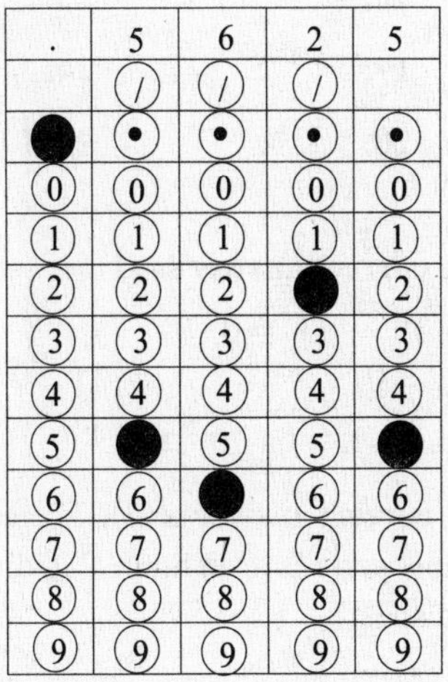

22. To begin, find the length of the missing side EF by using the Pythagorean theorem:

$(EF)^2 + 5^2 = 13^2$

$(EF)^2 + 25 = 169$

$(EF)^2 = 144$

$EF = 12.$

Since the tangent of an angle is the ratio $\dfrac{opposite}{adjacent}$, compute $\tan D = \dfrac{12}{5}$. Since this is equal to 2.4, this number may be coded onto your answer sheet as shown below.

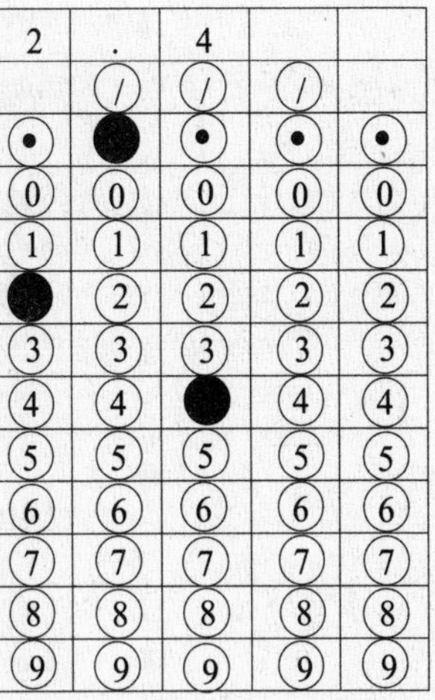

23. The formula for the slope of a line is $\dfrac{y_2 - y_1}{x_2 - x_1} = \dfrac{12 - 5}{6 - 4} = \dfrac{7}{2} = 3.5$. Therefore, 3.5 must be coded on your answer sheet, as shown below.

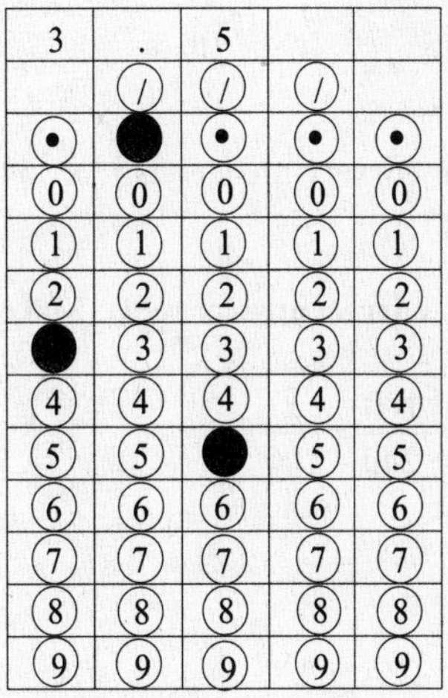

24. The correct answer is (3). The sale price is 80% of the list price. Let x = the list price. Then

$.80x = \$111.20$

$x = \dfrac{\$111.20}{.80}$

$x = \$139$

25. The correct answer is (3). $D = rt$

Since the two ships are traveling in opposite directions, we may imagine one ship to be standing still and the other moving away from it at their combined rates of 80 km/hr. In that case,

$D = rt$

$360 = 80t$

$t = 4\dfrac{1}{2}$ hours

Part 2

1. 3	**6.** 3	**11.** 0, -3	**16.** 15	**21.** 5
2. 4	**7.** 1	**12.** 4	**17.** 2	**22.** 4
3. 2	**8.** 2	**13.** 2	**18.** 4	**23.** 3
4. 3	**9.** 2	**14.** 3	**19.** 3	**24.** 4
5. 3	**10.** 5	**15.** .0264	**20.** 3	**25.** 1

1. The correct answer is (3). $5488.9 + 6.4 = 5495.3$

2. The correct answer is (4). Make a proportion:

$$\frac{180}{240} = \frac{x}{100}$$

$$\frac{3}{4} = \frac{x}{100}$$

$$4x = 300$$

$$x = 75\%$$

If the coat is selling for 75% of list, then it must have been discounted 25%.

3. The correct answer is (2). The Distributive Property says that $82(9 + 12) = 82(9) + 82(12)$

4. The correct answer is (3). Add 24 inches to the length and the width for the overlap. That makes the dimensions 76 inches × 94 inches. The perimeter of a rectangle is 2L + 2W, which in this case means that the perimeter is $2(76) + 2(94) = 340$ inches.

5. The correct answer is (3). The volume of a rectangular solid is given by the formula $V = L \times W \times H = 12 \text{ cm} \times 10 \text{ cm} \times 7 \text{ cm} = 840 \text{ cm}^3$.

6. The correct answer is (3). There are six surfaces on the rectangular solid. Two of them are 12 cm × 10cm, so both of these surfaces have areas of 120 cm². Two of the surfaces are 7 cm × 10 cm, so these two surfaces have areas of 70 cm². Finally, there are two 12 cm × 7 cm surfaces, with areas of 84 cm². Adding up the six areas gives us a total surface area of 548 cm².

7. The correct answer is (1). $x^2 - 9x - 22 = 0$

$(x + 2)(x - 11) = 0$

$x + 2 = 0$	$x - 11 = 0$
$x = -2$	$x = 11$

8. The correct answer is (2). There are twelve 5-minute periods in an hour. 240 divided by 12 = 20 dishes.

9. The correct answer is (2). To find the income available to the poorest $\frac{3}{5}$, add $5.4 + 12.0 + 17.6$ and get 35%. That is 6% less than the 41% available to the richest $\frac{1}{5}$.

10. The correct answer is (5). The poorest $\frac{4}{5}$ received 59% while the richest $\frac{1}{5}$ got 41%. In that order, the ratio is 59:41,

11. The given equation, $y = 7x - 3$, is already in the slope-intercept form. Therefore, the y-intercept is at –3, that is, at the point (0, -3). The grid should be filled in as shown below.

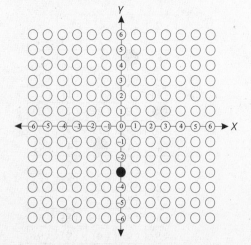

12. **The correct answer is (4).** If Mrs. Cogswell's chain has S sausages, then her neighbor's has $S + 3$. Since there are 57 sausages in total, $S + (S + 3) = 57$.

13. **The correct answer is (2).** Christine begins with $84. At the end of one year, she has earned 6% interest. The amount of money in the account at the end of one year, therefore, would be 84×1.06. After two years, she would have $84 \times 1.06 \times 1.06 = 84 \times (1.06)^2$. Following the pattern, after 4 years, she would have $84(1.06)^4$ in her bank account.

14. **The correct answer is (3).** Given that every cubic foot weighs 62.4 pounds, the total number of pounds for $18\frac{1}{4}$ cubic feet would simply be 18.25×62.4.

15. Since $0.03 - 0.0036 = 0.0264$, this number must be coded on the answer grid, as shown below.

16. One dimension of the garden is 3 feet. Half again as much as 3 is $3 + \frac{1}{2}(3) = 4\frac{1}{2}$. The perimeter of the garden, therefore, would be $3 + 3 + 4\frac{1}{2} + 4\frac{1}{2} = 15$ feet. Thus, this number must be coded in on the answer grid, as shown below.

17. **The correct answer is (2).** The gap between the two lines is greatest on Wednesday.

18. **The correct answer is (4).** Keeping a straight-edge square with the vertical axis and moving it until the "B" line just touches its right end, we find it comes out between the 40 and 50 marking. Call it 45. But since the sales volume is graphed in terms of thousands of dollars, we must multiply the 45 by 1000: $45 \times 1000 = \$45,000$.

19. **The correct answer is (3).** The graph of "A" dips from Wednesday to Thursday. All other choices are for intervals when the graph dips and rises or rises only.

20. **The correct answer is (3).** Store A did $80,000 in sales on Saturday and $40,000 on Monday: $80,000 - \$40,000 = \$40,000$.

21. The correct answer is (5). The graph shows sales for one week in store A and store B. There is no information given about number of shoppers.

22. The correct answer is (4). If there are Q quarters in the bank account, there are $25 - Q$ dimes. The total amount of money in the bank would be $.10(25 - Q) + .25Q$. Since there is a total of \$3.85 in the bank, $.10(25 - Q) + .25Q = 3.85$.

23. The correct answer is (3). The distance formula tells us that

$$D = \sqrt{(x_2 - x_1)^2 + (y_2 - y_1)^2} \cdot$$

By substitution, we get $D =$

$$\sqrt{(9 - 3)^2 + (-4 - 7)^2} = \sqrt{6^2 + (-11)^2} = \sqrt{36 + 121} = \sqrt{157}$$

24. The correct answer is (4). The x-coordinate of the midpoint is obtained by averaging the x-coordinates of the endpoints; the y-coordinate of the midpoint is found by averaging the y-coordinates of the endpoints. Therefore, the midpoint is $\left(\left(\frac{3+9}{2}\right),\left(\frac{7+-4}{2}\right)\right) = \left(\frac{12}{2}, \frac{3}{2}\right) = (6, 1\frac{1}{2})$

25. The correct answer is (1). If the number is represented by N, 9 added to the number is $N + 9$, twice the number is $2N$, and twice the number diminished by 6 is $2N - 6$. Therefore, $N + 9 = 2N - 6$.

NOTES

NOTES

NOTES

NOTES

NOTES

NOTES

Peterson's
Book Satisfaction Survey

Give Us Your Feedback

Thank you for choosing Peterson's as your source for personalized solutions for your education and career achievement. Please take a few minutes to answer the following questions. Your answers will go a long way in helping us to produce the most user-friendly and comprehensive resources to meet your individual needs.

When completed, please tear out this page and mail it to us at:

> Publishing Department
> Peterson's, a Nelnet company
> 2000 Lenox Drive
> Lawrenceville, NJ 08648

You can also complete this survey online at **www.petersons.com/booksurvey.**

1. **What is the ISBN of the book you have purchased? (The ISBN can be found on the book's back cover in the lower right-hand corner.)** _____

2. **Where did you purchase this book?**
   - ❑ Retailer, such as Barnes & Noble
   - ❑ Online reseller, such as Amazon.com
   - ❑ Petersons.com
   - ❑ Other (please specify) _____

3. **If you purchased this book on Petersons.com, please rate the following aspects of your online purchasing experience on a scale of 4 to 1 (4 = Excellent and 1 = Poor).**

	4	3	2	1
Comprehensiveness of Peterson's Online Bookstore page	❑	❑	❑	❑
Overall online customer experience	❑	❑	❑	❑

4. **Which category best describes you?**
   - ❑ High school student
   - ❑ Parent of high school student
   - ❑ College student
   - ❑ Graduate/professional student
   - ❑ Returning adult student
   - ❑ Teacher
   - ❑ Counselor
   - ❑ Working professional/military
   - ❑ Other (please specify) _____

5. **Rate your overall satisfaction with this book.**

Extremely Satisfied	Satisfied	Not Satisfied
❑	❑	❑

6. Rate each of the following aspects of this book on a scale of 4 to 1 (4 = Excellent and 1 = Poor).

	4	3	2	1
Comprehensiveness of the information	❏	❏	❏	❏
Accuracy of the information	❏	❏	❏	❏
Usability	❏	❏	❏	❏
Cover design	❏	❏	❏	❏
Book layout	❏	❏	❏	❏
Special features (e.g., CD, flashcards, charts, etc.)	❏	❏	❏	❏
Value for the money	❏	❏	❏	❏

7. This book was recommended by:
- ❏ Guidance counselor
- ❏ Parent/guardian
- ❏ Family member/relative
- ❏ Friend
- ❏ Teacher
- ❏ Not recommended by anyone—I found the book on my own
- ❏ Other (please specify) _____

8. Would you recommend this book to others?

Yes	Not Sure	No
❏	❏	❏

9. Please provide any additional comments.

_____

_____

_____

_____

_____

Remember, you can tear out this page and mail it to us at:

Publishing Department
Peterson's, a Nelnet company
2000 Lenox Drive
Lawrenceville, NJ 08648

or you can complete the survey online at **www.petersons.com/booksurvey.**

Your feedback is important to us at Peterson's, and we thank you for your time!

If you would like us to keep in touch with you about new products and services, please include your e-mail address here: _____